Ordinary Differential Equations

Ordinary Differential Equations

Emilio O. Roxin

The University of Rhode Island

Wadsworth Publishing Company, Inc.
Belmont, California

ISBN-0-534-00009-6

L. C. Cat. Card No. -74-174666

Printed in the United States of America

1 2 3 4 5 6 7 8 9 10——76 75 74 73 72

Portions of this book were originally published in 1968 in Spanish by EUDEBA, Buenos Aires, Argentina.

Preface

This book is intended to bridge the gap between a rather elementary course in which differential equations have simply been introduced and a more advanced course in which a strong mathematical background is a necessary prerequisite. Addressed to a variety of students, it introduces both higher mathematical concepts as well as the physical motivations behind them. In this way, the student with an abstract mathematical background will discover the intuitive bases of many mathematical concepts, while the student with an engineering or physical sciences background will discover that many mathematical abstractions are refined, sophisticated and, I would say, even beautiful versions of concepts rooted in simple physical intuition. This insight is important on the one hand because, as history shows, the physical sciences are an inexhaustible source of the best mathematical problems. On the other hand, there can be no successful analysis of many practical problems without the use of good, advanced mathematics.

Most parts of this book require only a thorough knowledge of calculus. In general, the student will have had an introductory course in differential equations, but this is not mandatory, since the subject is covered from the very beginning. For a more complete understanding of all subjects covered in this text, a familiarity with advanced calculus, linear algebra, and even topology and complex variables, would be helpful. But, wherever concepts of higher mathematics are mentioned, a brief explanation is included. This device is intended to awaken the interest of the student to such concepts as fixed-point theorems, differentiable manifolds, and so on. The same effect—in reverse—can be achieved through the detailed discussions of such topics as impulsive functions, which brings the abstract mathematician into contact with the "practical" world.

Even using this interrelated approach, it was simply impossible to cover all major areas pertaining to the theory of ordinary differential equations in this one text. Numerical methods, series expansions, boundary-value problems and other subjects were left out. On the other hand, such areas of current interest as topological properties of solutions of differential equations, global behavior of nonlinear systems, dynamical systems, and the direct method of Liapunov concerning the theory of stability have been presented, with the hope of arousing the student's interest in modern research. If this intention is even partially accomplished, a major goal of this book will have been achieved.

Finally, I want to express my gratitude to all those colleagues who encouraged me to write this book and helped me along the way. The reviewers, Profs. T. A. Burton, David Sanchez, and David V. V. Wend, did a particularly thorough job in going over the whole manuscript and in suggesting the many changes that have enhanced both the value and the appeal of this book.

Contents

Chapter 1 Elementary Methods 1

1. Ordinary Differential Equations 1
2. First-Order Equations: Geometrical Interpretation 5
3. Elementary Methods of Integration 9
4. Implicit Differential Equations 19
5. Equations of Higher Order and Systems of Equations 25
6. Approximate Methods of Solution 31
7. Qualitative Methods 32

References 33

Chapter 2 Existence and Uniqueness Theorems 34

1. Statement of the Problems 34
2. Existence Theorem 37
3. Uniqueness 43
4. Successive Approximations 48
5. Continuation of Solutions 52
6. Dependence on Initial Data and Parameters 54
7. Systems of Differential Equations 59
8. Additional General Theorems 63

References 68

Chapter 3 Linear Differential Equations 70

1. Matrices 70
2. Linear Homogeneous Differential Equation 82
3. Associated Matrix Equation 87
4. Adjoint Equation 94
5. Nonhomogeneous Linear Differential Equation 96
6. Linear Equation of Order n 99
7. Differential Equations in the Complex Domain 103

References 105

Chapter 4 Linear Systems with Constant Coefficients 106

1. Direct Method of Solution 106
2. Solution in Matrix Form 118
3. Operational Method 126

References 136

Chapter 5 Applications 137

1. Systems 137
2. Rest Points and Critical Points 144
3. Oscillations 147
4. Frequency Response 157

References 168

Chapter 6 Geometric Theory and Phase Space 169

1. Phase Space—Autonomous and Periodic Equations 169
2. Critical Points and Periodic Solutions 173
3. Topological Properties of Trajectories 180
4. Dynamical Systems 188
5. Limit Sets 194

References 200

Chapter 7 Autonomous Two-dimensional Systems—Critical Points 201

1. Linear Systems 201
2. Homogeneous Systems 211

3. Analytic Systems 217
4. Perturbed Linear Systems 229
References 240

Chapter 8 Autonomous Two-dimensional Systems—Global Study 241

1. Differential Equations on Manifolds 241
2. The Index 257
3. Trajectories on the Projective Plane 267
4. Limit Sets in the Plane 274
References 280

Chapter 9 Second-order Equations 282

1. Conservative Systems 282
2. Nonconservative Systems 290
3. Relaxation Oscillations 295
References 300

Chapter 10 Stability 301

1. Stability of Rest Points 301
2. Stability of Solutions 311
3. Direct Method of Liapunov 317
4. Stability of Particular Systems 326
References 336

Answers 337

Index 351

1

Elementary Methods

1 Ordinary Differential Equations

1.1 General definition

An ordinary differential equation is a functional relationship of the type

$$F(t, x, x', x'', \ldots, x^{(n)}) = 0, \tag{1.1.1}$$

between an independent variable t, an unknown function x(t), and a finite number of its derivatives $x'(t), \ldots, x^{(n)}(t)$.

1.2 General and particular solutions

To solve such a differential equation means to find a function $x(t)$ which satisfies the relation (1.1.1) identically for all values of t in some (possibly infinite) interval. For example, the differential equation

$$x + x' = 0 \tag{1.1.2}$$

is satisfied by the function

$$x(t) = ce^{-t} \tag{1.1.3}$$

for any constant value of c. Conversely, in Chapter 2 we will prove that every solution can be written in this form. The expression (1.1.3) is therefore called the *general solution* of equation (1.1.2).

The differential equation

$$x^2 + x'^2 = 1 \tag{1.1.4}$$

is satisfied by the function

$$x(t) = \sin t \tag{1.1.5}$$

and also by

$$x(t) = 1. \tag{1.1.6}$$

Thus (1.1.5) and (1.1.6) are called *particular solutions* of equation (1.1.4).

1.3 Order and degree

The *order* of a differential equation (1.1.1) is the maximal order of differentiation that appears in that expression. In case F is a polynomial in $x, x', \ldots, x^{(n)}$, the highest power of $x^{(n)}$ is called the *degree* of the equation. If F involves transcendental functions of $x, x', \ldots, x^{(n)}$, the differential equation is also called *transcendental.* For example, the equations

$$xx'^2 - t^3 + 1 = 0, \qquad x'^2 - tx = 0$$

are of first order and second degree. The equation

$$x' = \ln x$$

is transcendental.

1.4 Systems

There may be various unknown functions $x(t), y(t), \ldots$ for whose determination we have several differential equations of the type

$$F(t, x, x', \ldots, y, y', \ldots) = 0. \tag{1.1.7}$$

In general, if there are n unknown functions, there will also be n relationships like

(1.1.7); in this case we say that we have a *system* of n differential equations. A *solution* of such a system will be a set of n functions $x(t)$, $y(t)$, ... which satisfy all of these equations identically on some interval.

1.5 Explicit equation

The easiest case is when equation (1.1.1) can be solved for the derivative of highest order, thus obtaining

$$x^{(n)} = f(t, x, x', \ldots, x^{(n-1)}). \tag{1.1.8}$$

This form is called *explicit*. It is impossible, however, to put equation (1.1.1) in the explicit form if there is no value of $x^{(n)}$ satisfying equation (1.1.1) or if there are several such values. For example, from equation (1.1.4), we can obtain

$$x' = \pm\sqrt{1 - x^2};$$

hence equation (1.1.4) is satisfied by the solutions of either of the two equations

$$x' = +\sqrt{1 - x^2}$$

and

$$x' = -\sqrt{1 - x^2}.$$

1.6 Explicit form of a system

Given a system of differential equations, we may try to solve it for the derivatives of highest order of the unknown functions. Moreover, it is always possible to reduce such a system (including the case of a single differential equation) to that in which only derivatives of the first order appear. This can be done by introducing new unknown functions. For example, writing $x_1 = x$, $x_2 = x'$, ..., $x_n = x^{(n-1)}$, the differential equation (1.1.1) becomes the system

$$\begin{cases} x_1' = x_2 \\ x_2' = x_3 \\ \cdots\cdots\cdots \\ x_{n-1}' = x_n \\ F(t, x_1, x_2, \ldots, x_n, x_n') = 0. \end{cases}$$

If the last expression can be solved for x_n', then the last equation will be of the form

$$x_n' = f(t, x_1, \ldots, x_n).$$

The following, somewhat more general form is called the *explicit form of the system*:

$$\left.\begin{aligned} x_1' &= f_1(t, x_1, x_2, \dots, x_n) \\ x_2' &= f_2(t, x_1, x_2, \dots, x_n) \\ &\cdots\cdots\cdots\cdots\cdots\cdots \\ x_n' &= f_n(t, x_1, x_2, \dots x_n). \end{aligned}\right\} \tag{1.1.9}$$

Only first-order derivatives appear, and the system is solved for them. In this form the x_i can be thought of as components of a vector $\mathbf{x}$, and in agreement with the notation used in vector analysis, we may write

$$\mathbf{x}' = \mathbf{f}(t, \mathbf{x}), \tag{1.1.10}$$

where $\mathbf{f}$ is a vector with the same number of components as $\mathbf{x}$.

Solving the equation for the derivative of highest order is, in general, an analytical problem. A great part of the theory of differential equations is usually developed under the assumption that the differential equation or system is already in the explicit form (1.1.8) or (1.1.9), which is the most important case.

1.7 Throughout this book it will be assumed that the independent variable t is real (in symbols, $t \in R$). In most cases we will also assume that the unknown function x is real. However, in some cases it is convenient to consider functions $x(t)$ assuming complex values. In the linear case this actually makes the theory simpler. Applications of complex solutions to physical problems will be given in Chapter 5.

Problems

1. Verify that $x = \sin 2t$ satisfies the differential equation
$$(x'' + 2x)^2 + x'^2 = 4.$$
2. Verify that $x = t^{1/2}$ satisfies the differential equation
$$2x't - x = 0.$$
3. Verify that for every c, $x = te^{(t+c)}$ satisfies the differential equation
$$x' = (t + 1)/t.$$
4. Verify that every solution of $x + x' = 0$ and that every solution of $x' - x = 0$ satisfies the differential equation
$$x''x' - x''x + x'^2 - x'x = 0.$$
5. Verify that if $x(t)$ satisfies $x' = 1 + x^2$, then $y(t) = 1/x(t)$ satisfies $y' = -(1 + y^2)$.
6. Find a polynomial $x(t) = at^2 + bt + c$ which satisfies $x'' + 2x' + x = t^2 - 4t + 1$.
7. Find a polynomial $x(t) = at^3 + bt^2 + ct + d$ which satisfies $x'' - 6x + 6t^3 = 0$.

8. Find constants a and b such that $x(t) = a \sin t + b \cos t$ satisfies $x' + x = 2 \sin t$.
9. In the differential equation $x' = 2tx$, substitute $x(t) = e^{y(t)}$; then solve the differential equation for $y(t)$ and substitute back, obtaining $x(t)$.
10. In the differential equation $x' - x = 3t^2e^t$, substitute $x(t) = y(t)e^t$; then solve the differential equation for $y(t)$ and substitute back, obtaining $x(t)$.

2 First-order Equations: Geometrical Interpretation

2.1 Direction fields

Consider the differential equation

$$x' = f(t, x), \tag{1.2.1}$$

where t, x are real variables and $f(t, x)$ is a real function defined on a given domain of the t, x plane. Here, and in the following sections, we will define a *domain* as any subset of the coordinate space having the following properties: (a) it is open—that is, if P is a point of the domain, there is some neighborhood of P contained in the domain; (b) it is connected—that is, if P and Q are two points of the domain, they can be joined by a continuous curve lying totally within the domain.

At each point of that domain, a direction is given by the line having slope $f(t, x)$, and the differential equation tells us that the graph of a solution $x(t)$ should be tangent at each of its points $(t, x(t))$ to the corresponding line with slope $f(t, x(t))$. This graph of a solution is sometimes referred to as a *solution curve*.

Equation (1.2.1) determines, therefore, what is called a *field of directions* on the given domain. One can visualize this field of directions by taking many points and drawing through each such point (t, x) a segment of a line with slope $f(t, x)$. Thus Figures 1.1 and 1.2 show the fields corresponding to the equations

$$x' = \frac{x}{t} \quad \text{and} \quad x' = -\frac{t}{x},$$

respectively. It is quite evident that the curves

$$x = ct, \quad c = \text{const} \quad \text{and} \quad x^2 + t^2 = c, \quad c = \text{const}$$

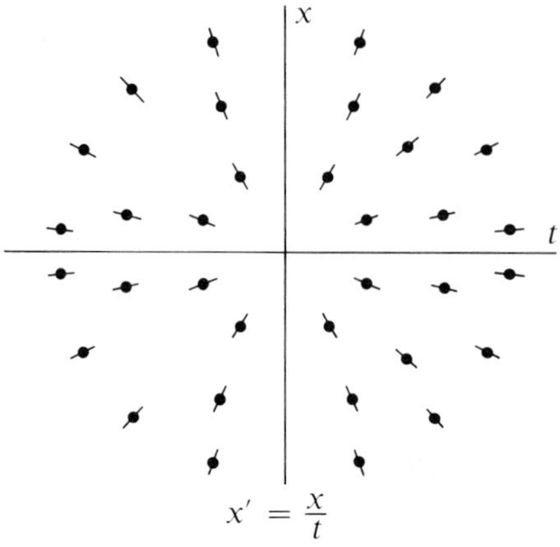

Figure 1.1

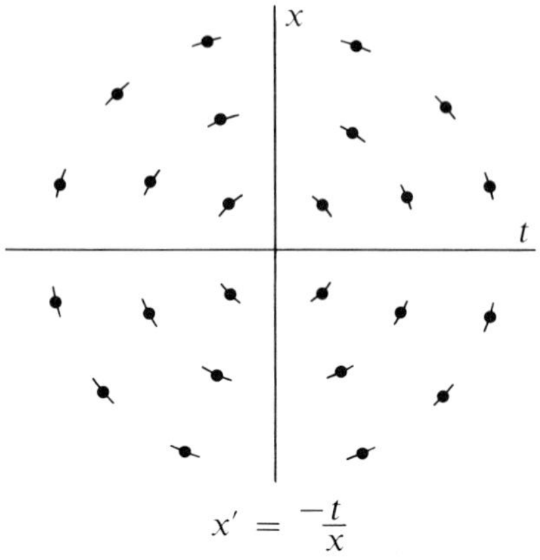

Figure 1.2

satisfy the condition of tangency and are therefore solution curves of the corresponding differential equations.

This method of visually guessing the solutions obviously lacks mathematical precision and can even lead to gross errors; nevertheless, it can be a good first step in acquiring an idea of the behavior of the solutions. In order to obtain a graphical representation, one can also draw the curves $f(t, x) = \text{constant}$, called *isoclines* because they correspond to points of equal inclination. As an example, Figure 1.3 illustrates the construction of the "towing" curve defined by the property of the segment of the tangent line, between the point of tangency and the intersection of the tangent line with the t-axis, to be of constant length l. The corresponding

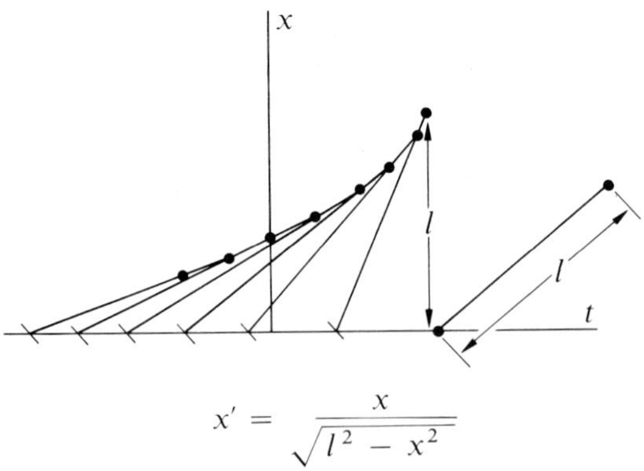

Figure 1.3

.ntial equation is

$$x' = \frac{x}{\sqrt{l^2 - x^2}},$$

nd the value of x' is independent of t. The isoclines are therefore horizontal lines.

2.2 The following corollaries are an immediate consequence of a geometric interpretation of the differential equation:

(1) If at some point (t, x) the derivative x' has value infinity, this does not necessarily imply a strange behavior on the part of the solution curve but simply that the tangent is vertical. Taking x as an independent variable and $t(x)$ as a function, the graph of this function can be perfectly regular, with the slope of the tangent line $dt/dx = 0$. It may certainly happen that x is not a (single-valued) function of t; for example, in the case $t = x^2$ for $t > 0$, two functions $x(t)$ are defined ($x = +\sqrt{t}$ and $x = -\sqrt{t}$), while for $t < 0$ no real function $x(t)$ is defined.

(2) At some points, $f(t, x)$ may become indeterminate; for example, it may be given by a quotient which takes the form 0/0. At such a point, the differential equation is not defined, but it is important to study the solutions in a neighborhood of the point. [Good examples are the origins in Figures 1.1 and 1.2.]

(3) If certain general conditions of regularity, which will be presented later, are satisfied, there will be infinitely many solutions $x(t)$ of the differential equation, and these solutions will cover some domain in the t, x plane in such a way that through each point (t, x) of this domain, there is one and only one solution. These curves can then be represented in the form

$$\varphi(t, x, c) = 0, \tag{1.2.2}$$

where for each constant value of c, we get one solution curve.

Conversely, given any family of curves defined by an equation of the form (1.2.2), at any point (t, x) the slope of the corresponding curve is determined from

$$\frac{\partial\varphi}{\partial x}\,dx + \frac{\partial\varphi}{\partial t}\,dt = 0$$

as

$$x' = -\frac{\partial\varphi/\partial t}{\partial\varphi/\partial x}. \tag{1.2.3}$$

In this last expression the constant c still appears; if we eliminate it between equations (1.2.2) and (1.2.3), we obtain the differential equation of the family (1.2.2).

If, in particular, this latter equation is

$$\varphi(t, x) = c,$$

we obtain the differential equation from

$$\varphi_x \, dx + \varphi_t \, dt = 0$$

as

$$x' = -\frac{\varphi_t}{\varphi_x}.$$

(4) In order to determine a particular solution it is necessary to give not only the differential equation but also an additional condition; for example, the condition might be a value of $x(t)$ for a particular value t_0 of t, thereby giving the point $(t_0, x(t_0))$ in the plane through which the solution should pass.

Interpreting the value t_0 as *initial instant*, the corresponding value $x_0 = x(t_0)$ is called the *initial value*, and one says that the solution $x(t)$ is determined by the *initial condition* $x(t_0) = x_0$.

Problems

As an exercise for the reader, in the following Problems we use coordinates x, y, with $y = y(x)$, $y' = dy/dx$.

1. Draw the direction field of the differential equation, using isoclines, and guess the shape of the solutions of

   (a) $y' = +\sqrt{y/x}$, (b) $y' = e^y$,
   (c) $y' = x^2y^2$, (d) $y' = (x^2 + y^2)^{-1}$.

2. Find the differential equations corresponding to the following families of curves; then find the points (x, y), where y' is indeterminate, and sketch the graph of the curves.

   (a) $y = c \sin x$. (b) $y^2 - x^2 = c$.
   (c) $y^3 - x^3 = c$. (d) $y = cx^4$.
   (e) $y = x^c$. (f) $y^2 + (x - c)^2 = c^2$.
   (g) $\dfrac{y^2 + (x + 1)^2}{y^2 + (x - 1)^2} = c$. (h) $\dfrac{y^2 + (x + a)^2}{y^2 + (x - a)^2} = c$ for fixed a.

3. All conics with foci at $x = \pm 1$, $y = 0$ are given by $(x^2/c) + y^2/(c - 1) = 1$. [$c > 1$ gives an ellipse and $0 < c < 1$ gives a hyperbola.] Show that the correspondent differential equation is

   $$y'^2xy + y'(x^2 - y^2 - 1) - xy = 0.$$

   Hence through each point (x, y) there are two conics of this family (one ellipse and one hyperbola).

4. Substituting y' for $-1/y'$ in the last differential equation, the equation is also satisfied. Infer that the family of curves is self-orthogonal, that is, the two curves passing through any given point (x, y) meet at a right angle.
5. Show that the family of parabolas $y^2 - 2cx - c^2 = 0$ is self-orthogonal.
6. Find the orthogonal trajectories of

   (a) the hyperbolas $y = c/x$, (b) the exponentials $y = ce^x$.

3 *Elementary Methods of Integration*

3.1 In some cases, the differential equation can be "solved" (i.e., the general solution can be found) by simple analytical methods. The most common examples will be considered in this section.

3.2 *Separable variables*

If the differential equation can be written in the form

$$\frac{dx}{dt} = \frac{f(t)}{g(x)},$$

then the variables t, x can be "separated" by writing

$$g(x)\,dx = f(t)\,dt,$$

from which

$$\int g(x)\,dx = \int f(t)\,dt + \text{const.}$$

Example 1

Continuous interest is accumulated when the interest is added continuously to the capital. If $x(t)$ is the capital at time t and p is the interest rate, then the interest added to the capital in the time interval dt is

$$dx = p \cdot x \cdot dt.$$

From this we obtain

$$\int_{x_0}^{x} \frac{dx}{x} = \ln x - \ln x_0 = p(t - t_0)$$

and finally

$$x(t) = x(t_0)e^{p(t-t_0)}.$$

The quantity $x(t_0)$ can be interpreted as the original capital invested at time t_0.

Example 2

A point moves along the x-axis with a velocity depending only on its position:

$$\frac{dx}{dt} = f(x).$$

In this case we obtain

$$\int_{x_0}^{x} \frac{dx}{f(x)} = t - t_0.$$

Sometimes the separation of variables can be accomplished by making a preliminary substitution (change of variables), as seen in the following example.

Example 3

$$\frac{dx}{dt} = f(x + t).$$

The substitution $x + t = u$, $du/dt = dx/dt + 1$ gives $du/dt = f(u) + 1$, and the solution

$$t - t_0 = \int_{u(t_0)}^{u(t)} \frac{du}{f(u) + 1}$$

is quickly obtained. Naturally, we should still replace u with $x + t$ in the solution.

3.3 Homogeneous equation

If x' depends only on the ratio x/t, that is, if

$$x' = f\left(\frac{x}{t}\right),$$

the substitution $x = x/t$ allows us to separate the variables. Indeed, differentiating $x = tz$, we obtain

$$x' = z + tz' = f(z);$$

hence

$$z' = \frac{f(z) - z}{t},$$

where the variables are separable.

Example 4

Let $x' = x/t + t/x$. With the indicated substitution, we obtain

$$z' = \frac{z + 1/z - z}{t} = \frac{1}{tz},$$

from which

$$\frac{z^2}{2} = \ln |t| + \text{const} = \ln |ct|$$

and finally

$$x^2 = t^2 \ln (c^2 t^2)$$

are obtained.

It should be noted that when x or t is zero, the above procedure breaks down and a more careful analysis is needed. Indeed, in this case x' becomes infinite. [See Figure 1.4.]

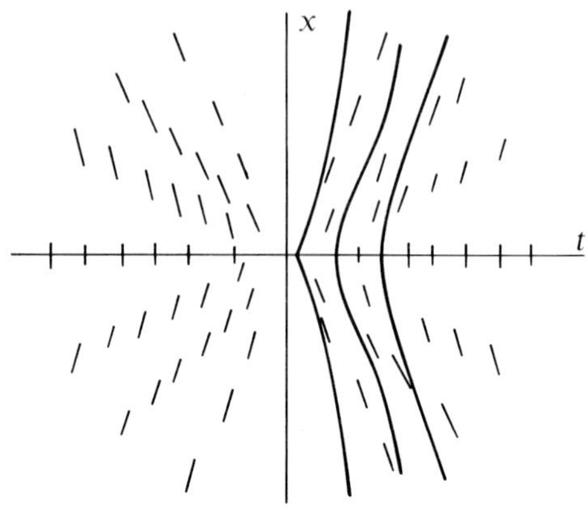

$$x' = \frac{x}{t} + \frac{t}{x}$$

Figure 1.4

3.4 Linear equation

A linear first-order differential equation is one of the form

$$x' + p(t)x = q(t). \tag{1.3.1}$$

It is called *homogeneous* if $q(t) \equiv 0$. In this case

$$x' + p(t)x = 0,$$

and the variables are separable

$$\frac{dx}{x} = -p(t)\,dt.$$

Let $P(t)$ be any integral of $p(t)$; the solution of this equation is then

$$x = ce^{-P(t)}$$

with c an arbitrary constant.

For the *nonhomogeneous* equation (1.3.1), we make the substitution $x(t) = u(t)v(t)$, choosing for $u(t)$ any solution of the corresponding homogeneous equation $u = e^{-P(t)}$, so that $u' + p(t)u = 0$. Replacing these expressions in (1.3.1), we obtain

$$v(u' + pu) + v'u = v'u = q.$$

Hence

$$v = \int \frac{q(t)}{u(t)} \, dt$$

and finally

$$x(t) = e^{-P(t)}\left[\int q(t)e^{P(t)} \, dt + c\right], \tag{1.3.2}$$

where $P(t) = \int p(t) \, dt$ and c is an arbitrary constant.

We will see later (Chapter 3, formula 3.5.3) that this is a particular case of a more general formula concerning systems of linear differential equations.

Example 5

Given the equation

$$x' + tx = t$$

and then applying the above formula, we obtain

$$x(t) = e^{-t^2/2}\left[\int te^{t^2/2} \, dt + c\right] = 1 + ce^{-t^2/2}$$

These solutions are represented in Figure 1.5.

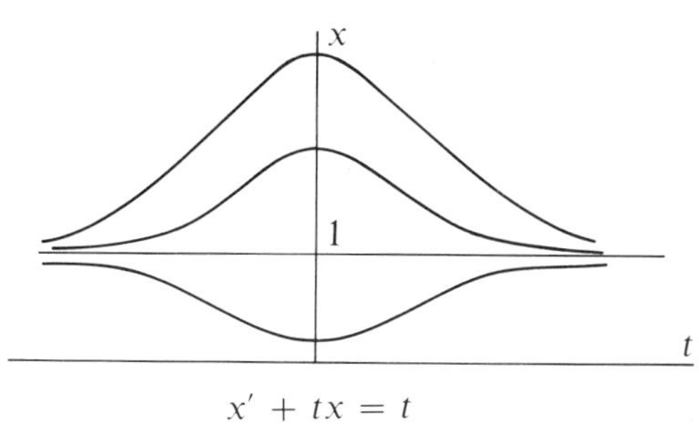

Figure 1.5

3.5 Bernoulli's equation

The equation

$$x' + f(t)x + g(t)x^n = 0 \quad (n \text{ is any real number})$$

is called *Bernoulli's equation*. The substitution $x = y^m$, $x' = my^{m-1}y'$ gives

$$my^{m-1}y' + f(t)y^m + g(t)y^{mn} = 0.$$

Dividing by y^{m-1} and choosing m such that $mn - m + 1 = 0$ (thus $m = -1/(n-1)$), the linear equation

$$y' + \frac{1}{m}[f(t)y + g(t)] = 0$$

is obtained.

Example 6

Consider the equation

$$x' + x + \frac{1}{x} = 0.$$

Applying the method given above with $n = -1$, $m = 1/2$, the equation for y is

$$y' + 2y + 2 = 0.$$

Solving this equation and substituting back the x variable, we get the solutions

$$x = \pm\sqrt{ce^{-2t} - 1}.$$

3.6 Riccati's equation

The equation

$$x' = f(t)x^2 + g(t)x + h(t)$$

is known as the *generalized Riccati equation*. [The proper Riccati equation has $f(t) = -1$.] This seems the simplest nonlinear equation, yet it cannot, in general, be integrated "elementarily." By this we mean that there is no formula in terms of $f(t)$, $g(t)$, and $h(t)$ which determines the solution $x(t)$, with a finite number of arithmetical operations ($+$, $-$, $\times$, $\div$, $\sqrt[n]{\ }$), using the elementary transcendental functions (e^x, $\ln x$, $\sin x$, $\cos x$) and also using possibly definite integrals. It should be noted that elementary transcendentals can also be defined by integrals; hence its use is irrelevant as a matter of principle. In this way the idea of *elementarily integrable* is more properly expressed as *reducible to quadratures* (definite integrals). So, we would consider a differential equation solved if we can express its solution in terms of integrals, even if these are not obtainable in terms of the elementary transcendentals (the elliptic functions, for example). Of course, the solution $x(t)$ of the differential equation may certainly exist (computable, for example, by power series), even in the case where no finite formula for $x(t)$ exists.

3.7 Exact equations

The differential equation $x' = f(t, x)$ can be written in the form

$$P(t, x)\, dt + Q(t, x)\, dx = 0, \tag{1.3.3}$$

and this can be done in many different ways.

We know from calculus that if P and Q are continuously differentiable and if the "crossed derivatives" are equal,

$$\frac{\partial P}{\partial x} = \frac{\partial Q}{\partial t}, \tag{1.3.4}$$

then there exists a *potential function* $F(t, x)$ such that $P = \partial F/\partial t$ and $Q = \partial F/\partial x$. Hence equation (1.3.3) can be written

$$\frac{\partial F}{\partial t}\, dt + \frac{\partial F}{\partial x}\, dx = 0.$$

The solutions are therefore

$$F(t, x) = \text{const},$$

where we assume that $P(t, x)$ and $Q(t, x)$ are not simultaneously zero.

When condition (1.3.4) is satisfied, equation (1.3.3) is called *exact*.

Example 7

Let

$$(1 + 2t \sin x)\, dt + (1 + t^2 \cos x)\, dx = 0.$$

Here

$$P(t, x) = 1 + 2t \sin x, \qquad \frac{\partial P}{\partial x} = 2t \cos x,$$

$$Q(t, x) = 1 + t^2 \cos x, \qquad \frac{\partial Q}{\partial t} = 2t \cos x.$$

Condition (1.3.4) is therefore satisfied, and we can find $F(t, x)$ by the following method.

If $\partial F/\partial t = 1 + 2t \sin x$, then by partial integration with respect to t, we obtain

$$F(t, x) = t + t^2 \sin x + g(x),$$

where $g(x)$ is the constant of integration (with respect to t).

Differentiating now with respect to x,

$$\frac{\partial F}{\partial x} = t^2 \cos x + g'(x) = Q(t, x) = 1 + t^2 \cos x,$$

from which $g'(t) = 1$ and $g(x) = x + c$ is obtained. The solution is therefore

$$F(t, x) = t + t^2 \sin x + x = \text{const.}$$

In general, condition (1.3.4) will not be satisfied; we can, nevertheless, try to find an *integrating factor* $M(t, x)$ such that if we multiply equation (1.3.3) by $M(t, x)$, the resulting differential equation

$$P(t, x)M(t, x)\, dt + Q(t, x)M(t, x)\, dx = 0$$

is exact. The condition of the crossed derivatives is now

$$\frac{\partial(MP)}{\partial x} = \frac{\partial(MQ)}{\partial t},$$

or

$$\frac{p\partial M}{\partial x} - \frac{Q\partial M}{\partial t} + M\left(\frac{\partial P}{\partial x} - \frac{\partial Q}{\partial t}\right) = 0.$$

This is a partial differential equation in the unknown function $M(t, x)$, but we do not need to find its general solution. Indeed, any particular solution will do the desired job. By exercising some ingenuity, we should be able to find such a particular solution.

Example 8

Let

$$x\,dt - t\,dx = 0.$$

Here $P = x$ and $Q = -t$, and condition (1.3.4) is not satisfied. For the integrating factor $M(t, x)$, we get the partial differential equation

$$x\frac{\partial M}{\partial x} + t\frac{\partial M}{\partial t} + 2M = 0.$$

If, for example, we restrict our attention to an expression of the form $M = M(t)$, so that $\partial M/\partial x = 0$, $\partial M/\partial t = M'$, this equation becomes

$$tM' + 2M = 0,$$

from which $M'/M = -2t$, $\ln M = c - \ln t^2$, and $M = 1/t^2$ is a solution. Multiplying the equation by this factor, we obtain

$$\frac{x}{t^2}\,dt - \frac{dx}{t} = 0.$$

The potential function is $F(t, x) = -x/t$. Obviously, in this trivial case the variables could have been separated; it is interesting to note that this separation is equivalent to multiplying by the different integrating factor $1/xt$.

Problems

1. Solve the differential equations

(a) $y' = k$, (b) $y' = ky$,
(c) $y' = ky/x$, (d) $y' = kx/y$.

In the last two problems, note the particular cases $k = 1, 2, -1$.

(e) $y' = \dfrac{y-a}{x-b}$, (f) $y' = \dfrac{\sin x}{\sin y}$.

2. Solve $y' = x + y$. [*Hint*: use the substitution $u = x + y$.]
3. Solve $y' = (4x - y)^2$. [*Hint*: use the substitution $u = 4x - y$.]
4. Solve

$$y' = \frac{(x-y)^2 - (x+y)}{(x-y)^2 + (x+y)}.$$

[*Hint*: use the substitution $\xi = x + y$, $\eta = x - y$, find the differential equation for $\eta(\xi)$, solve it, and substitute back.]

5. Show that

$$y' = \frac{a_1 x + b_1 y + c_1}{a_2 x + b_2 y + c_2}$$

is transformed into a homogeneous equation by the substitution $x = \xi + \alpha$, $y = \eta + \beta$, where α, β are the solutions of the linear algebraic equations $a_i \alpha + b_i \beta + c_i = 0$, $i = 1, 2$, provided the determinant of this linear system is different from zero. If the determinant is zero, then the substitution $u = a_1 x + b_1 y$ makes the differential equation solvable by separation of variables.

6. Solve the differential equation

$$y' = \frac{x^2 + xy + y^2}{x^2}.$$

7. Solve the differential equation

$$y' = \frac{2y^2 + x^2}{yx}.$$

8. Solve the following differential equations:
   (a) $y' - \dfrac{3x^2}{x^3 - 1} y = 3x^2$,
   (b) $y' + y \tan x = \cos x$,
   (c) $y^2\, dx + (3xy - 1)\, dy = 0$.
9. Solve
   (a) $y' + xy - y^2 \sin x = 0$,
   (b) $y' + 2y - 2(\cos x + \sin x)\sqrt{y} = 0$.
10. Prove that if, in the Riccati equation

$$y' = p(x) y^2 + q(x) y + r(x),$$

a particular solution $y_1(x)$ is known, the substitution $y = y_1 + z^{-1}$ leads to a linear differential equation for $z(x)$, and hence the general solution $y(x)$ may be obtained in this way.

11. Applying this method, find the general solution of

$$y' = y^2 + (1 - 2x^2) y + x^4 - x^2 + 2x,$$

for which $y_1 = x^2$ is a particular solution.

12. Show that

$$(a_1x + b_1 y)\, dy - (a_2 x + b_2 y)\, dx = 0$$

is exact if and only if $a_1 + b_2 = 0$. Assuming this condition, integrate.

13. Show that $(\cos x)^{-2}$ is an integrating factor for

$$\sin x \cos x \cos y\, dy + (\sin y + \cos^2 x)\, dx = 0.$$

Integrate.

14. Show that e^{xy} is an integrating factor for

$$(x - e^{-xy} \sin y)\, dy + (y + e^{-xy} \cos x)\, dx = 0.$$

Integrate.

4 *Implicit Differential Equations*

4.1 Implicit equations

A differential equation is in the general implicit form if it is given by a function

$$F(t, x, x') = 0. \tag{1.4.1}$$

In general, for each pair t, x in some domain of the (t, x) plane, there is a set of values of x' satisfying the equation (1.4.1). In particular cases this set of values may be empty. Sometimes every value of x' will satisfy the equation. Ordinarily, however, there are a finite number of values x' satisfying (1.4.1).

4.2. Application of the implicit function theorem

The implicit function theorem in calculus states: if the triple (t_0, x_0, x_0') satisfies equation (1.4.1) and, in some neighborhood of this triple, $F(t, x, x')$ is continuous and has continuous partial derivatives, and if $\partial F/\partial x' \neq 0$, then a function $x' = f(t, x)$ exists which is continuous and differentiable in some neighborhood of (t_0, x_0) and which satisfies identically (1.4.1), such that $f(t_0, x_0) = x_0'$.

From an abstract point of view, one can "solve" equation (1.4.1) for x' as long as the conditions of this theorem are satisfied, and can obtain as many functions $f(t, x)$ as "roots" x_0' exist, thereby satisfying (1.4.1).

If the function $F(t, x, p)$ is a polynomial in p, for example, there will be, in general, n roots $p_1, \ldots, p_n$, which will change continuously with (t, x). Writing these roots as $p_i(t, x)$, the differential equation (1.4.1) will separate into the n differential equation [recall Section 4.1]

$$x' = p_i(t, x) \quad i = 1, 2, \ldots, n. \tag{1.4.1a}$$

Every solution of any one of the equations (1.4.1a) satisfies equation (1.4.1) because $F(t, x, p_i) = 0$ is precisely the condition defining $p_i(t, x)$.

Example 9

Let

$$x'^2 = 1.$$

This equation separates into

$$x' = 1 \quad \text{and} \quad x' = -1.$$

The respective solutions are

$$x = c + t \quad \text{and} \quad x = c - t.$$

This result can be generalized immediately to all equations of the type

$$F(x') = 0.$$

If p_i are the roots of the equation $F(p) = 0$, the differential equation is decomposed into the equations

$$x' = p_i = \text{const},$$

and solutions are

$$x = p_i t + c.$$

These solutions can be written in the following elegant but less clear manner. Solving the last equation for p_i, we get $p_i = (x - c)/t$, and remembering that p_i satisfies $F(p_i) = 0$, we may write

$$F\left(\frac{x - c}{t}\right) = 0$$

as a general expression for the solutions. In the last example this would be

$$\left(\frac{c - x}{t}\right)^2 = 1$$

If the initial condition here is $t_0 = 0$, $x_0 = 0$, the solutions through this point are given by

$$x^2 = t^2,$$

which is precisely the equation of the two lines of slopes ± 1 through the origin.

4.3 *A geometrical interpretation*

In the three-dimensional (t, x, p) space, we may consider the surface

$$F(t, x, p) = 0. \tag{1.4.2}$$

All solutions of the differential equation may be conceived as three-dimensional curves on the surface (1.4.2), for which $x' = p$. If the vertical line $t = t_0$, $x = x_0$ intersects the surface at the points $p = p_i$ and $\partial F/\partial p \neq 0$ at these points, then they are regular points of the surface, and the tangent plane at each of these points is not vertical. In this case, in some neighborhood of (t_0, x_0), a differential equation $x' = f(t, x)$ is defined on each sheet of the surface and then determines a family of curves lying on that surface.

Of particular interest are those points (t, x, p), where both equations (1.4.2) and

$$\frac{\partial F}{\partial p} = 0 \tag{1.4.3}$$

are satisfied. Assuming that these points constitute a curve, its projection on the (t, x) plane is called the *p-discriminant* curve, $D(t, x) = 0$, of (1.4.2). Those triples $(t, x, p = x')$, which satisfy (1.4.2) and (1.4.3), are called *singular*. A *singular solution* of the differential equation (1.4.1) is a solution for which all triples (t, x, x') are singular. In other words, along the singular solution $x(t)$, (1.4.3) is satisfied identically for $p = x'$. Hence every singular solution is contained in the p-discriminant curve. The converse is false; the p-discriminant curve does not necessarily satisfy the differential equation, although it is a good candidate for a solution.

Example 10

The differential equation

$$x^2 + x'^2 = 1$$

can be imagined as representing curves on the surface

$$x^2 + p^2 = 1, \tag{1.4.4}$$

which is a cylinder in the (t, x, p) space. The solutions are therefore restricted to the region $|x| \leqq 1$. At every interior point of this region, two values of x' are determined by the differential equation. At the boundary points, where $|x| = 1$, there is only one value of x', and for the exterior of this region, there is no value of x' satisfying the equation, hence no solution.

For $|x| < 1$, the equation resolves into

$$x' = +\sqrt{1 - x^2} \quad \text{and} \quad x' = -\sqrt{1 - x^2}.$$

The solutions are obtained by the separation of variables as $x = \sin(t + c)$ with $c =$ arbitrary constant.

Through each point (t_0, x_0) with $|x_0| < 1$, there are two such solutions. [See Figure 1.6.] All those curves are tangent to the lines $x = \pm 1$ which, as envelopes of solutions, are solutions themselves; in fact, they are singular solutions of the differential equation.

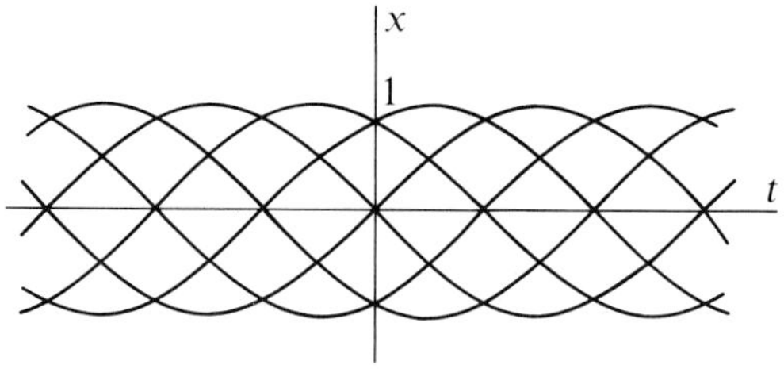

$x'^2 + x^2 = 1$

Figure 1.6

Note that on the surface (1.4.4) the solution curves are helices. Indeed, the equation of these curves in (t, x, p) space is obtained by differentiating (1.4.4)

$$xx' + pp' = 0,$$

with the condition that $p = x'$, $p' = -x$ holds whenever $p \neq 0$. From these last two expressions, we obtain

$$x'' = -x,$$

the equation of the harmonic oscillator. The solutions in the (t, x, p) space are, therefore,

$$x = \sin(t + c), \qquad p = \cos(t + c),$$

which are the above mentioned helices. Their projections on the (t, x) plane, obtained by setting $p = 0$, are the solutions of the original equation. The singular solutions appear as envelopes of these projections and have no counterpart on the surface (1.4.4).

4.4 Clairaut's equation

The equation

$$x = tx' + f(x') \tag{1.4.5}$$

is known as *Clairaut's equation*. Obviously, for $x' = c$ and $c =$ constant, the line

$$x = tc + f(c)$$

satisfies the equation. In the (t, x, p) space we may consider the surface

$$x = tp + f(p), \tag{1.4.6}$$

which, together with $x' = p$, is equivalent to (1.4.5). Differentiating (1.4.6), we obtain

$$p = p + p'(t + f'(p)),$$

or

$$p'(t + f'(p)) = 0.$$

There are two possibilities: either $p' = 0$, which implies $p = x' =$ constant (this gives the above mentioned straight lines), or

$$t + f'(p) = 0.$$

This last equation is precisely the condition satisfied by the envelopes of the lines (1.4.6), with $p =$ constant. Indeed, we know that in calculus such an envelope is obtained by eliminating the parameter p between equation (1.4.6) and the equation obtained by differentiating the latter with respect to p. This envelope of solutions is therefore the singular solution.

Example 11

Let

$$x = tx'.$$

The straight line solutions are

$$x = tc.$$

The singular solution is obtained by differentiating with respect to c as $0 = t$; hence $x = tc = 0$, and the envelope reduces to the origin.

Example 12

Let

$$x = 2tx' - x'^2.$$

The general solution is

$$x = 2ct - c^2.$$

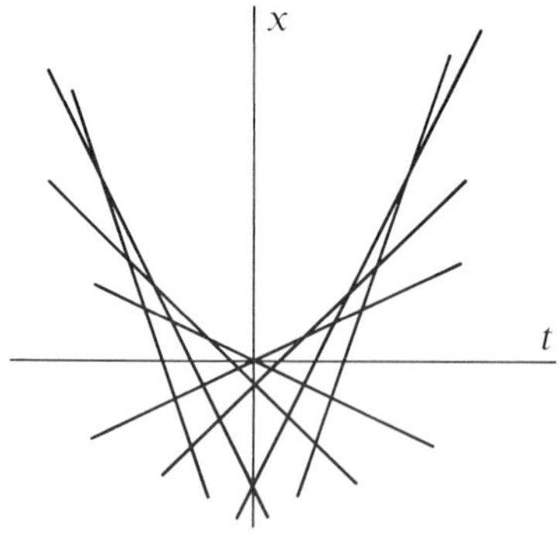

$x = 2tx' - x'^2$

Figure 1.7

To obtain the singular solution, we differentiate with respect to the parameter c

$$0 = 2t - 2c,$$

and eliminating c, we get the parabola $x = t^2$. [See Figure 1.7.]

Problems

1. Analyze the following differential equations and indicate
   (a) the set of points (x, y) where there is no solution,
   (b) the set of points (x, y) where there are several solutions (how many?).
   Then,
   (c) find the general solution of the differential equation,
   (d) find the discriminant curve,
   (e) see if the discriminant curve is or contains a singular solution.
   (a) $y'^2 = 1 - y$,
   (b) $y'^2 = (1 - y)^{-1}$,
   (c) $y'^2 = \dfrac{1}{y^2} - 1.$
2. Solve the following Clairaut's equations, finding the general and singular solutions:
   (a) $y = y'x - \frac{2}{3}y'^{3/2}$,
   (b) $y = y'x + \sqrt{1 + y'^2}$,
   (c) $y = y'x - x\sqrt{y'}$.

---

5 *Equations of Higher Order and Systems of Equations*

5.1 Special methods

The solution of the second-order differential equation

$$F(t, x, x', x'') = 0 \tag{1.5.1}$$

can be accomplished in an elementary fashion in only a few special cases, even when the equation is given in the explicit form

$$x'' = f(t, x, x'). \tag{1.5.2}$$

In this section we shall present some examples in which the solution can be obtained by special devices.

It should be noted that, theoretically, the problem of integrating a second-order differential equation is considered solved if it is reduced to solve a first-order differential equation—just as a first-order equation is considered solved if it is reduced to quadratures.

For a second-order equation, the general solution will contain two arbitrary constants. [The general solution of a first-order equation contained one such constant.] Here, the *initial value problem* is: for $t = t_0$, the values of $x(t_0) = x_0$ and $x'(t_0) = y_0$ are given, and these are precisely the two arbitrary constants determining the particular solution.

5.2 *Equation without* x

If the differential equation is

$$F(t, x', x'') = 0,$$

we take $p(t) = x'(t)$ as a new unknown function for which

$$F(t, p, p') = 0.$$

Solving this, we get $p = \varphi(t, c_1)$ and finally

$$x = \int \varphi(t, c_1)\, dt + c_2 .$$

5.3 *Equation without t*

If

$$F(x, x', x'') = 0,$$

we write $x' = p$, and if we imagine p as a function $p(x)$ of x, we obtain

$$x'' = \frac{dx'}{dt} = \frac{dp}{dx} \cdot \frac{dx}{dt} = \frac{dp}{dx} p.$$

The equation is thus transformed into a new one:

$$F_1(x, p, dp/dx) = 0.$$

This equation is of the first order; hence if $p = p(x, c)$ is its solution, then

$$\frac{dx}{dt} = p(x, c),$$

which can be integrated by the separation of variables.

Example 13

A point $x(t)$ describes a straight line with an acceleration which is a function of position alone

$$x'' = f(x).$$

With $p = dx/dt$, we obtain

$$p\,\frac{dp}{dx} = f(x).$$

By separation of variables,

$$\frac{p^2}{2} = \int f(x)\,dx + c,$$

or

$$\frac{dx}{dt} = \pm\sqrt{2\int f(x)\,dx + c}.$$

Separating variables again, we obtain

$$t = \int \frac{dx}{\sqrt{2\int f(x)\,dx + c}} + c'.$$

c and c' are the two arbitrary constants.

5.4 *Equation homogeneous in x, x', x''*

Let

$$F(t, x, x', x'') = 0$$

be such that $F(t, kx, kx', kx'') = k^n F(t, x, x', x'')$ for some real n and all t and k.

F is then said to be homogeneous of degree n in x, x', x''. We introduce a new unknown function $y(t)$ by the relation

$$x = \exp \int y(t)\, dt,$$

so that

$$x' = y(t) \exp \int y(t)\, dt$$

and

$$x'' = (y' + y^2) \exp \int y\, dt.$$

Substituting in the differential equation, we obtain

$$\left[\exp \int y\, dt\right]^n F(t, 1, y, y' + y^2) = 0.$$

Since the exponential function is never zero, we may disregard the first factor, leaving an implicit differential equation of the first order.

5.5 *Systems of differential equations*

We mentioned in Section 1.5 that every differential equation of order n can be reduced to a system of n equations of the first order by means of the substitution

$$x = x_1, \qquad x' = x_2, \qquad x'' = x_3, \ldots, x^{(n-1)} = x_n,$$

which transforms the equation

$$x^{(n)} = f(t, x, x', x'', \ldots, x^{(n-1)})$$

into the system

$$\begin{cases} x_1' = x_2 \\ x_2' = x_3 \\ \cdots\cdots\cdots \\ x_n' = f(t, x_1, x_2, \ldots, x_n). \end{cases}$$

If we disregard algebraic difficulties, we may state that conversely, every system of n first-order differential equations can be reduced to a single equation of order n. Given the n differential equations with n unknown functions,

$$F_i(t, x_1, x_2, \ldots, x_n, x_1', \ldots, x_n') = 0 \quad i = 1, 2, \ldots, n,$$

and differentiating each one $n - 1$ times, we obtain $n(n - 1)$ relations of the form

$$\Phi_j(t, x_1, x_1', \ldots, x_1^{(n)}, x_2, x_2', \ldots, x_2^{(n)}, x_3, \ldots) = 0.$$

Along with the original n equations, these constitute an algebraic system of n^2 equations. By purely algebraic operations, we can, in general, eliminate $n^2 - 1$ variables, obtaining one equation without them. If we choose to eliminate the variables

$$x_2, x_2', \ldots, x_2^{(n)}, x_3, x_3', \ldots, x_3^{(n)}, \ldots, x_n^{(n)},$$

which are exactly $(n - 1)(n + 1) = n^2 - 1$ in number, the resulting equation is of the form

$$G(t, x_1, x_1', \ldots, x_1^{(n)}) = 0,$$

and therefore the desired differential equation. Once this equation is integrated, the remaining functions $x_i(t)$ can be found by purely algebraic methods. In order to get x_2, for example, we should eliminate $x_2', x_2'', \ldots, x_2^{(n)}, x_3, x_3', \ldots, x_3^{(n)}, \ldots, x_n^{(n)}$, obtaining a relation

$$H(t, x_1, x_1', \ldots, x_1^{(n)}, x_2) = 0,$$

which we then solve for x_2.

All these processes of algebraic elimination of variables between several relations can be performed *in general*. By this we mean that if the conditions of a suitable implicit function theorem are satisfied, those resulting functions exist. The actual algebraic computation may certainly be prohibitive in length, or they may not be solvable in terms of finite expressions. If the conditions of the implicit function theorem are not satisfied, the given system is "degenerate," and there may be incompatibilities or indeterminacies. In such cases, the system should first be carefully analyzed from an algebraic viewpoint.

If the functions $F_i(t, x_1, \ldots, x_n)$ are not continuously differentiable, then even from the theoretical viewpoint the equivalence between the system and the single equation of order n is no longer true.

In physics and engineering applications, the problems are often formulated in such a way that there is only one unknown function, for whose determination there is an nth-order differential equation. Hence single equations are very important.

On the other hand, in the theoretical development of differential equations, it is usually simpler to assume the equations in the form of a system of first-order equations, which we may even assume to be in the explicit form. The system

$$x_i' = f_i(t, x_1, x_2, \ldots, x_n), \quad i = 1, 2, \ldots, n$$

can then be written in vector notation as

$$\mathbf{x}' = \mathbf{f}(\mathbf{t}, \mathbf{x}),$$

where the letter $\mathbf{x}$ now stands for the vector $(x_1, x_2, \ldots, x_n)$, similarly $\mathbf{f} = (f_1, f_2, \ldots, f_n)$. This vector notation is not only a shorthand which saves writing, but it also has the intrinsic vector meaning which appeals strongly to the intuition. For this reason we prefer, in general, to use the vector notation when dealing with systems.

Problems

In the following problems we will return to our standard notation: t is the independent variable.

1. Integrate these differential equations.
   (a) $x'' = g - kx'$ [differential equation of a falling body, g = acceleration of gravity, k = coefficient of friction due to air resistance].
   (b) $x'' = \dfrac{x'}{t}$.
   (c) $x'' = \dfrac{t}{x'}$.
   (d) $x'' = -2tx'$.
   (e) $x'' = -k \sin x$ [differential equation of the frictionless pendulum]. The second integration cannot be expressed in terms of elementary functions, so leave the solution as an indefinite integral, i.e., reduce the problem to a quadrature.
   (f) $x'' = x$.
   (g) $x'' = x^{-3}$.
2. Reduce this differential system for $x(t)$, $y(t)$, ... to a single differential equation of higher order:
   (a) $x' = x - y, \quad y' = x + y$.
   (b) $x' = y - z, \quad y' = z - x, \quad z' = x - y$.
   (c) $x' = x^2 - y, \quad y' = x^2 + y$.
   (d) $x' = e^y, \quad y' = e^x$.
3. The radioactive element X disintegrates in such a way that if $x(t)$ atoms are present at time t, in the time interval dt, $k_1 x$ atoms disappear, thus producing $k_1 x$ atoms of element Y. Element Y also disintegrates, so that if $y(t)$ atoms are present, $k_2 y$ atoms disintegrate in the time interval dt. Write the two differential equations for $x(t)$, $y(t)$ and solve them for $x(0) = x_0$, $y(0) = 0$. What will be the maximum amount of element Y present at any time? Generalize for a disintegration chain of more than two elements.

6 *Approximate Methods of Solution*

6.1 In only a few cases can the solutions of a differential equation be written in closed form, that is, written as a finite expression involving only elementary functions with a finite number of algebraic operations and integrations. In fact, it can be proved that no such closed-form solution exists for many differential equations of great applications importance.

Thus we are led to consider computational methods, which give the numerical values of a (particular) solution, or analytical methods, which use infinite algorithms such as power-series expansions, trigonometric series, integral transforms, and so on. All these methods are of paramount importance in the applied sciences.

6.2 Numerical methods

In many applications we are only interested in finding the particular solution of a given differential equation determined by initial or equivalent conditions. This is true in astronomy, for example, when we want to chart the motion of the planets or predict eclipses. In order to know the function $x(t)$, in most cases it will be enough to know the numerical values of this function for a sufficiently numerous set $t = t_1, t_2, t_3, \ldots$. This is, by the way, how we know values of the logarithmic or trigonometric functions from tables. Numerical methods aim to obtain such a sequence of values with the best possible accuracy and the least computational effort, taking into consideration the available computer hardware.

In the last century numerical methods were already developed and widely used in connection with "hand" computation. The recent availability of electronic digital computers created added interest in numerical methods; many calculations which were formerly considered too lengthy and therefore disregarded are now within easy reach. However, it is not within the scope of this book to go into the details of numerical methods.

6.3 Graphical methods

Graphical methods replace the numerical computation of a solution by the geometrical construction of the corresponding solution curve. The degree of accuracy is, in general, notably lower. This is why graphical methods are mostly disregarded in "serious" mathematical books, but they enjoy some popularity in engineering and the applied sciences. The relatively low degree of approximation of these methods is just right for many practical problems. The method becomes especially attractive if the integration of the differential equation is just one step within a completely graphical procedure for solving an entire problem. Graphical methods are therefore best studied in relation to specific applications.

6.4 Physical methods—analog computers

Just as the graphical methods make use of geometry in order to approximate solution curves of a differential equation, the laws of mechanics or physics can be applied in general. Closely related are the mechanical instruments used in finding the integral of a given function. And by incorporating devices for addition, subtraction, and multiplication in an appropriate way, we may obtain a physical system whose behavior is described by a given differential equation or system of differential equations. Hence, by measuring the behavior of the system (recording, for example, x as a function of time on a graph), it is possible to obtain the particular solution $x(t)$.

A set of blocks, each able to perform a given operation (addition, multiplication, integration, etc.), and interconnecting devices appropriate for coupling these blocks together is usually called an *analog computer*. Analog computers acting mechanically are curiosities, but electronic analog computers are effectively used in actual computations. Analog computers are usually constructed to solve a restricted class of differential equations. The design and use of a particular computer, therefore, is intimately related to the field of its applicability. It can be extremely helpful in studying the behavior of a certain system (described by some differential equation) under different initial conditions, under slight changes of some coefficients, or under some other perturbations.

7 *Qualitative Methods*

7.1 If an expression for the general solution of a differential equation is known, practically all interesting questions concerning the behavior of the solutions can be studied with the aid of this expression. Such problems are usually the boundedness of solutions, existence of constant solutions (rest points), stability of such rest points, existence of periodic solutions, and so on. Even so, when the expression for the general solution becomes very complicated, the study of such properties from this expression may be very cumbersome, and sometimes it is easier to obtain some properties of solutions from a study of the differential equation itself.

In fact, as we have already noted, most equations do not have general solutions expressible in closed form. Hence it is very important to be able to derive properties of the solutions of a differential equation without having to "solve" it.

It should be noted that numerical methods, of paramount importance when our interest lies in only one particular solution (defined, for example, by initial conditions), are of relatively little value when it comes to ascertaining properties *of all* solutions (such as boundedness). In order to get even a faint idea of the general behavior of all solutions, we would have to compute a large number of particular solutions, and even then, we could only formulate a good guess concerning the behavior of solutions in general.

Historically, two names are closely related to the birth and development of the ideas leading to the qualitative methods, *Henri Poincaré* and *A. M. Liapunov*. Poincaré developed, within the theory of differential equations, some problems related to topology. We will examine such typical problems in Chapters 6, 7, and 8. Liapunov gave the solid foundation of a theory of stability, which we will study in Chapter 10.

References

More details on direct methods of integration of particular differential equations, together with the general theory, can be found in the classic

E. L. Ince. *Ordinary Differential Equations*. New York: Dover, 1926 (1st ed.), 1956.

The most encyclopedic collection of particular differential equations with their solutions and/or known properties, preceded by a thorough theoretical introduction, is given in

E. Kamke. *Differentialgleichungen, Lösungsmethoden, und Lösungen*, Volume I of *Gewöhnliche Differentialgleichungen*. Leipzig: Akademischer Verlag, 1959 (6th ed.).

Numerical and graphical solutions can be found in

F. A. Willers. *Practical Analysis, Graphical and Numerical Methods*. New York: Dover, 1948.

For numerical methods in particular, a good reference is

L. Collatz. *Numerical Treatment of Differential Equations*, in the series *Grundlehren der Mathematikche Wissenschaft*. Berlin: Springer, 1960.

For analog computers and similar devices, we refer the reader to

W. W. Soroka. *Analog Methods in Computation and Simulation*. New York: McGraw-Hill, 1954.

2

Existence and Uniqueness Theorems

1 Statement of the Problems

1.1 In order to avoid purely technical difficulties, we will assume that the differential equation is given in the explicit form (1.1.8) or (1.1.9), for a system. Initially, we will consider the simplest case of a first-order equation

$$x' = f(t, x). \tag{2.1.1}$$

As noted in Section 1.1.5, an nth-order equation can be written as a system of first-order equations, and such a system can, in turn, be written in the vector form (1.1.10)

$$\mathbf{x}' = \mathbf{f}(t, \mathbf{x}),$$

which has the same formal appearance as (2.1.1). Once the questions of existence and uniqueness of solutions for equation (2.1.1) are examined, the theorems and methods of proof can be easily applied to the vector equation representing a system of differential equations. Hence we will concentrate most of our attention on the first-order equation (2.1.1) and only later, in Sections 7 and 8, consider systems.

1.2 Differential and integral equations

Given the first-order differential equation (2.1.1)

$$x' = f(t, x)$$

with the initial condition

$$x(t_0) = x_0, \tag{2.1.2}$$

the basic problem is to find a solution, that is, a function $x(t)$ defined on some interval containing t_0, satisfying (2.1.2), such that its derivative exists and satisfies (2.1.1) at every point of this interval.

Obviously, this is not always possible. For example, if we take for $f(t, x)$ in (2.1.1) the *Dirichlet function*, defined as 1 if t is rational and 0 if t is irrational, there is no hope of finding $x(t)$ satisfying (2.1.1). [The theory of the Lebesgue integral shows that the indefinite integral of the Dirichlet function is a constant. By again differentiating this integral, we obtain derivative zero, which coincides with the requirement at the irrational points only.]

Let us assume that $f(t, x)$ is defined and continuous on some domain D of the t, x plane. Assume also that (t_0, x_0) belongs to D. We will show that in this case the existence question can be answered in the affirmative: there exists a function $x(t)$, defined on some open t interval including the point t_0, such that the derivative $x'(t)$ exists and is continuous, and relations (2.1.1) and (2.1.2) are satisfied.

This solution $x(t)$ therefore satisfies the integral relation

$$x(t) = x_0 + \int_{t_0}^{t} f(s, x(s))\, ds \tag{2.1.3}$$

on the given t interval.

Conversely, any continuously differentiable function $x(t)$ satisfying (2.1.3) also satisfies both relations (2.1.1) and (2.1.2). The integral equation (2.1.3) is therefore equivalent to the differential equation (2.1.1) plus the initial condition (2.1.2).

1.3 Generalized solutions of a differential equation

The integral equation (2.1.3) can be satisfied by a function $x(t)$ defined on a t interval, which is not differentiable at each point of the interval; an example would be a function whose graph has a corner (a point of discontinuity of the derivative). From an applications viewpoint, there is nothing wrong in admitting such functions $x(t)$. In fact, it is necessary to do so if we want some practical problems to have solutions. This leads us to introduce the idea of generalized solutions.

We will call $x(t)$ a *generalized solution* of (2.1.1) and (2.1.2) if it satisfies the integral equation (2.1.3). It can be shown that this is equivalent to requiring that $x(t)$ be absolutely continuous and satisfy, besides condition (2.1.2), the differential equation (2.1.1) at "almost all" points t of the interval.

Note 1: The expression "for almost all values of t" means for all values of t except possibly a set of (Lebesgue) measure zero. A set on the line has (Lebesgue) measure zero if it can be covered by a collection of intervals having an arbitrarily small total length. It should be mentioned that changing the values of a function at the points t belonging to a set of measure zero does not affect the integrability of the function nor does it change the value of its integral. An example of a set of measure zero is any countable set of points.

Note 2: A function $x(t)$ is absolutely continuous on a finite interval I, if for every $\varepsilon > 0$ there is a $\delta > 0$ such that for any finite set $I_1, I_2, \ldots, I_n$ of nonoverlapping intervals of I, where the total length $(I_1 + I_2 + \cdots + I_n)$ is less than δ, the sum of the oscillations of $x(t)$ in the intervals I_i is not greater than ε. Hence

$$\sum_{i=1}^{n} \max \{x(t_i') - x(t_i'') \,|\, t_i', t_i'' \in I_i\} \leqq \varepsilon.$$

It can be proved that every absolutely continuous function has a derivative at all points t in the interval I, with the possible exception of a set of points of (Lebesgue) measure zero. Furthermore, the function $x(t)$ is equal to the integral of its derivative.*

The importance of considering generalized solutions will appear when we consider the case where $f(t, x)$ in (2.1.1) is discontinuous (a discontinuous differential equation). As long as $f(t, x)$ is continuous, however, the derivative $x'(t)$ must also be continuous, with solutions and generalized solutions being identical.

1.4 Uniqueness, continuation, and dependence of solutions on initial conditions and parameters

Once the problem of existence of a solution is settled, the next thing to consider is its uniqueness. In practical terms, if we know that some given physical system is described by a differential equation but that this equation has several solutions, we still do not know which one will represent the true behavior of the physical system.

Another problem concerning a solution is its *interval of existence*. A classical existence theorem asserts that a solution exists in some neighborhood of the initial point t_0, that is, in some interval (t_1, t_2), with $t_1 < t_0 < t_2$. Nothing is said, however, about the actual length of this interval. It is therefore important to give criteria for how far a solution can be extended, while always satisfying the differential equation, of course.

Finally, in practical problems the initial conditions and also some parameters of the differential equation (its coefficients, for example) are subjected to small measurement errors. In order for the solutions to have a practical meaning, a slight change in the initial data or parameters should result in a correspondingly small change in the solution. This means the solution should depend continuously on those data.

---

* For proof and more details, see the standard textbooks for functions of a real variable, particularly H. L. Royden, *Real Analysis*, New York: Macmillan, 1968 (2nd ed.), Chapters 3–5.

Problems

1. Write the differential equations and initial conditions equivalent to
   (a) $x(t) = 1 + \int_0^t sx^2(s)\,ds,$ (b) $x(t) = a + bt + \int_1^t x(s)\,ds,$

   (c) $tx(t) = 2 + \displaystyle\int_1^t \frac{x(s)}{s}\,ds \quad t > 0,$ (d) $\displaystyle\frac{x(t)}{t} = \int_2^t \frac{x(s)}{s^2}\,ds \quad t > 0.$

2. Prove that the second-order differential equation $x'' = f(t, x)$, with $x(0) = a$, $x'(0) = b$, is equivalent to

$$x(t) = a + \int_0^t \left\{ b + \int_0^s f(r, x(r))\,dr \right\} ds,$$

   and hence $x(t) = a + bt + \int_0^t (t - r) f(r, x(r))\,dr$.

2 *Existence Theorem*

2.1 Cauchy-Peano Theorem

If $f(t, x)$ is defined and continuous on the rectangle

$$\mathscr{D}: |t - t_0| \leqq a, \qquad |x - x_0| \leqq b, \tag{2.2.1}$$

and if in this rectangle the inequality

$$|f(t, x)| \leqq b/a \tag{2.2.2}$$

is satisfied, then there exists a function $x(t)$, defined in the interval $\mathscr{I}: |t - t_0| \leqq a$, with a continuous derivative, satisfying the differential equation (2.1.1) *and the initial condition* (2.1.2).

Condition (2.2.2) prevents the solution $x(t)$ from leaving the rectangle $\mathscr{D}$ through its upper or lower edge, thereby penetrating into a region where the differential equation is not defined. In this case the solution $x(t)$ would not be defined in the whole interval $\mathscr{I}$. This follows from the fact that $f(t, x)$ is the slope of $x(t)$.

The given statement of the existence theorem is equivalent to the following:

If $f(t, x)$ is defined, continuous, and bounded on the strip

$$\mathscr{D}': |t - t_0| \leqq a \qquad -\infty < x < +\infty, \tag{2.2.3}$$

then there exists a function $x(t)$, defined on the interval $\mathscr{I}: |t - t_0| \leqq a$, having a continuous derivative and satisfying (2.1.1) *and* (2.1.2).

In order to show the equivalence of both statements, let us assume that the first statement is true. In this case the second one holds also since to prove the existence of the solution under its assumptions requires only a restriction of one's attention to a rectangle of sufficient height. Because $f(t, x)$ is bounded, we may indeed write $|f(t, x)| \leqq M$, set $b/a = M$ (i.e., define $b = aM$), and the conclusion follows from the first statement.

If we assume the second statement to be true and want to prove the first one, we encounter immediate difficulty because by the first assumption, the function $f(t, x)$ is assumed to be defined only in the rectangle $\mathscr{D}$. In this case, we extend the definition of the function $f(t, x)$ to the whole strip $\mathscr{D}'$ by

$$f(t, x) = f(t, x_0 + b) \quad \text{if} \quad x > x_0 + b,$$

$$f(t, x) = f(t, x_0 - b) \quad \text{if} \quad x < x_0 - b.$$

The extended function $f(t, x)$ is still continuous and bounded. Hence we may apply the second statement proving the existence of the solution $x(t)$ in the whole interval $\mathscr{I}$. But from the assumption (2.2.2) of the first statement, this solution cannot leave the rectangle $\mathscr{D}$ on its upper or lower border; so it stays within this rectangle for all $t \in \mathscr{I}$, and the first statement is proved.

2.2 ε-approximate solutions

As an auxiliary tool for proving the existence of solutions and other theorems, we now introduce a new concept.

A function $u(t)$ is called an ε-approximate solution of equation (1.1.1) *in the interval $\mathscr{I}$ if it is absolutely continuous in this interval and if for almost all values of t (that is, for all values of t except possibly a set of (Lebesgue) measure zero), the condition*

$$|u'(t) - f(t, u(t))| \leqq \varepsilon \tag{2.2.4}$$

is satisfied.

As an example of ε-approximate solutions, we will construct a polygonal line approximating the graph of a solution, under the assumptions of the existence theorem.

As $f(t, x)$ is continuous on the compact domain $\mathscr{D}$, it is uniformly continuous and for every $\varepsilon > 0$ there is a $\delta > 0$ such that, as long as all considered values belong to $\mathscr{D}$,

$$|t - t_1| \leqq \delta, \qquad |x - x_1| \leqq \delta$$

imply

$$|f(t, x) - f(t_1, x_1)| < \varepsilon. \tag{2.2.5}$$

Again by compactness of $\mathscr{D}$, the continuous function $f(t, x)$ is bounded on $\mathscr{D}$: $|f(t, x)| \leqq M \leqq b/a$ by (2.2.2). Taking

$$0 < h \leqq \min(\delta, \delta/M),$$

the inequality (2.2.5) will be satisfied whenever

$$|t - t_1| \leqq h, \qquad |x - x_1| \leqq Mh.$$

In order to construct the ε-approximate solution, which we will call $u_\varepsilon(t)$, we now start subdividing the interval $\mathscr{I}$ with the points $\cdots < t_{-1} < t_0 < t_1 < t_2 \cdots$ in such a way that each subinterval has length no greater than h.

Starting at the point (t_0, x_0), we determine in each subinterval (t_{i-1}, t_i) a straight line segment (linear function) of slope $f(t_{i-1}, u_{i-1})$ if $i > 0$, or $f(t_i, u_i)$ if $i - 1 < 0$. In this way we obtain a polygonal line of vertices (t_i, u_i). This polygonal line cannot leave the rectangle $\mathscr{D}$ on its upper or lower border because its slope is never greater than M in absolute value. Hence the function $u_\varepsilon(t)$ satisfies (2.2.4). It is therefore an ε-approximate solution.

2.3 *Approximating sequence*

In order to prove the theorem, we will consider a sequence $\varepsilon_i \to 0$. For each ε_i consider an approximate solution $u_{\varepsilon_i}(t)$ determined by the method explained above.

In Figure 2.1 the approximations to the solution of $x' = e^{-x}$, which passes through the origin, are given for $h = 1, 1/2$. The exact solution $x = \ln(1 + t)$ is also indicated.

That the successive approximations do not always tend to the solutions monotonically is shown in Figure 2.2, where a similar graph is represented for the equation $x' = 1 - 2x$, with the exact solution $x = (1 - e^{-2t})/2$.

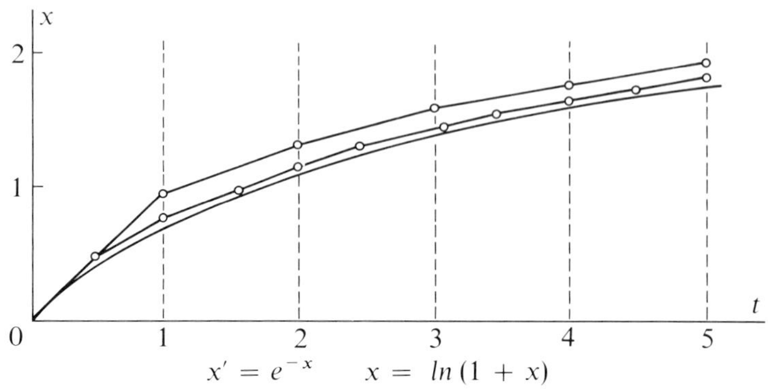

Figure 2.1

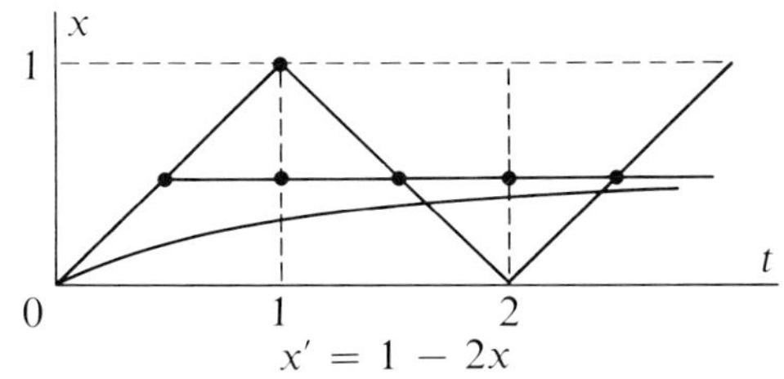

Figure 2.2

Once it is clear that such an approximating sequence of ε-solutions can be constructed, we should consider its possible limit for $\varepsilon_i \to 0$. In general it is not true that the sequence of approximating functions $u_{\varepsilon_i}(t)$ tends to a limit function, but we will see that this can be proved true if we restrict our attention to some properly chosen subsequence. This will enable us to obtain a limit function which can then be proved to satisfy the differential equation.

2.4 *Arzelá-Ascoli's Lemma*

Let

$$\{u_n(t) \mid n = 1, 2, \ldots\}$$

be a sequence of uniformly bounded and equicontinuous functions defined in the finite interval $\mathscr{I}$. Then there exists a subsequence

$$\{u_{n_i}(t) \mid i = 1, 2, \ldots\}$$

which, in the interval, converges uniformly to some limit function.

Note: That the first sequence is uniformly bounded means that there is a constant M such that for all $n = 1, 2, \ldots$ and all $t \in \mathscr{I}$ the inequality $|u_n(t)| \leqq M$ holds. That the sequence is equicontinuous means that for any $\varepsilon > 0$ there is a $\delta > 0$ such that $t_1, t_2 \in \mathscr{I}$, $|t_1 - t_2| < \delta$, imply $|u_n(t_1) - u_n(t_2)| < \varepsilon$ for all $n = 1, 2, \ldots$.

To prove this lemma, we must first take a sequence of values $t_0, t_1, t_2, \ldots$ belonging to $\mathscr{I}$, such that the set $\{t_i | i = 1, 2, \ldots\}$ is dense in $\mathscr{I}$. [This means that in every subinterval of $\mathscr{I}$, no matter how small, there is some t_i.] If the interval $\mathscr{I} = [0, 1]$, this sequence may be, for example, 0, 1, 1/2, 1/4, 3/4, 1/8, 3/8, ..., i.e., the set of binary rationals $p/2^q$ in [0, 1].

Next we consider the values of the functions at $t = t_0$, that is, $\{u_n(t_0)\}$. Because the functions are uniformly bounded, this is a bounded set. Therefore it has some limit point $U(t_0)$, and we can select a subsequence $u_{01}, u_{02}, u_{03}, \ldots$ such that

$$\lim_{i \to \infty} u_{0i}(t_0) = U(t_0).$$

For $t = t_1$, the values of these functions, $u_{0i}(t_1)$, constitute a bounded set, and we can again select a converging subsequence which we now call $u_{11}, u_{12}, u_{13}, \ldots$. For this new subsequence we have, therefore,

$$\lim_{i \to \infty} u_{1i}(t_0) = U(t_0), \qquad \lim_{i \to \infty} u_{1i}(t_1) = U(t_1);$$

the first of these limits follows from the fact that this sequence is a subsequence of the previous one.

Next we select a subsequence of the last one, which also converges at $t = t_2$, and using this procedure we obtain successive subsequences

$$\begin{array}{llll} u_{01}(t), & u_{02}(t), & u_{03}(t), & \cdots \\ u_{11}(t), & u_{12}(t), & u_{13}(t), & \ldots \\ u_{21}(t), & u_{22}(t), & u_{23}(t), & \ldots \\ \multicolumn{4}{c}{\cdots\cdots\cdots\cdots\cdots\cdots} \end{array}$$

From its construction it follows that the subsequence $u_{k1}(t), u_{k2}(t), u_{k3}(t), \ldots$ converges for $t = t_0, t_1, \ldots, t_k$ to the respective values $U(t_0), U(t_1), \ldots, U(t_k)$.

Now consider the *diagonal sequence*

$$u_{11}(t), \quad u_{22}(t), \quad u_{33}(t), \quad \ldots.$$

It converges for all values $t = t_0, t_1, t_2, \ldots$ because, except for a finite number of terms, it is a subsequence of every subsequence previously constructed. This method of proof, devised by Georg Cantor, is known as the *diagonal method.* Cantor used it to prove that the real numbers are not countable.

We have thus determined a subsequence (the diagonal one) which converges on the dense set $t_0, t_1, t_2, \ldots$. Renumbering the functions, we may call this subsequence simply $u_0(t), u_1(t), u_2(t), \ldots$. We will prove that this sequence not only converges for the mentioned values of t but that it also converges uniformly in the whole interval $\mathscr{I}$. To prove this we will use Cauchy's criterion which states: a necessary and sufficient condition for the sequence $\{u_n\}$ to be convergent is that for every $\varepsilon > 0$ there is an N such that $n > N$, $m > N$ imply $|u_n - u_m| < \varepsilon$.

Since the functions $u_n(t)$ are equicontinuous, for every $\varepsilon > 0$ there is a $\delta > 0$ such that if $\tau_1, \tau_2 \in \mathscr{I}$, then $|\tau_1 - \tau_2| < \delta$ implies

$$|u_n(\tau_1) - u_n(\tau_2)| < \varepsilon/3$$

for all values of n.

Now we subdivide the interval $\mathscr{I}$ in subintervals $\Delta_1, \Delta_2, \ldots, \Delta_p$ of length less than δ. In each Δ_k we choose a value t_k from the sequence $t_0, t_1, \ldots$, as determined above. This is possible because the sequence is dense in $\mathscr{I}$.

On this finite set $\{t_{k'}\}$ the sequence $u_n(t)$ converges; hence, given the value of $\varepsilon > 0$, there is an N such that $n > N$, $m > N$ imply $|u_n(t_{k'}) - u_m(t_{k'})| < \varepsilon/3$ for all $t_{k'}$, $k = 1, 2, 3, \ldots, p$.

Every value $t \in \mathscr{I}$ belongs to some interval Δ_k. We can therefore write for the corresponding $t_{k'}$

$$\begin{aligned}|u_n(t) - u_m(t)| \leqq |u_n(t) - u_n(t_{k'})| &+ |u_n(t_{k'}) - u_m(t_{k'})| \\ &+ |u_m(t_{k'}) - u_m(t)| \leqq 3\varepsilon/3 = \varepsilon.\end{aligned}$$

By Cauchy's criterion we see that the sequence $u_n(t)$ converges to a limit function $U(t)$ for all values of $t \in \mathscr{I}$. This convergence is uniform because in the last inequality we may pass to the limit as $m \to \infty$, thereby obtaining

$$\varepsilon \geqq \lim_{m \to \infty} |u_n(t) - u_m(t)| = |u_n(t) - U(t)|$$

uniformly for all $t \in \mathscr{I}$, as long as $n > N$.

The uniform convergence implies that the limit function $U(t)$ is continuous on the interval $\mathscr{I}$.

2.5 *Proof of the existence theorem*

Under the hypotheses of the theorem we constructed a sequence of ε_i-approximate solutions u_{ε_i} for a sequence $\varepsilon_i \to 0$. By Lemma 2.4 we can select a uniformly converging subsequence $u_i(t)$. Renumbering the subscripts, if necessary, we may assume that $u_i(t)$ is an ε_i-approximate solution, where $\varepsilon_i \to 0$ as $i \to \infty$. For each i, $u_i(t_0) = x_0$ and for all $t \in \mathscr{I}$,

$$u_i(t) = x_0 + \int_{t_0}^{t} [f(s, u_i(s)) + w(s)]\, ds,$$

where $|w(s)| \leqq \varepsilon_i$ for all s between t_0 and t. From this we obtain

$$|u_i(t) - x_0 - \int_{t_0}^{t} f(s, u_i(s))\, ds| \leqq \varepsilon_i |t - t_0|.$$

As the function $u_i(t)$ converges uniformly, the same is true for $f(t, u_i(t))$. We may therefore take the limit under the integral sign, thereby obtaining

$$U(t) = x_0 + \int_{t_0}^{t} f(s, U(s))\, ds.$$

We have already seen in Section 2.1.2 that this integral equation is equivalent to the differential equation at "almost all" points $t \in \mathscr{I}$. But then, by continuity, dU/dt exists at all points and is equal to $f(t, U(t))$ without exception. This concludes the proof of the existence theorem.

Problems

1. Prove that if the functions $f_n(t)$, $n = 1, 2, \ldots$, have a uniformly bounded derivative, $|f_n'(t)| < K$ for all t and all n, then this family of functions is equicontinuous.
2. Prove that the family $f_n(t) = 1/n \sin(n^2 t)$, $0 \leqq t \leqq 2\pi$, $n = 1, 2, \ldots$, is equicontinuous in spite of the fact that the functions $f_n(t)$ do not have uniformly bounded derivatives.
3. Prove that the sequence $f_n(t) = \sin nt$, $n = 1, 2, \ldots$, is bounded but does not have any pointwise convergent subsequence. Is it equicontinuous?
4. Prove that the family $f_n(t) = t^n$, $n = 1, 2, \ldots$, is equicontinuous in every t interval $[0, a]$, with $0 < a < 1$, but not in $[0, 1]$. What is the limit function in $[0, 1]$?

---

3 *Uniqueness*

3.1 In the proof of the existence theorem, an arbitrary selection of the approximating polygonal lines is made in order to obtain the converging subsequence. Additional arbitrary steps are used in its construction: the way the interval is subdivided in order to determine the polygonal approximations and certainly the actual method of choosing precisely polygonal lines as ε-approximations. It is reasonable to expect that if we change any of these arbitrary choices, we may be able to obtain a different limit function which satisfies the differential equation. In general this is indeed the case, and if we do not impose any further restrictions, there is the possibility of more than one solution satisfying both the differential equation and the initial condition.

Example 1

Consider the initial value problem

$$x' = \sqrt[3]{x}, \qquad x(0) = 0.$$

By separation of variables the solution $x = (2t/3)^{3/2}$ is easily obtained for $t \geqq 0$. Besides, the function $x(t) = 0$ is obviously a solution. Hence the solution is not unique. It is interesting to note that the polygonal method of approximating the solution of this initial value problem will yield the identically zero solution.

3.2 Lipschitz condition

The function $f(x)$ is said to satisfy a Lipschitz condition if there exists a constant k (called a Lipschitz constant) such that for any two values x_1, x_2 of the domain of definition of $f(x)$,

$$|f(x_1) - f(x_2)| \leqq k|x_1 - x_2| \tag{2.3.1}$$

holds.

The Lipschitz condition implies the continuity of $f(x)$ but not necessarily its differentiability. For example, the function defined by $f(x) = x \sin 1/x$ for $x \neq 0$, $f(0) = 0$ satisfies a Lipschitz condition, or is said to be "lipschitzian," but it is not differentiable at the origin.

The function $f(x) = +\sqrt{x}$ defined for $x \geqq 0$ is not lipschitzian at $x = 0$; neither is the function $\sqrt[3]{x}$ used in the last example.

3.3 Gronwall's Lemma

Let $z(t)$ be continuous in the interval $[t_0, t_1]$, where it satisfies the integral inequality

$$0 \leqq z(t) \leqq a + \int_{t_0}^{t} (kz(s) + b)\, ds, \tag{2.3.2}$$

a, b, k being nonnegative constants and $t_0 \leqq t \leqq t_1$. Then

$$z(t_1) \leqq [a + b(t_1 - t_0)]e^{k(t_1 - t_0)}. \tag{2.3.3}$$

Proof

Consider the function $g(t) = z(t)e^{-k(t-t_0)}$, which is continuous on the interval $[t_0, t_1]$. Let m be the maximum of $g(t)$ in this interval, and let $m = g(t')$ for some $t' \in [t_0, t_1]$. Then

$$\begin{aligned} z(t') &= g(t')e^{k(t'-t_0)} = me^{k(t'-t_0)} \\ &\leqq a + \int_{t_0}^{t'} (kme^{k(s-t_0)} + b)\, ds \\ &= a + b(t' - t_0) + m(e^{k(t'-t_0)} - 1). \end{aligned}$$

Subtracting $me^{k(t'-t_0)}$ from the third and last expression, we obtain

$$0 \leqq a + b(t' - t_0) - m,$$

or

$$m \leqq a + b(t' - t_0).$$

Finally,

$$\begin{aligned} z(t_1) &= g(t_1)e^{k(t_1-t_0)} \leqq me^{k(t_1-t_0)} \\ &\leqq [a + b(t_1 - t_0)]e^{k(t_1-t_0)}. \end{aligned}$$

3.4 Approximation Theorem

Assume that on the rectangle

$$\mathscr{D}: |t - t_0| \leqq a, \qquad |x - x_0| \leqq b,$$

the function $f(t, x)$ is defined, continuous, and satisfies a Lipschitz condition with respect to x; that is, there exists a constant k such that

$$|f(t, x_1) - f(t, x_2)| \leqq k|x_1 - x_2| \tag{2.3.4}$$

for every (t, x_1), $(t, x_2) \in \mathscr{D}$. Let $u_0(t)$ and $u_1(t)$ be functions defined in $|t - t_0| \leqq a$ with values in $|u - x_0| \leqq b$. Assume that $u_0(t)$ is continuously differentiable and satisfies the differential equation (2.1.1) *and* *the initial condition (2.1.2). Also assume that $u_1(t)$ is an ε-approximate solution; i.e., it is absolutely continuous and satisfies*

$$|u_1'(t) - f(t, u_1(t))| \leqq \varepsilon \tag{2.3.5}$$

for almost all values of t in the interval, together with the initial condition

$$|u_1(t_0) - x_0| \leqq \delta. \tag{2.3.6}$$

Then

$$|u_1(t) - u_0(t)| \leqq \delta e^{k|t-t_0|} + \varepsilon |t - t_0| e^{k|t-t_0|}. \tag{2.3.7}$$

Proof

We may assume that $t > t_0$; the other case is proved in a similar fashion. From the condition that $u_0(t)$ is an exact solution and $u_1(t)$ an ε-approximate one, we get

$$|u_1'(t) - f(t, u_1(t)) - u_0'(t) + f(t, u_0(t))| \leqq \varepsilon.$$

Integrating,we obtain

$$\left| u_1(t) - u_1(t_0) - u_0(t) + u_0(t_0) - \int_{t_0}^{t} [f(s, u_1(s)) - f(s, u_0(s))] \, ds \right|$$
$$\leq \varepsilon(t - t_0).$$

This follows from the well-known fact that for any integrable function $g(t)$ and for $t_0 < t$,

$$\left| \int_{t_0}^{t} g(t) \, dt \right| \leqq \int_{t_0}^{t} |g(t)| \, dt.$$

Define $z(t)$ by $z(t) = |u_1(t) - u_0(t)|$. Then $z(t)$ is nonnegative and satisfies the inequality

$$z(t) \leq z(t_0) + \int_{t_0}^{t} |f(s, u_1(s)) - f(s, u_0(s))| \, ds + \varepsilon(t - t_0).$$

Because $f(t, x)$ is lipschitzian,

$$z(t) \leq z(t_0) + \int_{t_0}^{t} kz(s) \, ds + \varepsilon(t - t_0) = z(t_0) + \int_{t_0}^{t} [kz(s) + \varepsilon] \, ds.$$

We may therefore apply Gronwall's lemma and obtain

$$z(t) \leq [z(t_0) + \varepsilon(t - t_0)]e^{k(t-t_0)},$$

from which (2.3.7) is obtained, if we substitute for $z(t)$ and $z(t_0)$ the values given above.

This approximation theorem is not only useful in proving the next theorem (uniqueness) but also for estimating the error in numerical computations of approximate solutions.

3.5 Uniqueness Theorem

Under the assumptions of the Cauchy-Peano theorem and assuming that in $\mathscr{D}$ the function $f(t, x)$ satisfies the Lipschitz condition (2.3.4), *the differential equation* (2.1.1) *with the initial condition* (2.1.2) *has one and only one solution in the interval*

$$|t - t_0| \leqq a.$$

Proof

The existence of the solution has already been given by the Cauchy-Peano theorem [see Section 2.2.1]; only the uniqueness remains to be proved.

Assume that there are two solutions $u_0(t)$ and $u_1(t)$ satisfying (2.1.1) and (2.1.2). We may then apply the approximation theorem with arbitrarily small $\varepsilon > 0$ and $\delta > 0$, from which it follows that the difference $|u_0(t) - u_1(t)|$ is arbitrarily small, hence zero. This proves the theorem.

In Example 2.3.2 the function $\sqrt[3]{x}$ does not satisfy a Lipschitz condition at $x = 0$, and this explains the possibility that the solution is not unique.

It must be noted that the Lipschitz condition is sufficient but not necessary for the uniqueness of the solution. There exist weaker conditions (some of which we will consider later) that are also sufficient to insure uniqueness.

Problems

1. Indicate for which initial points (t_0, x_0) the following differential equations admit a local solution, i.e., a solution defined in some neighborhood of t_0, and also where the solution may fail to be unique:

   (a) $x' = \dfrac{x^2}{t^2 - x^2}$, (b) $x' = +\sqrt{x(x-1)}$,

   (c) $x' = \sqrt[3]{x(x-1)}$, (d) $x' = \ln \sin x$,

   (e) $x' = |x|^t$.

2. Give an upper bound for the difference between the functions $x(t)$ and $y(t)$ defined by

   (a) $x' = 2x + t \quad x(0) = 5,$
   $y' = 2y + t \quad y(0) = 5.1,$

   (b) $x' = +\sqrt{x} \quad x(0) = 1,$
   $y' = +\sqrt{y + 0.01} \quad y(0) = 1.$

3. Show that the function $f(t, x) = tx$ satisfies a Lipschitz condition in x for each fixed value of t, on any interval $a \leqq t \leqq b$ but not on the whole t-axis.

4. $f(x)$ is said to satisfy a Lipschitz condition of order α, if

$$|f(x) - f(y)| \leqq k|x - y|^\alpha$$

   for some constant k. Prove that if $f(x)$ satisfies a Lipschitz condition of order $\alpha > 1$, then $f(x) = \text{constant}$.

4 Successive Approximations

4.1 Idea of the method

Given the differential equation (2.1.1) and the initial condition (2.1.2), we can try to obtain the solution by an iterative method known as *Picard's method of successive approximations*.

To begin we may take any reasonable function as an approximation of order zero, the constant $x(t) = x_0$, for example. In practical cases (this method is suitable for actual computations) we may achieve a better guess about the solution; starting with this better approximation, the convergence of the method will in general be quicker, which is certainly of practical value. For theoretical purposes we are satisfied to start with the constant mentioned above.

In order to obtain better and better approximations, we will use the integral equation (2.1.3), which is equivalent to the initial value problem (2.1.1), (2.1.2). We define the successive approximations $x_n(t)$ by

$$x_{n+1}(t) = x_0 + \int_{t_0}^{t} f(s, x_n(s))\, ds \quad n = 0, 1, 2, \ldots, \tag{2.4.1}$$
$$x_0(t) \equiv x_0 .$$

Since x_0 is the initial value, all approximations $x_n(t)$ satisfy the initial condition (2.1.2). Furthermore, it follows from (2.4.1) that if any approximation happens to be an exact solution of the differential equation, then all the following approximations will coincide with it.

It is a reasonable guess that under suitable conditions, the sequence $x_n(t)$ will converge to a limit function $x(t)$ which will satisfy the integral relation

$$x(t) = x_0 + \int_{t_0}^{t} f(s, x(s))\, ds, \tag{2.4.2}$$

obtained from (2.4.1) by letting $x_n(t) \to x(t)$ and $x_{n+1}(t) \to x(t)$. This would prove that we have found a solution $x(t)$ to our problem.

4.2 Convergence Theorem

Under the conditions of Theorem 3.5, *the successive approximations defined by* (2.4.1) *converge uniformly to the unique solution of the initial value problem* (2.1.1), (2.1.2).

Proof

Formula (2.4.1) can be written as

$$x_{n+1}(t) - x_n(t) = \int_{t_0}^{t} [f(s, x_n(s)) - f(s, x_{n-1}(s))]\, ds. \tag{2.4.3}$$

We will prove that the series

$$\lim_{n \to \infty} x_n(t) = x_0 + [x_1(t) - x_0] + [x_2(t) - x_1(t)] + \cdots \tag{2.4.4}$$

converges uniformly to a limit function $x(t)$. Then, by continuity of $f(s, x(s))$ in the rectangle $\mathscr{D}$ previously considered, the functions

$f(s, x_n(s))$ will also converge uniformly to the limit function $f(s, x(s))$. By a classical theorem of functions of a real variable, we may then take the limit under the integral sign in formula (2.4.1) and obtain (2.4.2), which is the desired result.

In order to prove the convergence of the series (2.4.4), we first note that all these functions $x_n(t)$ are actually defined on the whole interval $|t - t_0| \leqq a$. This is indeed the case because according to their definition (2.4.1), these functions $x_n(t)$ are continuous and their derivatives are given by $f(t, x_{n-1}(t))$. Hence, by assumption (2.2.2), the slope of the graph of $x_n(t)$ is bounded in absolute value in such a way that for $|t - t_0| \leqq a$, $|x_n(t) - x_0| \leqq b$, the curve remains in the rectangle $\mathscr{D}$ defined in (2.2.1).

Writing $M = b/a$ and taking into account the Lipschitz condition (2.3.4), it follows that for $t > t_0$,

$$|x_1(t) - x_0| \leqq \int_{t_0}^{t} |f(s, x_0)|\, ds \leqq M(t - t_0),$$

$$|x_2(t) - x_1(t)| \leqq \int_{t_0}^{t} |f(s, x_1(s)) - f(s, x_0)|\, ds$$

$$\leqq \int_{t_0}^{t} k|x_1(s) - x_0|\, ds \leqq kM \frac{(t - t_0)^2}{2!}.$$

It also follows by induction that, in general,

$$|x_{n+1}(t) - x_n(t)| \leqq \int_{t_0}^{t} |f(s, x_n(s)) - f(s, x_{n-1}(s))|\, ds$$

$$\leqq \int_{t_0}^{t} k|x_n(s) - x_{n-1}(s)|\, ds \leqq M \frac{k^n(t - t_0)^{n+1}}{(n + 1)!}.$$

For $t < t_0$ the above formulas are still valid, if we write $|t - t_0|$ instead of $t - t_0$, which is easy to ascertain.

The series (2.4.4) is thus majorized by the series

$$\frac{M}{k} \sum_{0}^{\infty} \frac{k^n |t - t_0|^n}{n!} = \frac{M}{k} e^{k|t - t_0|}$$

and therefore converges uniformly on the interval $|t - t_0| \leqq a$. This, along with our initial observations, concludes the proof of the theorem.

4.3 Functional analytic proof

The convergence theorem can be proved using the modern tools of functional analysis. This proof is interesting because it gives a deeper insight into the nature of the problem, which is really much more general than this application to differential equations.

Let us consider every continuous function $x(t)$, defined in the interval $\mathscr{I}: |t - t_0| \leqq a$, as an element of an abstract space. This space will be linear in the sense that these functions can be added and multiplied by numbers and that these operations obey the usual rules. Introducing the norm

$$\|x\| = \max \{|x(t)\|, t \in \mathscr{I}\},$$

we will have a *linear normed space*.

Define an operator T which assigns to each function $x(t)$ of this space another function in the space by the formula

$$T[x(t)] = x_0 + \int_{t_0}^{t} f(s, x(s))\, ds.$$

Then, for every two functions $x(t)$, $y(t)$, we have

$$T[x] - T[y] = \int_{t_0}^{t} [f(s, x(s)) - f(s, y(s))]\, ds$$

and, by the Lipschitz condition which $f(t, x)$ is assumed to satisfy,

$$\|T[x] - T[y]\| = \max \{|T[x] - T[y]|\}$$

$$\leqq k \int_{t_0}^{t} \max |x(s) - y(s)|\, ds \leqq ka\|x - y\|.$$

By taking the interval $\mathscr{I}$ (actually its half-length a) sufficiently small, we can achieve $ka < 1$; hence

$$\|T[x] - T[y]\| < \|x - y\|.$$

An operator satisfying this last inequality is called a *contraction operator* because the distance between two points shrinks when the operator is applied. It can be proved [see, for example, Kolmogorov and Fomin, *Introductory Real Analysis*, New York: Prentice-Hall, 1970] that under these conditions there exists a *fixed point* of the mapping determined by this operator, i.e., an element $x(t)$ such that $T[x] = x$. This fixed point is the solution of our problem.

4.4 An application of the method of successive approximations

As a trivial example, consider the differential equation

$$x' = x, \; x(0) = 1.$$

According to formula (2.4.3), we obtain the successive approximations

$$x_1(t) = 1 + \int_0^t 1 \, dt = 1 + t,$$

$$x_2(t) - x_1(t) = \int_0^t t \, dt = t^2/2,$$

$$x_3(t) - x_2(t) = \int_0^t (t^2/2) \, dt = t^3/3!,$$

$$\cdots\cdots\cdots\cdots\cdots\cdots\cdots$$

and the exponential series results

$$x(t) = 1 + t + t^2/2 + t^3/3! + \cdots.$$

Problem

1. Starting with $x_0(t) \equiv x_0$, find the successive approximations and, if possible, the limit function $x(t)$ for
   (a) $x' = x + 1 \quad x(0) = 0$, (b) $x' = x^2 \quad x(0) = 1$,
   (c) $x' = -2tx \quad x(0) = 1$.

5 Continuation of Solutions

5.1 Continuation Theorem

If $f(t, x)$ is defined and continuous on the closure $\mathscr{D}$ of a bounded domain of the t, x plane, and if the differential equation (2.1.1) has the solution $x(t)$ defined on the open interval (a, b) and its graph lies within $\mathscr{D}$, then the limits $x(a + 0)$ and $x(b - 0)$ exist. Furthermore, if either of the points $(a, x(a + 0))$ or $(b, x(b - 0))$ is an interior point of $\mathscr{D}$, then the solution $x(t)$ can be continued beyond the corresponding end point of the interval (a, b).

Proof

As $\mathscr{D}$ is bounded, $x(t)$ will also be bounded, and any sequence $x(t_i)$ with $t_i \to b$ from the left will have some limit value B_1. Let us prove that this limit value is unique; that is, if B_2 is another limit value of the sequence $x(t_i)$, it must coincide with B_1. Otherwise, the values of $x(t_i)$ will indeed oscillate indefinitely between neighborhoods of B_1 and B_2 as $t_i \to b$. But this is impossible because $f(t, x)$, which is the slope of the curve $x(t)$, is bounded.

For t tending to a from the right, the same happens; hence $x(t)$ is defined on the closed interval $[a, b]$. The possibility of continuing the solution outside this interval then becomes a straightforward application of the existence theorem of the solution in some interval around the initial point.

5.2 *Remarks*

According to the preceding theorem, the solution $x(t)$ is always continuable to the boundary of the domain $\mathscr{D}$. [It is not possible for the solution to "disappear" in the interior of $\mathscr{D}$ nor for it to behave as shown in Figure 2.3.]

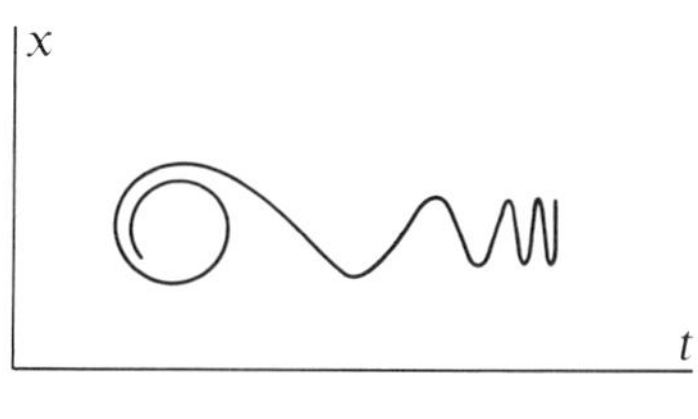

Figure 2.3

If the domain is not bounded, then it is certainly possible that the solution $x(t)$, defined in (a, b), may tend to infinity as $t \to a + 0$ or as $t \to b - 0$. In such cases the solution is not continuable.

5.3 *Interval of existence*

The maximal interval on which a solution $x(t)$ can be continued is called the *interval of existence* of the solution. If the differential equation is defined on some open set (the entire t, x plane, for example), then the interval of existence of any solution is always an open interval. Hence:

(a) for $x' = x$, $x(0) = 1$, the solution $x = e^t$ has $(-\infty, +\infty)$ as the interval of existence;

(b) for $x' = x^2$, $x(0) = 1$, the solution is $x = (1-t)^{-1}$ with the interval of existence $(-\infty, 1)$;

(c) for $x' = 1 + x^2$, $x(0) = 0$, the solution is $x = \tan t$ and the interval of existence is $\left(-\frac{\pi}{2}, +\frac{\pi}{2}\right)$.

5.4 Remarks

In the continuation theorem it is not necessary to assume the solution $x(t)$ to be unique. The assumption that $f(t, x)$ will be continuous may be replaced by any other condition insuring the existence of a solution through each point, together with the assumption that $f(t, x)$ is bounded. Later, in Section 2.8, we will show how some of these more general conditions insure the existence of a solution of the differential equation.

Problem

1. Find the interval of existence of the solutions of:

(a) $x' = -t/x$ $\quad x(0) = 1$, $\qquad$ (b) $x' = x^2 - 2x + 2$ $\quad x(0) = a$,
(c) $x' = x/(t^2 - t)$ $\quad x(-1) = 1$, $\qquad$ (d) $x' = x/(t^2 - t)$ $\quad x(0) = 1$,
(e) $x' = x/(t^2 - t)$ $\quad x(1) = 1$.

6 Dependence on Initial Data and Parameters

6.1 General expression of the solution

The solution $x(t)$ of a differential equation is really a function not only of the variable t but also of the initial data t_0, x_0. So, for example, any solution of $x' = x$ may be written as $x(t) = x_0 e^{t-t_0}$, a function of t, t_0 and x_0. In general, we will find it useful to express this functional dependence as

$$x = F(t, t_0, x_0). \tag{2.6.1}$$

If the solution through any initial point is unique, this function $F(t, t_0, x_0)$ is well-determined, and it is easy to prove that

$$x_1 = F(t_1, t_0, x_0), \qquad x_2 = F(t_2, t_1, x_1)$$

imply

$$x_2 = F(t_2, t_0, x_0). \tag{2.6.2}$$

This formula can also be written as

$$F(t_2, t_1, F(t_1, t_0, x_0)) = F(t_2, t_0, x_0).$$

In addition,

$$x_1 = F(t_1, t_0, x_0)$$

is equivalent to

$$x_0 = F(t_0, t_1, x_1). \tag{2.6.3}$$

Finally,

$$F(t_0, t_0, x_0) = x_0. \tag{2.6.4}$$

These relations (2.6.1) through (2.6.4) describe the *evolution* of a system.

In usual applications the differential equations may depend on certain *parameters*. In that case, obviously, the solution will also depend on those parameters. For example, the equation $x' = kx$ has the general solution $x = x_0 e^{k(t-t_0)}$, a function of t, t_0, x_0, and k.

In general, if the differential equation is

$$x' = f(t, x, p), \tag{2.6.5}$$

then the general solution should be written as

$$x = F(t, t_0, x_0, p). \tag{2.6.6}$$

There may be more than one parameter, but for the sake of simplicity we will consider only one.

6.2 *Theorem*

If on a region $\mathscr{D}$, which is the closure of a bounded domain, the function $f(t, x, p)$ is continuous in (t, x, p) and satisfies a Lipschitz condition in x, then the solution (2.6.6) *of the differential equation* (2.6.5) *is a continuous function of (t, t_0, x_0, p).*

Proof

Using approximation formula (2.3.7), the proof is fairly easy to construct; hence we will present only an outline.

In formula (2.3.7) the factor δ does indeed correspond to a variation in x_0; a variation of t_0 is equivalent, when multiplied by some upper bound of the absolute value of the slope, to a variation of x_0; and similarly, the variation of t may be related to the value $x(t)$. One should keep in mind that because $\mathscr{D}$ is closed and bounded, hence compact, the values of the slope $f(t, x, p)$ are bounded. As any function, continuous on a compact domain, is uniformly continuous there, a sufficiently small change in the parameter p will change the function $f(t, x, p)$ by less than ε, uniformly in $\mathscr{D}$. The new solution is therefore an ε-approximation of the first differential equation, and formula (2.3.7) shows the continuity of $F(t, t_0, x_0, p)$.

6.3 Remarks

It is interesting to observe that assuming only continuity of $f(t, x, p)$, which does not insure the uniqueness of the solution, is not enough to prove Theorem 6.2. [The Lipschitz condition was indeed used in the proof of Theorem 3.4.] A counterexample is given by the equation

$$x' = \begin{cases} +\sqrt{x} & \text{for } x \geqq 0, \\ 0 & \text{for } x \leqq 0. \end{cases}$$

The (nonunique) solutions are represented in Figure 2.4, where it is apparent that the function $F(t, t_0, x_0)$ is not only not continuous but is not even defined for $x_0 = 0$.

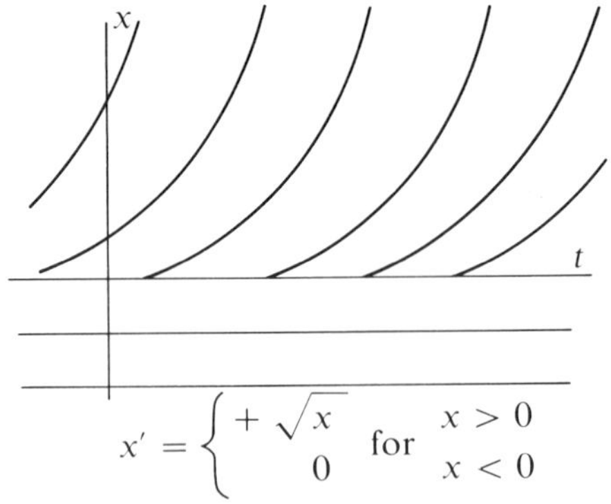

Figure 2.4

It should also be noted that for a function with a continuous derivative, the Lipschitz condition is equivalent to the boundedness of the derivative. Hence a function with a continuous derivative is Lipschitz continuous on any compact region. Therefore Theorem 6.2 can be restated substituting the Lipschitz condition in x with the requirement that $f(t, x, p)$ have a continuous partial derivative with respect to x. In this formulation, the theorem is slightly less general than in the form given above.

6.4 Theorem

If on the closed bounded domain $\mathscr{D}$ of the t, x plane, $f(t, x)$ is continuous in (t, x) and continuously differentiable in x, then the general solution $F(t, t_0, x_0)$ of (2.1.1) is also differentiable in x_0.

Proof

In order to simplify the formulas, let us assume $t_0 = 0$. This amounts to making a preliminary change of independent variable $\tau = t - t_0$ and does not restrict the generality of the proof.

Let $u(t) = F(t, 0, u_0)$ be a fixed solution of equation (2.1.1) and let $x(t) = F(t, 0, x_0)$ be a nearby solution. All we have to prove is that for $x_0 \to u_0$, the quotient

$$\frac{x(t) - u(t)}{x_0 - u_0}$$

has a limit, which is then the partial derivative $\partial F(t, 0, x_0)/\partial x_0$.

Using the integral expression (2.1.3) for the differential equation, we obtain

$$x(t) - u(t) = x_0 - u_0 + \int_0^t [f(s, x(s)) - f(s, u(s))]\, ds$$

and therefore

$$\frac{x(t) - u(t)}{x_0 - u_0} = 1 + \int_0^t \frac{f(s, x(s)) - f(s, u(s))}{x(s) - u(s)} \cdot \frac{x(s) - u(s)}{x_0 - u_0}\, ds. \tag{2.6.7}$$

From this we obtain

$$\left|\frac{x(t) - u(t)}{x_0 - u_0}\right| \leqq 1 + \int_0^t \left|\frac{f(s, x(s)) - f(s, u(s))}{x(s) - u(s)}\right| \cdot \left|\frac{x(s) - u(s)}{x_0 - u_0}\right| ds.$$

The first factor in the integrand can be expressed as the derivative $\partial f(s, x)/\partial x$ evaluated at some intermediate point; so in $\mathscr{D}$ it is bounded by some constant k independent of x_0.

Gronwall's lemma is applicable to $[x(t) - u(t)]/[x_0 - u_0]$; hence this quotient is bounded by a constant M.

Equation (2.6.7) can also be written

$$\frac{x(t) - u(t)}{x_0 - u_0} = 1 + \int_0^t \left\{ \left[\frac{\partial f(s, x)}{\partial x}\right]_{x=u(s)} + \eta \right\} \cdot \frac{x(s) - u(s)}{x_0 - u_0}\, ds. \quad (2.6.8)$$

Here η can be taken arbitrarily small because $\partial f(s, x)/\partial x$ is continuous, and by Theorem 2.6.2, the solution $x(s)$ tends to $u(s)$ as $x_0 \to u_0$.

Therefore, writing $\varepsilon = \eta M$, the quotient

$$y(t) = \frac{x(t) - u(t)}{x_0 - u_0}$$

is an ε-approximate solution of the differential equation

$$y' = [\partial f(t, x)/\partial x]_{x=u(t)}\, y \quad (2.6.9)$$

with the initial condition

$$y(0) = 1. \quad (2.6.10)$$

These two equations are equivalent to the integral equation (2.6.8) without the term in η.

As in the existence theorem, this ε-approximate solution converges to a limit function. [Different sequences with $\varepsilon \to 0$ cannot have distinct limits because the solution of the differential equation is unique.] This limit function is, therefore, the limit of the above quotient, that is, $\partial F(t, 0, x_0)/\partial x_0$. Hence we have proved both the existence of this partial derivative and the fact that as a function of t, it satisfies the differential equation (2.6.9).

Problem

1. Show that if $f(t, x)$ is continuous in (t, x) and continuously differentiable in x, and if $x = u(t)$ is a fixed solution of $x' = f(t, x)$, then the variational equation (2.6.9)

$$y' = [\partial f(t,x)/\partial x]_{x=u(t)}\, y,$$

with initial condition $y(0) = 1$, has a unique solution.

7 *Systems of Differential Equations*

7.1 Vector notation

The explicit form of a system of n differential equations with n unknown functions is

$$x^{i'} = f^i(t, x^1, x^2, \ldots, x^n), \quad i = 1, 2, \ldots, n. \tag{2.7.1}$$

It is convenient to consider the functions $x^i(t), f^i(t)$, respectively, as components of the *vector functions* $\mathbf{x}(t)$ and $\mathbf{f}(t)$. By definition, the derivative of such a vector function is obtained by taking the derivative of each component. Hence the system (2.7.1) can be written in vector form as

$$\mathbf{x}' = \mathbf{f}(t, \mathbf{x}). \tag{2.7.2}$$

We will see that most properties of the *scalar* first-order differential equations can be generalized to *vector equations.*

We will assume that the algebraic properties of the vectors are known. For vector calculus, the basic idea is the *limit*. In order to define the limit, it is convenient to have the idea of the *norm* of a vector. [The norm plays the role of a sort of absolute value or geometric length of a vector.]

Let $\mathbf{x} = (x^1, x^2, \ldots, x^n)$ and $\mathbf{y} = (y^1, y^2, \ldots, y^n)$ be n-dimensional vectors. *Equality* of vectors is defined by

$$\mathbf{x} = \mathbf{y} \quad \text{if and only if} \quad x^i = y^i$$

for all $i = 1, 2, \ldots, n$. *Addition* (and consequently *subtraction*) is defined componentwise:

$$\mathbf{x} + \mathbf{y} = (x^1 + y^1, x^2 + y^2, \ldots, x^n + y^n).$$

In order to *multiply* a *vector* $\mathbf{x}$ by a *scalar* (i.e., a number) k, each component should by multiplied by k. Thus

$$k\mathbf{x} = (kx^1, kx^2, \ldots, kx^n).$$

The *scalar* (or *inner*) *product* of two vectors is defined by

$$\mathbf{xy} = x^1y^1 + x^2y^2 + \cdots + x^ny^n = \sum_{i=1}^{n} x^iy^i.$$

The scalar product of two vectors is a scalar, i.e., a number, *not* a vector. If two vectors have scalar product equal to zero, they are said to be *orthogonal.*

Note: In vector formulas, super-indices should not be mistaken for exponents. When we have to write powers of components of a vector, we will write them as $(x^i)^2$, $(y^j)^3$. Powers of vectors themselves never appear because there is no true product between vectors. For the scalar product of a vector with itself, we write $\mathbf{xx}$.

The *euclidean norm* of a vector is defined by the nonnegative quantity $\|\mathbf{x}\|$ such that

$$\|\mathbf{x}\|^2 = \sum_{i=1}^{n} (x^i)^2.$$

Its square can also be written as the scalar product $\|\mathbf{x}\|^2 = \mathbf{xx}$ and represents the square of the length of the vector $\mathbf{x}$ in cartesian coordinates.

Sometimes another norm, defined by

$$\|\mathbf{x}\|_1 = \sum_{i=1}^{n} |x^i|,$$

makes calculations simpler and is therefore used. It should be mentioned that both definitions of norm are equivalent in their applications to calculus, as we will see.

The *limit* of a vector (both for the case of sequences and for a continuous vector function $\mathbf{x}(t)$) is defined by

$$\begin{aligned} \mathbf{x} \to 0 \quad &\text{if and only if} \quad \|\mathbf{x}\| \to 0, \\ \mathbf{x} \to \mathbf{x}_0 \quad &\text{if and only if} \quad \|\mathbf{x} - \mathbf{x}_0\| \to 0. \end{aligned}$$

The two different norms given above are equivalent for defining the limit. Indeed, if the first norm of a sequence of vectors tends to zero, so does the second norm, and conversely. Hence every sequence of vectors which is convergent according to the first norm is also convergent according to the second norm and has the same limit, and conversely. Finally, the limit can also be defined componentwise:

$$\mathbf{x} \to \mathbf{x}_0 \quad \text{if and only if} \quad x^i \to x_0^i$$

for all $i = 1, 2, \ldots, n$, and this definition is again equivalent to the former one.

The *derivative* and the *integral*, being limits, are also taken componentwise:

$$\frac{d\mathbf{x}}{dt} = \left(\frac{dx^1}{dt}, \frac{dx^2}{dt}, \ldots, \frac{dx^n}{dt}\right),$$

$$\int_a^b \mathbf{x}(t)\, dt = \left(\int_a^b x^1(t)\, dt, \int_a^b x^2(t)\, dt, \ldots, \int_a^b x^n(t)\, dt\right).$$

In the sequel we state the basic theorems for vector differential equations. Their proofs are, in most cases, literal repetitions of the corresponding proofs for the scalar case. We will therefore give only very brief proofs, allowing the reader to work the details as an exercise.

In vector notation, the initial condition for a vector differential equation is

$$\mathbf{x}(t_0) = \mathbf{x}_0\,. \tag{2.7.3}$$

7.2 *Existence Theorem*

If the vector function $\mathbf{f}(t, \mathbf{x})$ *is defined and continuous on the cylinder*

$$\mathscr{D}: |t - t_0| \leqq a, \qquad \|\mathbf{x} - \mathbf{x}_0\| \leqq b,$$

where it satisfies

$$\|\mathbf{f}(t, \mathbf{x})\| \leqq b/a,$$

then there exists a vector function $\mathbf{x}(t)$, *defined on the interval* $|t - t_0| = a$, *having a continuous derivative and satisfying the differential equation* (2.7.2) *with the initial condition* (2.7.3).

For the proof, the reader should go back to Theorem 2.1; this proof is valid by simply "translating" the absolute values $|x|$, $|x - x_0|$, and so on, by the norms $\|\mathbf{x}\|$, $\|\mathbf{x} - \mathbf{x}_0\|$, etc. For example, an ε-approximate solution will now be defined as a vector function $\mathbf{u}(t)$ such that

$$\|\mathbf{u}'(t) - \mathbf{f}(t, \mathbf{u}(t))\| \leqq \varepsilon.$$

Also, a formula similar to (2.2.5) can be established: by continuity of $\mathbf{f}(t, \mathbf{x})$ on the compact region $\mathscr{D}$, a value $\delta > 0$ can be found such that for $|t - t_1| < \delta$, $\|\mathbf{x} - \mathbf{x}_1\| < \delta$, we have

$$\|\mathbf{f}(t, \mathbf{x}) - \mathbf{f}(t_1, \mathbf{x}_1)\| < \varepsilon.$$

The selection process of a convergent subsequence, as found in the Arzelá-Ascoli lemma, should proceed in the following way. Given the sequence of vector functions $\mathbf{x}_i(t) = (x_i^1(t), \ldots, x_i^n(t))$, $i = 1, 2, 3, \ldots,$ we consider initially the first components $x_i^1(t)$. If, as in the Arzelá-Ascoli lemma, we are able to find a subsequence $x_{i_j}^1(t)$ which is convergent, we disregard the other members of the sequence and renumber the sequence so that now the vector functions $\mathbf{x}_i(t)$ have their first component $x_i^1(t)$ convergent. If the second component $x_i^2(t)$ is still not convergent (as will generally be the case), we should select a subsequence so that the second component also converges. By doing this process n times, we finally obtain the sequence of (again renumbered) vector functions $\mathbf{x}_i(t)$ such that all components converge to corresponding limit functions. Hence this sequence converges in the vector sense (componentwise).

We note finally that for vector functions (and $b > a$)

$$\left\| \int_a^b \mathbf{x}(t)\, dt \right\| \leqq \int_a^b \|\mathbf{x}(t)\|\, dt.$$

Using these statements, the reader should be able to reformulate the proof of Theorem 7.2 and those that follow.

7.3 Uniqueness Theorem

If the conditions of the preceding theorem are satisfied and if there exists a constant k such that throughout $\mathscr{D}$, the Lipschitz condition in $\mathbf{x}$

$$\|\mathbf{f}(t, \mathbf{x}_1) - \mathbf{f}(t, \mathbf{x}_2)\| \leqq k\|\mathbf{x}_1 - \mathbf{x}_2\|$$

holds, then the solution of (2.7.2) *and* (2.7.3) *is unique.*

7.4 Successive approximations

Under the conditions of the preceding uniqueness theorem, the method of successive approximations [see Section 2.4.1] *is applicable and converges to the solution of the initial value problem* (2.7.2–2.7.3).

7.5 Continuation of the solution

If, on the closed bounded domain $\mathscr{D}$ of the $t, \mathbf{x}$ space, the function $\mathbf{f}(t, \mathbf{x})$ is continuous, then every solution of equation (2.7.2) *is continuable up to the boundary of $\mathscr{D}$.*

7.6 *Dependence on initial data and parameters*

If, on the closed bounded domain $\mathscr{D}$ of the t, $\mathbf{x}$ space and for $p \in [p_1, p_2]$, the function $\mathbf{f}(t, \mathbf{x}, p)$ is continuous in $(t, \mathbf{x}, p)$ and satisfies a Lipschitz condition in $\mathbf{x}$, then the general solution $\mathbf{F}(t, t_0, \mathbf{x}_0, p)$ is a continuous function of $(t, t_0, \mathbf{x}_0, p)$.

In Theorem 7.3 the parameter p may also be a vector $\mathbf{p} = (p^1, p^2, \ldots, p^m)$, which means that there are really m scalar parameters appearing in the equation. The dimension m does not need to be equal to n.

Problems

1. Write in vector notation the system of differential equations
$$x' = x^2 + y^2, \qquad y' = x^2 - y^2, \qquad x(0) = 1, \qquad y(0) = 0.$$
[*Hint*: write $\mathbf{x} = \begin{pmatrix} x \\ y \end{pmatrix}$, $\mathbf{f}(t, \mathbf{x}) = \begin{pmatrix} f \\ g \end{pmatrix}$ and give the expressions of f and g.] Also give the initial condition.
2. Write as a system, i.e., componentwise, the vector differential equation
$$\mathbf{x}' = \mathbf{x}/\|\mathbf{x}\|,$$
where $\|x\|$ is the euclidean norm.
3. Carry out the details of the proof for
(a) Theorem 7.2, (b) Theorem 7.3,
(c) Theorem 7.4, (d) Theorem 7.5.
4. Write the second-order differential equation as a system and solve by successive approximations.
(a) $x'' = x \qquad x(0) = 1, \quad x'(0) = 0.$
(b) $x'' = 2x^3 \qquad x(0) = 1, \quad x'(0) = 0.$
(c) $x'' = -2x' - 2x \qquad x(0) = 1, \quad x'(0) = -1.$

8 *Additional General Theorems*

8.1 *Need for a more general theory*

Up to this point, we have discussed the differential equation $\mathbf{x}' = \mathbf{f}(t, \mathbf{x})$ under the assumption that $\mathbf{f}(t, \mathbf{x})$ is continuous in both variables. This seems proper within the context of classical mathematics. [Indeed it is almost unavoidable if we want the solutions $\mathbf{x}(t)$ to have a derivative everywhere without exceptions.] Nevertheless, it proves to be too restricted for some important practical applications. For example, in many automatic electric circuits (and these may be incorporated into larger mechanical systems) there are relays (switches) which have only two positions: open or closed. As they operate, the differential equation of the system changes discontinuously. If we want to always describe the behavior of the system with the same differential equation, we have to consider the function $\mathbf{f}(t, \mathbf{x})$ as discontinuous. This type of *relay control* (also known as *bang-bang* control) is commonly used in many, very simple regulating devices, and from a theoretical point of view, it has frequently been proved to give an "optimal" control.

We will see that allowing the function $\mathbf{f}(t, \mathbf{x})$ to be discontinuous in t does not change the theory very greatly. On the other hand, allowing $\mathbf{f}(t, \mathbf{x})$ to be discontinuous in $\mathbf{x}$ introduces extremely difficult problems; for example, in very simple instances one may obtain infinitely many solutions, no solutions at all, solutions to one side only, and other kinds of strange behavior.

In all applications in which $\mathbf{f}(t, \mathbf{x})$ is allowed to be discontinuous, we must take the concept of solution of the differential equation in its generalized sense [see Section 2.1.3]. Indeed, the solution $\mathbf{x}(t)$ will generally have a discontinuous derivative (e.g., the solution may have "corners"), so that the differential equation cannot hold everywhere. For $\mathbf{x}(t)$ to be a generalized solution, the differential equation has to be satisfied *almost everywhere*, that is, for all t except possibly a set of t values of (Lebesgue) measure zero. On the other hand, continuity of $\mathbf{x}(t)$ is not enough; the solutions have to be absolutely continuous [see Section 2.1.3].

8.2 Carathéodory's Theorem

If on the $n + 1$-dimensional cylinder

$$\mathscr{D}: |t - t_0| \leqq a, \qquad \|\mathbf{x} - \mathbf{x}_0\| \leqq b,$$

the n-dimensional vector function $\mathbf{f}(t, \mathbf{x})$ is (Lebesgue) measurable in t for every fixed $\mathbf{x}$ and continuous in $\mathbf{x}$ for every fixed t, and if there is a (Lebesgue) integrable function $m(t)$ majorizing $\|\mathbf{f}(t, \mathbf{x})\|$,

$$\|\mathbf{f}(t, x)\| \leqq m(t) \quad \text{for all } (t, \mathbf{x}) \in \mathscr{D}, \tag{2.8.1}$$

then the differential equation

$$\mathbf{x}' = \mathbf{f}(t, \mathbf{x}) \tag{2.8.2}$$

with the initial condition

$$\mathbf{x}(t_0) = \mathbf{x}_0, \quad (t_0, \mathbf{x}_0) \in \mathscr{D} \tag{2.8.2'}$$

has a generalized solution which exists in some interval $|t - t_0| \leqq c$ for some positive c.

The following statement is equivalent to this existence theorem.* *If* $\mathbf{f}(t, \mathbf{x})$ *is defined on the slab* $\mathscr{D}'$: $|t - t_0| \leqq a$ *with* $\mathbf{x}$ *arbitrary, if it is (Lebesgue) measurable in t for every fixed* $\mathbf{x}$ *and continuous in* $\mathbf{x}$ *for every fixed t, and if there is a (Lebesgue) measurable function* $m(t)$ *majorizing* $\|\mathbf{f}(t, \mathbf{x})\|$, *i.e., satisfying* (2.8.1), *then the differential equation* (2.8.2) *with the initial condition* (2.8.2′) *has a generalized solution which exists in the whole interval* $|t - t_0| \leqq a$.

Note: In this theorem neither the functions $\mathbf{f}(t, \mathbf{x})$ nor $m(t)$ need to be bounded, but $m(t)$ needs to be *integrable*, which means that the integral $\int_{t_0}^{t} |m(s)|\, ds$ must be finite for every t in $|t - t_0| \leqq a$.

Since for any solution, $\mathbf{x}(t) = \mathbf{x}_0 + \int_{t_0}^{t} \mathbf{f}(s, \mathbf{x}(s))\, ds$, and this integral is majorized, in norm, by $\int_{t_0}^{t} m(t)\, dt$, $\mathbf{x}(t)$ cannot escape to infinity in finite time.

8.3 Continuation Theorem†

Under the conditions of Theorem 8.2, *all solutions can be continued to the boundary of* $\mathscr{D}$.

8.4 Uniqueness Theorem

In addition to the assumptions of Theorem 8.2, *let* $\mathbf{f}(t, \mathbf{x})$ *satisfy the condition*

$$\|\mathbf{f}(t, \mathbf{x}_1) - \mathbf{f}(t, \mathbf{x}_2)\| \leqq \psi(|t - t_0|, \|\mathbf{x}_1 - \mathbf{x}_2\|),$$

where the function $\psi(\tau, \rho)$ *is nonnegative, defined on* $0 < \tau < a$, $0 \leqq \rho \leqq 2b$, *measurable in t for fixed* ρ, *and continuous and nondecreasing in* ρ *for fixed t; let a majorizing function* $m(t) \geqq \psi(t, \rho)$ *exist which is (Lebesgue) integrable in the interval* $\alpha \leqq t \leqq a$ *for every* $0 < \alpha < a$. *Furthermore, let* $u(t) \equiv 0$ *be the only absolutely continuous function in* $[0, a]$ *with* $u(0) = u'(0) = 0$ *satisfying* $u'(t) = \psi(t, u(t))$ *almost everywhere in* $[0, a]$. *Then the initial value problem* (2.8.2), (2.8.2′) *has a unique solution.*

---

* The original proof for this theorem can be found in C. Carathéodory, *Vorlesungen über reelle Funktionen*, London: Chelsea Publishing Co., 1968 (3rd ed.), p. 582. It can also be studied in Coddington and Levinson, *Theory of Ordinary Differential Equations*, New York: McGraw-Hill, 1955, Chapter 2.

† For the proof of this theorem, as well as Theorems 8.4 and 8.5 which follow, we refer the reader to Coddington and Levinson (1955).

Note: As a particular case, the above condition is satisfied if the following generalized Lipschitz condition holds:

$$\|\mathbf{f}(t, \mathbf{x}_1) - \mathbf{f}(t, \mathbf{x}_2)\| \leqq \Phi(|t - t_0|) \cdot \Psi(\|x_1 - x_2\|)$$

with

$$\int_0^a \Phi(\tau)\, d\tau < \infty \quad \text{and} \quad \int_0^\delta \frac{d\rho}{\Psi(\rho)} = \infty$$

for every $\delta > 0$.

8.5 Dependence on initial conditions and parameters

Let $\mathscr{D}$ be the cylinder considered in Theorem 8.2, $\mathscr{I}_\mathbf{p}$ the interval $\|\mathbf{p} - \mathbf{p}_0\| \leqq c$, and $\mathscr{D}_\mathbf{p}$ the cartesian product $\mathscr{D} \times \mathscr{I}_\mathbf{p}$. [Here we assume $\mathbf{p}$ to be an m vector.] Let $\mathbf{f}(t, \mathbf{x}, \mathbf{p})$ be defined on $\mathscr{D}_\mathbf{p}$, where it is measurable in t for fixed $(\mathbf{x}, \mathbf{p})$, continuous in $(\mathbf{x}, \mathbf{p})$ for $\mathbf{p} = \mathbf{p}_0$, $\mathbf{x}$ arbitrary, and t fixed; let $\|\mathbf{f}(t, \mathbf{x}, \mathbf{p})\| \leqq m(t)$ and $|\int_{t_0}^{t_0 \pm a} m(t)\, dt| < b$, where $m(t)$ is (Lebesgue) integrable in $|t - t_0| \leqq a$. Assume, furthermore, that for $\mathbf{p} = \mathbf{p}_0$ the initial value problem

$$\mathbf{x}' = \mathbf{f}(t, \mathbf{x}, \mathbf{p}), \qquad \mathbf{x}(t_0) = \mathbf{x}_0 \tag{2.8.3}$$

has a unique solution $\mathbf{u}_{\mathbf{p}_0}(t)$ in the interval $|t - t_0| \leqq a$.

Then there is a $\delta > 0$ such that for

$$|t_0 - t_0'| < \delta, \qquad \|\mathbf{x}_0 - \mathbf{x}_0'\| < \delta, \qquad \|\mathbf{p} - \mathbf{p}_0\| < \delta,$$

every solution $\mathbf{u}_\mathbf{p}(t)$ of equation (2.8.3) with initial condition $\mathbf{x}(t_0') = \mathbf{x}_0'$ is continuable over the entire interval $|t - t_0| \leqq a$ and tends uniformly to $\mathbf{u}_{\mathbf{p}_0}(t)$ in this interval, as $t_0' \to t_0$, $\mathbf{x}_0' \to \mathbf{x}_0$, $\mathbf{p} \to \mathbf{p}_0$.

Note 1: The great generality of the assumptions should be noticed. For $\mathbf{p} \neq \mathbf{p}_0$ the solution *does not* need to be unique. Nevertheless, *any* sequence of solutions $\mathbf{u}_\mathbf{p}(t)$ tends uniformly to the (unique) solution $\mathbf{u}_{\mathbf{p}_0}(t)$.

Note 2: With respect to differentiability of the solution, it can be proved that if $\mathbf{f}(t, \mathbf{x}, \mathbf{p})$ is of class C^r in $(t, \mathbf{x}, \mathbf{p})$ (i.e., if all partial derivatives up to the order r exist and are continuous), then the solution $\mathbf{F}(t, t_0, \mathbf{x}_0, \mathbf{p})$ is of class C^r in $(t, t_0, \mathbf{x}_0, \mathbf{p})$ and of class C^{r+1} in t.

8.6 *Maximal and minimal solutions*

If solutions of a differential equation are not unique, it is possible to prove the following for the one-dimensional case, where **x** is a scalar.

If the differential equation satisfies the conditions of the existence theorem, there are in general many solutions through each initial point (t_0, x_0). Among them all, there is one *maximal solution* $x_M(t)$ and one *minimal solution* $x_m(t)$, such that for every other solution $x(t)$ of the initial value problem (2.8.2), (2.8.2′),

$$x_m(t) \leqq x(t) \leqq x_M(t) \tag{2.8.4}$$

in the common interval of existence of $x_m(t)$ and $x_M(t)$.

Conversely, for each t_1 in this common interval of existence and for every intermediate value α, $x_m(t_1) < \alpha < x_M(t_1)$, there is necessarily one solution $x(t)$ of the initial value problem such that $x(t_1) = \alpha$. This means that the solutions of the initial value problem fill completely the space between $x_m(t)$ and $x_M(t)$.

8.7 *Kneser's theorem*

The generalization of this property of maximal and minimal solutions to the case where **x** is a vector of dimension n is not trivial. It was given by H. Kneser and can be best explained with the concept of attainable set.

Given the initial value problem $\mathbf{x}' = \mathbf{f}(t, \mathbf{x})$, $\mathbf{x}(t_0) = \mathbf{x}_0$, consider all possible solutions $\mathbf{x}(t)$. For a given fixed value $t = \tau$, the set of all values of these solutions, i.e., the set $\{\mathbf{x}(\tau)\}$, is called the *attainable set* from $(t_0, \mathbf{x}_0)$ for $t = \tau$. Then Kneser's theorem* can be stated as follows:

> *Under the assumptions of the Theorem* 8.2, *the attainable set from* $(t_0, \mathbf{x}_0)$ *for any* $t = \tau$ *in the interval* $|\tau - t_0| \leqq a$, *is a closed and connected set.*

In the one-dimensional case, this connected set is the interval bounded by the minimal and maximal solutions.

Problems

1. Let $u(t)$ and $v(t)$ be two solutions of the differential equation $x' = f(t, x)$, where $f(t, x)$ is continuous. Prove that the two functions $m(t) = \min\{u(t), v(t)\}$ and $M(t) = \max\{u(t), v(t)\}$ also satisfy the same differential equation. [*Hint*: consider at any given value of t, the cases $u(t) = v(t)$, $u(t) > v(t)$, $u(t) < v(t)$.

---

* The proof of this theorem can be found in P. Hartman, *Ordinary Differential Equations*, New York: Wiley, 1964, p. 15.

2. Let $u(t) > 1$ be continuously differentiable on $[a, a + \varepsilon]$. Prove that if $u'(t) \leqq ku(t) \ln u(t)$, then

$$\ln u(t) \leqq e^{k(t-a)} \ln u(a), \quad a \leqq t \leqq a + \varepsilon.$$

3. Prove that if $f(t, x)$ is continuous and $|f(t, x) - f(t, y)| \leqq k(t)|x - y|$ for $t \in [0, a]$, and if $\int_0^a k(t)\,dt < \infty$, then $x' = f(t, x)$, $x(0) = c$ has a unique solution in $[0, a]$.

4. Integrate the following discontinuous differential equations and discuss the singularities. The function sgn x is defined by

$$\operatorname{sgn} x = \begin{cases} +1 & \text{if } x > 0 \\ 0 & \text{if } x = 0 \\ -1 & \text{if } x < 0. \end{cases}$$

(a) $x' = y, \quad y' = -x + \operatorname{sgn} y.$
(b) $x' = 2y/t, \quad y' = 1/t.$
(c) $x' = \operatorname{sgn} y, \quad y' = \operatorname{sgn} x.$
(d) $x' = \operatorname{sgn} y, \quad y' = -\operatorname{sgn} x.$
(e) $x' = \operatorname{sgn} x, \quad y' = -\operatorname{sgn} y.$

5. For the following differential equations: (i) Where is the solution not unique? Give maximal and minimal solutions. (ii) For which (t, t_0, x_0) is the general solution $F(t, t_0, x_0)$ continuous in x_0?

(a) $x' = \begin{cases} +\sqrt{x} & \text{for } x \geqq 0 \\ -\sqrt{x} & \text{for } x \leqq 0. \end{cases}$

(b) $x' = \begin{cases} -\sqrt{x} & \text{for } x \geqq 0 \\ +\sqrt{x} & \text{for } x \leqq 0. \end{cases}$

(c) $x' = -\sqrt[3]{x}.$

(d) $x' = f(t, x) = \begin{cases} 0 & \text{if } x \leqq 0 \\ 2x/t & \text{if } 0 < x < t^2 \\ 2t & \text{if } x \geqq t^2. \end{cases}$

[Verify that $f(t, x)$ is continuous.]

References

For answers to questions concerning the theory of functions of a real variable, Lebesgue integration, discontinuous functions, and so forth, we refer the reader to

H. L. Royden. *Real Analysis*. New York: Macmillan, 1968 (2nd ed.).

For additional details and proofs concerning general topics in this chapter, see

E. Coddington and N. Levinson. *Theory of Ordinary Differential Equations*, Chapters 1 and 2. New York: McGraw-Hill, 1955.

A very extensive reference for existence and uniqueness theorems and properties of non-unique solutions is

P. Hartman. *Ordinary Differential Equations*. New York: Wiley, 1964.

A specific reference for differential inequalities of the type of Gronwall's inequality is

V. Lakshmikantham and S. Leela. *Differential and Integral Inequalities*, Volumes I and II. New York: Academic Press, 1969.

Original proofs used in this chapter can be found in

G. Peano. Sull' integrabilitá delle equazione differenziali di primo ordine. *Atti Accad. Torino* **21** (1885–1856): 677–685.

G. Peano. Démonstration de l'integrabilité des équations différentielles ordinaires. *Math. Annalen* **37** (1890): 182–228.

C. Arzelá. Funzioni di linee. *Atti R. Accad. Lincei Rend.* (4) **5**, (1889): 342–348.

C. Arzelá. Sulle funzioni di linee. *Mem. R. Accad. Bologna* (5) **5**, (1895): 225–244.

H. Kneser. Über die Lösungen eines Systems Gewöhnlicher Differentialgleichungen . . . . *Stz. B. Preuss. Akad. Wiss. Phys.-Math. Klasse* (1923): 171–174.

E. Kamke. Zur Theorie der Systeme Gewöhnlicher Differentialgleichungen, II. *Acta Math.* **58** (1932): 57–85. [This paper is concerned with Kneser's theorem.]

3

Linear Differential Equations

1 Matrices

1.1 Matrix algebra

An $(n \times m)$ matrix is a rectangular array of numbers called its elements:

$$A = (a^1_j) = \begin{pmatrix} a^1_1 & a^1_2 & \cdots & a^1_m \\ a^2_1 & a^2_2 & \cdots & a^2_m \\ a^n_1 & a^n_2 & \cdots & a^n_m \end{pmatrix}. \tag{3.1.1}$$

The matrix is said to be *real* if all its elements a^i_j are real; it is said to be *complex* if all its elements a^i_j are complex. The positive integers n, m correspond, respectively, to the number of rows and the number of columns of the matrix. If $n = m$, then A is said to be a *square matrix*; if $n \neq m$, then A is said to be a *rectangular matrix*. $(n \times m)$ is sometimes called the *dimension* of the matrix. For $m = 1$, the matrix has only one column and is therefore called a *column vector*. Similarly, for $n = 1$, the matrix is called a *row vector*.

Two matrices are defined to be equal if they have the same dimension and their correspondent elements are equal. Thus

$$A = (a^i_j) = B = (b^i_j) \quad \text{if and only if} \quad a^i_j = b^i_j \quad \text{for} \begin{cases} 1 = 1, 2, \ldots, n \\ j = 1, 2, \ldots, m. \end{cases}$$

The *sum* of two matrices having the same number of rows and columns is defined by

$$A + B = (a^i_j) + (b^i_j) = (a^i_j + b^i_j). \tag{3.1.2}$$

The *product* of two matrices, such that the number of columns of the first factor is equal to the number of rows of the second factor, is defined by multiplying rows by columns. Hence

$$A \cdot B = (a^i_j) \cdot (b^j_k) = \left(\sum_{j=1}^{m} a^i_j \, b^j_k \right). \tag{3.1.3}$$

These operations are possible only if the dimensions of the matrices are properly chosen. They are always possible, however, when we operate within the class of square matrices of dimension $(n \times n)$, with fixed n. Most of our examples will refer to operations between square matrices of the same dimension, but some of their properties will be clearly shown to be generalizable to the case of rectangular matrices.

The *zero matrix* is a matrix all of whose elements are zero. It has the property that, when added to any other matrix of the same dimension, it leaves the latter invariant. The zero matrix is usually designated by 0; its dimension is generally obvious from the context.

Given any matrix A, the *opposite matrix* $-A$ is the one that, when added to A, gives 0. Obviously, it is obtained by changing the sign of all elements of A.

The *unit matrix* is defined as a matrix that, when multiplied by any other matrix, leaves the latter invariant. This definition requires that we operate within the set of square $(n \times n)$ matrices for some fixed n; so the unit matrix must also be square. It is usually denoted by E and is composed of ones in the main diagonal and zeros elsewhere. Hence

$$E = (\delta^i_j) \quad \text{where} \quad \delta^i_j = \begin{cases} 0 & \text{if } i \neq j \\ 1 & \text{if } i = j. \end{cases} \tag{3.1.4}$$

When we think of the unit matrix as an identity operator (explained later), it is usually called an *identity matrix* and denoted by I.

Note: There are infinitely many zero matrices and unit matrices, depending upon their dimensions. In most cases, however, it will be unnecessary to distinguish the various possible dimensions of these matrices by special notation.

The *product of a matrix by a number* is defined by multiplying every element of the matrix by this number.

It is easy to see that both the operations of addition and multiplication of matrices and the multiplication of a matrix by a number obey the usual laws of arithmetic;

the only exception is the commutative law which does not usually hold true for the product of two matrices. Hence, in general, $A \cdot B \neq B \cdot A$. [If A is an $(n \times m)$ matrix and B is an $(m \times p)$ matrix, the product $B \cdot A$ is not even feasible!]

Given the matrix A, the *transpose* A^T is obtained by interchanging rows and columns.

The (complex) *conjugate matrix* $\bar{A}$ is obtained by replacing each element of A by its complex conjugate.

The *inverse matrix* of A, denoted A^{-1}, is defined by the relation

$$A \cdot A^{-1} = A^{-1} \cdot A = E. \tag{3.1.5}$$

[It can be shown that any "right" inverse is also a "left" inverse and conversely; also, the inverse is unique.]

The inverse of a product is given by

$$(A \cdot B)^{-1} = B^{-1} \cdot A^{-1}.$$

The *determinant* of a square matrix A will be denoted by det A. According to the rule of multiplication of determinants, $\det (A \cdot B) = (\det A)(\det B)$. Since $\det E = 1$, it follows that $\det (A^{-1}) = (\det A)^{-1}$. It also follows that if $\det A = 0$, then the matrix A cannot have an inverse and is therefore said to be *singular*. Conversely, it can be shown that if $\det A \neq 0$, then A has an inverse and is said to be *nonsingular*.

1.2 Norm of a matrix

We define the *norm* of the matrix $A = (a_j^i)$ by

$$\|A\| = \sum_{i,j} |a_j^i|. \tag{3.1.6}$$

Here, if A is an $(n \times m)$ matrix, then the summation index i (superindex) runs from 1 to n, while j (subindex) runs from 1 to m. When there is no danger of confusion, the ranges of the summation indices will be omitted. Occasionally the summation indices themselves will be omitted, retaining only the $\sum$ symbol.

This definition of norm leads to the following properties, which are easy to check:

$$\left.\begin{array}{ll} \text{(a)} & \|A + B\| \leqq \|A\| + \|B\|, \\ \text{(b)} & \|A \cdot B\| \leqq \|A\| \cdot \|B\|, \\ \text{(c)} & \|A \cdot \mathbf{x}\| \leqq \|A\| \cdot \|\mathbf{x}\|. \end{array}\right\} \tag{3.1.7}$$

In this last inequality, $\mathbf{x}$ is a column vector, that is, $m = 1$.

Note: *Vectors as matrices.* n-dimensional vectors can be written either as column vectors (i.e., $(n \times 1)$ matrices) or as row vectors (i.e., $(1 \times n)$ matrices). The transpose of a row vector is a column vector and conversely. Hence, if $\mathbf{x}$ is an n-dimensional column vector and $\mathbf{y}$ an n-dimensional row vector, then the product $\mathbf{y} \cdot \mathbf{x}$ is a number (scalar), while the product $\mathbf{x} \cdot \mathbf{y}$ is an $(n \times n)$ matrix. [In Section 2.7.1 we mentioned that the norm of a vector may be defined by (3.1.6).]

The scalar or inner product of two column vectors $\mathbf{x}$, $\mathbf{z}$ is defined in matrix notation as $\mathbf{x}^T \cdot \mathbf{z} = \mathbf{z}^T \cdot \mathbf{x}$ (for real vectors). The scalar product of two row vectors is similarly $\mathbf{y} \cdot \mathbf{v}^T = \mathbf{v} \cdot \mathbf{y}^T$.

1.3 Linear transformations—eigenvalues

The product

$$A\mathbf{x} = \mathbf{y},$$

where A is an $(n \times m)$ matrix and $\mathbf{x}$ is an m-dimensional column vector, is well-defined, and $\mathbf{y}$ turns out to be an n-dimensional column vector. This equation defines a mapping or transformation of the space of m-dimensional vectors into the space of n-dimensional vectors.

Since the matrix product is distributive, this transformation is linear. Hence

$$A(\mathbf{x}_1 + \mathbf{x}_2) = A\mathbf{x}_1 + A\mathbf{x}_2 .$$

The matrix A represents, in some sense, the transformation itself or the "operator" performing the mapping.

The most important case is when A is square, that is, $(n \times n)$. This corresponds to the mapping of the n-dimensional vector space into itself. When A is nonsingular, the inverse matrix corresponds to the inverse transformation. Thus

$$\mathbf{x} = A^{-1}\mathbf{y}.$$

For many applications, one basic problem is the following: given an $(n \times n)$ matrix A, find a vector $\mathbf{u}$ and a number λ such that

$$A\mathbf{u} = \lambda\mathbf{u}. \tag{3.1.8}$$

This condition is equivalent to

$$A\mathbf{u} - \lambda\mathbf{u} = 0$$

or

$$(A - \lambda E)\mathbf{u} = 0.$$

In order for a vector $\mathbf{u}$ to exist, which satisfies this condition, it is necessary and sufficient that

$$\det(A - \lambda E) = 0. \tag{3.1.9}$$

In developing this determinant, equation (3.1.9) becomes an algebraic equation of degree n in λ, known as the *characteristic equation* of the matrix A. It is easy to see how this algebraic equation is really of degree n: its coefficient of the highest degree term is always 1 and therefore never vanishes. Hence the characteristic equation always has n roots (real or complex), which are not necessarily all distinct. For each one of these *characteristic roots* λ_k, there exists at least one vector $\mathbf{u}_k$ satisfying (3.1.8). Traditionally, $1/\lambda_k$ is called an *eigenvalue* of the matrix A. [λ_k itself is usually called an eigenvalue.] $\mathbf{u}_k$ is the corresponding *eigenvector*.

Example 1

(a) The matrix

$$A = \begin{pmatrix} 1 & 1 \\ -2 & 4 \end{pmatrix}$$

has the characteristic equation

$$\det\begin{pmatrix} 1-\lambda & 1 \\ -2 & 4-\lambda \end{pmatrix} = \lambda^2 - 5\lambda + 6 = 0.$$

The characteristic roots are $\lambda_1 = 2$, $\lambda_2 = 3$. Corresponding eigenvectors are

$$\mathbf{u}_1 = \begin{pmatrix} 1 \\ 1 \end{pmatrix} \quad \text{and} \quad \mathbf{u}_2 = \begin{pmatrix} 1 \\ 2 \end{pmatrix}.$$

(b) The matrix

$$A = \begin{pmatrix} 0 & 1 \\ 1 & 0 \end{pmatrix}$$

has the characteristic equation $\lambda^2 - 1 = 0$, with the characteristic roots $\lambda_1 = 1$, $\lambda_2 = -1$. Corresponding eigenvectors are

$$\mathbf{u}_1 = \begin{pmatrix} 1 \\ 1 \end{pmatrix} \quad \text{and} \quad \mathbf{u}_2 = \begin{pmatrix} 1 \\ -1 \end{pmatrix}.$$

(c) The matrix

$$A = \begin{pmatrix} 0 & -1 \\ 1 & 0 \end{pmatrix}$$

has the characteristic roots $\lambda_1 = i$, $\lambda_2 = -i$ (i being the imaginary unit) and therefore no real eigenvectors.

If A is *symmetric* ($A^T = A$), then any two eigenvectors corresponding to different characteristic roots are orthogonal.

This can be proved by letting

$$A\mathbf{u}_h = \lambda_h \mathbf{u}_h, \qquad A\mathbf{u}_k = \lambda_k \mathbf{u}_k.$$

Then

$$\mathbf{u}_h^T A \mathbf{u}_k = \mathbf{u}_h^T \lambda_k \mathbf{u}_k = \lambda_k \mathbf{u}_h^T \mathbf{u}_k;$$
$$\mathbf{u}_h^T A \mathbf{u}_k = \mathbf{u}_h^T A^T \mathbf{u}_k = (A\mathbf{u}_h)^T \mathbf{u}_k = \lambda_h \mathbf{u}_h^T \mathbf{u}_k.$$

Hence $(\lambda_h - \lambda_k)\mathbf{u}_h^T \mathbf{u}_k = 0$ and the scalar product $\mathbf{u}_h^T \mathbf{u}_k$ must be zero.

1.4 Similar matrices

Consider an n-dimensional vector space whose elements $\mathbf{x}$, $\mathbf{y}$, ... will be written as column vectors. A *linear coordinate transformation* in this space is given by

$$\mathbf{y} = P\mathbf{x}, \tag{3.1.10}$$

where P is an $(n \times n)$ matrix. P must be nonsingular, that is, it must have an inverse P^{-1}, so that

$$\mathbf{x} = P^{-1}\mathbf{y} \tag{3.1.10'}$$

gives the inverse transformation. According to our interpretation of (3.1.10–3.1.10′) as coordinate changes, we think of $\mathbf{x}$ and $\mathbf{y}$ as representing the *same* geometrical vector.

The *general linear transformation* from an n-dimensional vector space into another n-dimensional vector space, which may be the same vector space, is also given by an expression of the type

$$\bar{\mathbf{x}} = A\mathbf{x}.$$

We now think of $\mathbf{x}$ and $\bar{\mathbf{x}}$ as *different* vectors ($\bar{\mathbf{x}}$ being the image of $\mathbf{x}$). Here the matrix A, which represents the linear transformation, does not need to have an inverse. [If A has no inverse, the transformation is called degenerate.]

Take A to represent a linear transformation from a vector space into itself. If, in this vector space, we change coordinates according to (3.1.10–3.1.10′), we get

$$\bar{\mathbf{y}} = P\bar{\mathbf{x}} = PA\mathbf{x} = PAP^{-1}\mathbf{y}.$$

Hence the same linear transformation is now represented, in the new $\mathbf{y}$-coordinates, by the matrix

$$B = PAP^{-1}. \tag{3.1.11}$$

Two matrices A, B, such that (3.1.11) holds for some nonsingular matrix P, are called *similar*. They represent the *same* linear transformation in different coordinate systems.

The similarity of matrices, written $A \sim B$, is an equivalence relation:

(a) $A \sim A$ for every matrix A,

(b) $A \sim B$ implies $B \sim A$,

(c) $A \sim B$ and $B \sim C$ imply $A \sim C$.

The proof is left to the reader.

Similar matrices have the same characteristic equation and therefore the same characteristic roots. The proof follows.

$$\begin{aligned}\det(B - \lambda E) &= \det(PAP^{-1} - \lambda E)\\ &= \det(PAP^{-1} - P\lambda EP^{-1})\\ &= \det[P(A - \lambda E)P^{-1}]\\ &= \det P \cdot \det(A - \lambda E) \cdot \det P^{-1}\\ &= \det(A - \lambda E).\end{aligned}$$

The converse is not true: matrices with the same characteristic equation may not be similar. An example is given by the (2×2) matrices

$$A = \begin{pmatrix} 0 & 1 \\ 0 & 0 \end{pmatrix} \quad \text{and} \quad B = \begin{pmatrix} 0 & 0 \\ 0 & 0 \end{pmatrix},$$

with characteristic equation $\lambda^2 = 0$. Both matrices cannot be similar because the linear transformations they represent are essentially different; that is, their difference could not be only a matter of a coordinate change.

1.5 Canonical form

Similar matrices represent the same linear transformation in different coordinate systems, as shown above. A common trick in analytic geometry is to choose the most convenient coordinate system for studying a given problem. Similarly, it is also reasonable to ask for the coordinate system in which a given linear transformation is represented by the simplest type of matrix. In other words, given a matrix, find a similar matrix that is as simple as possible.

It is possible to show that given any $(n \times n)$ matrix A, it can be transformed by a similarity transformation (3.1.11) into a matrix with the following structure (called the *Jordan canonical form*):

$$J = \begin{pmatrix} J_0 & 0 & 0 & \cdots & 0 \\ 0 & J_1 & 0 & \cdots & 0 \\ \cdots & \cdots & \cdots & \cdots & \cdots \\ 0 & 0 & 0 & \cdots & J_p \end{pmatrix}. \tag{3.1.12}$$

Here the zeros are, in general, rectangular submatrices. J_0 is a square submatrix of the type

$$J_0 = \begin{pmatrix} \lambda_1 & 0 & 0 & \cdots & 0 \\ 0 & \lambda_2 & 0 & \cdots & 0 \\ \cdots & \cdots & \cdots & \cdots & \cdots \\ 0 & 0 & 0 & \cdots & \lambda_r \end{pmatrix}, \tag{3.1.13}$$

while the remaining J_i's are of the form

$$J_i = \begin{pmatrix} \lambda_{r+i} & 1 & 0 & \cdots & 0 \\ 0 & \lambda_{r+i} & 1 & \cdots & 0 \\ \cdots & \cdots & \cdots & \cdots & \cdots \\ 0 & 0 & 0 & \cdots & \lambda_{r+i} \end{pmatrix}. \tag{3.1.14}$$

[Note that J_0 is a diagonal matrix, while for $i > 0$, J_i has ones in the parallel above the main diagonal.]

If we write out the characteristic equation of the matrix J (3.1.12), we see immediately that the only nonzero term in the development of the determinant is the diagonal; hence the numbers $\lambda_1, \lambda_2, \ldots, \lambda_{r+p}$ are the characteristic roots of J. All these numbers do not need to be distinct. Of course, some characteristic roots may also be complex. [A particularly interesting case is that in which all the n characteristic roots of J are distinct. In this case, the matrix J reduces to its diagonal part J_0.]

Given a matrix A, a basic problem is to transform it by a similarity transformation into its canonical form J. As seen in the preceding section, both A and J have the same characteristic roots. Hence the first step is to find and solve the characteristic equation of A. If all the characteristic roots are distinct, we can immediately write down J: it must reduce to its diagonal part J_0, and the numbers λ_i are the characteristic roots.

We should also mention that canonical matrices J, which differ only by the way their characteristic roots are ordered, are considered essentially the same. With this understanding, the canonical form of a matrix A is uniquely determined; that is, no two essentially different canonical matrices are similar.

When multiple roots appear, the knowledge of the characteristic roots of a matrix is not enough for writing down the canonical form. This becomes evident if we look, for example, at all possible canonical forms of (2×2) matrices:

(a) distinct characteristic roots λ_1, λ_2

$$\begin{pmatrix} \lambda_1 & 0 \\ 0 & \lambda_2 \end{pmatrix};$$

(b) two equal roots $\lambda_1 = \lambda_2 = \lambda$

$$\begin{pmatrix} \lambda & 0 \\ 0 & \lambda \end{pmatrix} \quad \text{and} \quad \begin{pmatrix} \lambda & 1 \\ 0 & \lambda \end{pmatrix}.$$

For $n = 3$, there are six possible canonical forms:

(a) distinct characteristic roots

$$\begin{pmatrix} \lambda_1 & 0 & 0 \\ 0 & \lambda_2 & 0 \\ 0 & 0 & \lambda_3 \end{pmatrix};$$

(b) two roots equal: $\lambda_1 \neq \lambda_2 = \lambda_3 = \lambda$

$$\begin{pmatrix} \lambda_1 & 0 & 0 \\ 0 & \lambda & 0 \\ 0 & 0 & \lambda \end{pmatrix} \quad \text{and} \quad \begin{pmatrix} \lambda_1 & 0 & 0 \\ 0 & \lambda & 1 \\ 0 & 0 & \lambda \end{pmatrix};$$

(c) three roots equal: $\lambda_1 = \lambda_2 = \lambda_3 = \lambda$

$$\begin{pmatrix} \lambda & 0 & 0 \\ 0 & \lambda & 0 \\ 0 & 0 & \lambda \end{pmatrix}, \quad \begin{pmatrix} \lambda & 0 & 0 \\ 0 & \lambda & 1 \\ 0 & 0 & \lambda \end{pmatrix}, \quad \text{and} \quad \begin{pmatrix} \lambda & 1 & 0 \\ 0 & \lambda & 1 \\ 0 & 0 & \lambda \end{pmatrix}.$$

Finally, note that a characteristic root may be zero. For the case of complex roots, it is possible to transform the Jordan canonical matrix into another real form, as we will see in the applications to differential equations.

1.6 Functions of matrices

Once we have defined the basic arithmetic operations with matrices, we can define the natural powers of a matrix

$$A^r = \overbrace{A \cdot A \cdot \dots \cdot A}^{r \text{ times}}.$$

Different powers of the same matrix commute:

$$A^r \cdot A^s = A^s \cdot A^r = A^{r+s}.$$

We also can define polynomials of a matrix. Hence

$$P(A) = c_0 E + c_1 A + c_2 A^2 + \cdots + c_n A^n,$$

where $c_0, \ldots, c_n$ are scalars (constants).

Changing a matrix by a similarity transformation $A = QBQ^{-1}$, the powers A^r are changed into

$$\begin{aligned} A^r &= (QBQ^{-1})^r = (QBQ^{-1})(QBQ^{-1}) \cdots (QBQ^{-1}) \\ &= QB^rQ^{-1}. \end{aligned}$$

Similarly, for a polynomial $P(A)$,

$$\begin{aligned} P(A) = P(QBQ^{-1}) &= c_0 E + c_1 QBQ^{-1} + c_2 QB^2Q^{-1} \\ &\quad + \cdots + c_n QB^nQ^{-1} \\ &= Q(c_0 E + c_1 B + c_2 B^2 + \cdots + c_n B^n)Q^{-1} \\ &= QP(A)Q^{-1}. \end{aligned} \tag{3.1.15}$$

The limit of a matrix is defined by means of the norm (3.1.6). Hence, for a sequence $A_1, A_2, A_3, \ldots$

$$A = \lim_{i \to \infty} A_i \quad \text{if and only if} \quad \lim_{i \to \infty} \|A - A_i\| = 0. \tag{3.1.16}$$

It is therefore meaningful to write series of matrices. For example, one can define the exponential of a matrix by the series

$$e^A = E + \frac{A}{1} + \frac{A^2}{2} + \frac{A^3}{3!} + \cdots + \frac{A^n}{n!} + \cdots. \tag{3.1.17}$$

This function will certainly be defined only for those matrices for which this series is convergent. It is surprisingly easy to prove that the exponential series converges for every matrix A.

1.7 Theorem

The exponential series (3.1.17) *converges for every matrix A.*

The definition (3.1.16) of convergence is equivalent to the condition that each element $a^h_{k(i)}$ of the matrix A_i should tend, for $i \to \infty$, to the corresponding element a^h_k of the matrix A. By Cauchy's convergence criterion, a necessary and sufficient condition for the sequence $a^h_{k(i)}$ to be convergent is that $[a^h_{k(i)} - a^h_{k(j)}]$ tend to zero as $i \to \infty, j \to \infty$ (independently). If this is true for every element of the matrix, then $\|A_i - A_j\| \to 0$ as $i \to \infty$, $j \to \infty$, and conversely. Cauchy's criterion can therefore be applied to matrices.

It follows that the condition for convergence of the series (3.1.17) is

$$\lim_{\substack{p \to \infty \\ q > p}} \left\| \sum_{n=p}^{q} \frac{A^n}{n!} \right\| = 0. \tag{3.1.18}$$

But

$$\left\| \sum_{n=p}^{q} \frac{A^n}{n!} \right\| \leqq \sum_{n=p}^{q} \left\| \frac{A^n}{n!} \right\| \leqq \sum_{n=p}^{q} \frac{\|A\|^n}{n!}.$$

The latter is the correspondent partial sum of the usual numerical exponential series for $e^{\|A\|}$, which is convergent for every $\|A\|$; hence (3.1.18) is satisfied.

1.8 Diagonal matrices—logarithms

If A is diagonal

$$A = \begin{pmatrix} a_1 & & & \\ & a_2 & & \\ & & \ddots & \\ & & & a_n \end{pmatrix}$$

(all nonwritten elements are zero), it is easy to see that

$$e^A = \begin{pmatrix} a^{a_1} & & & \\ & e^{a_2} & & \\ & & \ddots & \\ & & & e^{a_n} \end{pmatrix}. \tag{3.1.19}$$

The same formula holds if the a_i's are submatrices.

If A is diagonal, then the matrix

$$B = \begin{pmatrix} \ln a_1 & & & \\ & \ln a_2 & & \\ & & \ddots & \\ & & & \ln a_n \end{pmatrix} \tag{3.1.20}$$

has the property that $e^B = A$. It can be shown that for every nonsingular matrix A, there is some B satisfying this relation, and such a matrix B is therefore a logarithm of A. This logarithm matrix is not uniquely defined. [Even in the field of complex numbers, the logarithm is not unique.]

Problems

1. Prove formulas (3.1.7).
2. If the norm of a matrix is defined by $\|A\| = \max |a_j^i|$ and then if A is an $(n \times m)$ and B an $(m \times r)$ matrix, prove that

$$\|AB\| \leq m \|A\| \|B\|.$$

3. If for vectors $x = (x^1, \ldots, x^n)$ we define two norms

$$\|x\|_1 = \sum |x^i|, \qquad \|x\|_2 = \sqrt{\sum (x^i)^2},$$

prove that $\|x\|_1 \to 0 \Leftrightarrow \|x\|_2 \to 0$.

4. (a) Show that if A is an $(n \times n)$ matrix and $\lambda = 0$ is one of its characteristic roots, then A is singular. (b) Also show that in this case the linear transformation $R^n \to R^n$, represented by A, maps R^n into a proper linear subspace of R^n.
5. Prove the converse of the last statement (b) in Problem 4.
6. Show that for an upper triangular matrix (a_j^i) [upper triangular means that $a_j^i = 0$ for $i > j$], the characteristic roots are the elements a_j^i of the main diagonal.
7. Show that any skew-symmetric $(n \times n)$ matrix A (i.e., $A = -A^T$) with n odd is singular.
8. Find the (2×2) matrix whose correspondent linear transformation is
   (a) a rotation through an angle α [notice the cases $\alpha = \pi/2$, $\alpha = \pi$],
   (b) a symmetry with respect to the x-axis,
   (c) a symmetry with respect to the line $x = y$.
9. Write all canonical (4×4) matrices with all characteristic roots equal to 2.
10. Find the characteristic roots and eigenvectors of the rotation matrix

$$A = \begin{pmatrix} \cos\alpha & -\sin\alpha \\ \sin\alpha & \cos\alpha \end{pmatrix}.$$

For which values of α are the eigenvectors real?

11. Prove that every $(n \times n)$ matrix with n odd has at least one real eigenvector. Show that this is not the case for n even.

12. Prove that

$$\begin{pmatrix} 3 & 2 \\ 1 & -1 \end{pmatrix} \quad \text{and} \quad \begin{pmatrix} 1 & 3 \\ 2 & 1 \end{pmatrix}$$

are similar.

13. Prove that any matrix A and its transpose A^T are similar. [*Hint*: find P in $A^T = PAP^{-1}$.]

14. Find the characteristic roots and eigenvectors of

(a) $\begin{pmatrix} -1 & -1 & 3 \\ -4 & 1 & 4 \\ -4 & -1 & 6 \end{pmatrix}$, (b) $\begin{pmatrix} 5 & 0 & 1 \\ 6 & 2 & -4 \\ 6 & 0 & -2 \end{pmatrix}$,

(c) $\begin{pmatrix} 1 & -1 & 1 \\ 0 & 0 & 1 \\ 0 & -1 & 2 \end{pmatrix}$, (d) $\begin{pmatrix} 0 & -1 & 2 \\ 0 & 0 & 1 \\ -1 & -1 & 3 \end{pmatrix}$.

15. Find the canonical forms of the matrices

(a) $\begin{pmatrix} -4 & 4 \\ -3 & 4 \end{pmatrix}$, (b) $\begin{pmatrix} 23 & -12 \\ 36 & -19 \end{pmatrix}$, (c) $\begin{pmatrix} 21 & -12 \\ 36 & -21 \end{pmatrix}$,

(d) $\begin{pmatrix} 16 & -10 \\ 26 & -16 \end{pmatrix}$, (e) $\begin{pmatrix} 19 & -10 \\ 26 & -13 \end{pmatrix}$, (f) $\begin{pmatrix} -7 & 5 \\ -13 & 9 \end{pmatrix}$.

16. Find the canonical forms of the matrices in Problem 14.

17. Prove that if the nonsingular matrix P commutes with A, then A is invariant with respect to the coordinate transformation represented by P. Prove the converse.

18. If A is symmetric and $X' = AX$, prove that

$$(X')^T X = X^T X'.$$

19. If A is skew-symmetric ($A^T = -A$) and $X' = AX$, prove that

$$X^T X = \text{const}.$$

Here X' is the derivative of X.

2 *Linear Homogeneous Differential Equation*

2.1 *Existence and uniqueness of the solution*

The vector differential equation

$$\mathbf{x}' = A(t)\mathbf{x}, \tag{3.2.1}$$

where $\mathbf{x}$ is an n vector and $A(t)$ is an $(n \times n)$ matrix whose elements are functions of t, is called *linear homogeneous*. Written componentwise, (3.2.1) is

$$x^{i'} = \sum_{j=1}^{n} a_j^i(t)x^j \quad i = 1, 2, \ldots, n,$$

from which it is clear that (3.2.1) represents a system of scalar differential equations. Equation (3.2.1) is sometimes referred to as vector-matrix differential equation.

We will assume that the elements $a_j^i(t)$ are measurable and integrable functions of t in the interval $[a, b]$ or, briefly, that $A(t)$ has this property. Then, given the initial condition

$$\mathbf{x}(t_0) = \mathbf{x}_0 \tag{3.2.2}$$

or

$$x^i(t_0) = x_0{}^i \qquad i = 1, 2, \ldots, n,$$

where $t_0 \in [a, b]$, there exists a unique solution in some neighborhood of t_0 by virtue of the existence theorem. Theorem 2.7.3 does not, however, assure the existence of the solution in the whole interval $[a, b]$. This will be proved in the following theorem.

2.2 Theorem

If $A(t)$ is measurable and integrable on $[a, b]$, then (3.2.1–3.2.2) has a unique solution, which exists in the whole interval $[a, b]$.

Proof

Since all elements of $A(t)$ are integrable in $[a, b]$, there is a positive function $k(t)$, integrable in $[a, b]$, such that

$$\|A(t)\| \leqq k(t).$$

In every bounded region of x space, the right-hand side of (3.2.1) is majorized in norm by $Mk(t)$ for a suitable M; hence Theorem 2.7.3 assures the existence and uniqueness of the solution. It remains only to be proven that it cannot tend to infinity in finite time.

We may write, for $t_1 > t$ [with $\mathbf{x} = \mathbf{x}(t)$, $\mathbf{x}_1 = \mathbf{x}(t_1)$],

$$\frac{d}{dt}\|\mathbf{x}\| = \lim \frac{\|\mathbf{x}_1\| - \|\mathbf{x}\|}{t_1 - t} \leqq \lim \frac{\|\mathbf{x}_1 - \mathbf{x}\|}{t_1 - t}$$

$$= \left\|\frac{dx}{dt}\right\| = \|A(t)\mathbf{x}\| \leqq \|A(t)\| \; \|\mathbf{x}\| \leqq k(t)\|\mathbf{x}\|.$$

Hence $\|\mathbf{x}(t)\|$ is majorized by the solution of the equation $z' = k(t)z$, namely,

$$z(t) = z(t_0) \exp \{\textstyle\int_t^{t_0} k(s)\, ds\}.$$

This proves that in the interval $[t_0, b]$ the solution $\mathbf{x}(t)$ cannot escape to infinity and therefore exists up to $t = b$. For the interval $[a, t_0]$, the proof is similar.

2.3 Trivial solution. Linear independence

The function $\mathbf{x}(t) \equiv 0$ always satisfies equation (3.2.1) and is called the *trivial solution.* [In most cases, when speaking about any solution of this equation, a nontrivial solution is inferred.]

The functions $\mathbf{x}_1(t), \mathbf{x}_2(t), \ldots, \mathbf{x}_n(t)$ *are called linearly independent in an interval* $[a, b]$ *if the only linear combination*

$$\mathbf{x}_1(t)c^1 + \mathbf{x}_2(t)c^2 + \cdots + \mathbf{x}_n(t)c^n,$$

with constant coefficients c^i, *which is identically zero in the interval* $[a, b]$, *is the one corresponding to* $c^1 = c^2 = \cdots = c^n = 0$. *Otherwise, they are linearly dependent.*

2.4 Linearity Theorem

The set of all solutions of equation (3.2.1), *where* (3.2.1) *is assumed to satisfy the existence and uniqueness conditions of Theorem* 3.2.2, *forms a vector space of dimension n (equal to the dimension of the vector* $\mathbf{x}$).

The conclusion can also be stated as follows: (i) the sum and difference of two solutions are again solutions, (ii) a constant multiple of a solution is a solution, (iii) there is some set of exactly n linearly independent solutions such that every other solution is a linear combination of these.

Proof

It is easy to check that if $\mathbf{x}(t)$ and $\mathbf{y}(t)$ satisfy equation (3.2.1), then so do $\mathbf{x}(t) + \mathbf{y}(t)$, $\mathbf{x}(t) - \mathbf{y}(t)$, and $k\mathbf{x}(t)$, for k a constant.

In order to show that there exists a base of n linearly independent solutions, let us define the solutions

$$\mathbf{x}_1 = \begin{pmatrix} x_1^1 \\ x_1^2 \\ \vdots \\ x_1^n \end{pmatrix}, \qquad \mathbf{x}_2 = \begin{pmatrix} x_2^1 \\ x_2^2 \\ \vdots \\ x_2^n \end{pmatrix}, \qquad \ldots, \qquad \mathbf{x}_n = \begin{pmatrix} x_n^1 \\ x_n^2 \\ \vdots \\ x_n^n \end{pmatrix} \tag{3.2.3}$$

by the differential equation and the following initial conditions:

$$\mathbf{x}_1(t_0) = \begin{pmatrix} 1 \\ 0 \\ 0 \\ \vdots \\ 0 \end{pmatrix}, \qquad \mathbf{x}_2(t_0) = \begin{pmatrix} 0 \\ 1 \\ 0 \\ \vdots \\ 0 \end{pmatrix}, \qquad \ldots, \qquad \mathbf{x}_n(t_0) = \begin{pmatrix} 0 \\ 0 \\ 0 \\ \vdots \\ 1 \end{pmatrix}. \tag{3.2.4}$$

Assuming that for some $t = t_1 \in [a, b]$,

$$c_1 \mathbf{x}_1(t_1) + c_2 \mathbf{x}_2(t_1) + \cdots + c_n \mathbf{x}_n(t) = 0,$$

it follows that the function

$$\mathbf{x}(t) = c_1 \mathbf{x}_1(t) + c_2 \mathbf{x}_2(t) + \cdots + c_n \mathbf{x}_n(t)$$

satisfies the differential equation and the condition $\mathbf{x}(t_1) = 0$. Since the solution is unique, it must therefore coincide with the trivial solution. But, for $t = t_0$,

$$\mathbf{x}(t_0) = \begin{pmatrix} c_1 \\ c_2 \\ \vdots \\ c_n \end{pmatrix} = 0;$$

hence $c_1 = c_2 = \cdots = c_n = 0$.

Finally, let us prove that all solutions are linear combinations of the solutions defined above. All we need note is that any initial condition

$$\mathbf{x}(t_0) = \begin{pmatrix} x^1(t_0) \\ x^2(t_0) \\ \vdots \\ x^n(t_0) \end{pmatrix}$$

can be represented as the linear combination

$$\mathbf{x}(t_0) = c_1 \mathbf{x}_1(t_0) + c_2 \mathbf{x}_2(t_0) + \cdots + c_n \mathbf{x}_n(t_0),$$

with $c_i = x^i(t_0)$. Hence, by the uniqueness of the solution, for all values of $t \in [a, b]$, it will be true that

$$\mathbf{x}(t) = c_1\mathbf{x}_1(t) + c_2\mathbf{x}_2(t) + \cdots + c_n\mathbf{x}_n(t),$$

which completes the proof.

Problems

1. Indicate in which intervals the existence and uniqueness theorem (Theorem 3.2.2) applies to the vector differential equation $\mathbf{x}' = A(t)\mathbf{x}$, where

(a) $A = \begin{pmatrix} 2 & t^2 \\ t & t^3 \end{pmatrix}$, (b) $A = \begin{pmatrix} t & \cot t \\ 1 & \sin t \end{pmatrix}$,

(c) $A = \begin{pmatrix} \sin t & \cos t \\ 1 - t & 1 + t \end{pmatrix}$, (d) $A = \begin{pmatrix} t^{-1} & 1 \\ 1 & t \end{pmatrix}$,

(e) $A = \begin{pmatrix} (1 + t^2)^{-1} & (2 - t^2)^{-1} \\ t^{-1} & t^{-2} \end{pmatrix}$, (f) $A = \begin{pmatrix} \cot 2t & t \\ \tan 2t & 1 \end{pmatrix}$.

2. Given the differential equation $\mathbf{x}' = A(t)\mathbf{x}$ with the initial condition $\mathbf{x}^T(0) = (1, 0, 1)$ [$\mathbf{x}$ is a column vector, so $\mathbf{x}^T$ is a row vector] and knowing that $\|A(t)\| < 5$ in the interval $[0, 2]$, estimate $\|\mathbf{x}(2)\|$ according to the inequality appearing in the proof of Theorem 3.2.2.

3. Indicate if the following (scalar) functions $x_1(t)$, $x_2(t)$ are linearly independent in the indicated intervals:

(a) $x_1(t) = \begin{cases} 0 & \text{for } t < 0 \\ t & \text{for } t \geqq 0 \end{cases}$ $\quad x_2(t) = \begin{cases} 0 & \text{for } t < 1 \\ t - 1 & \text{for } t \geqq 1 \end{cases}$

in the intervals $[-1, 0]$, $[-1, +1]$, $[0, 1]$, $[1, 2]$;

(b) $x_1(t) = \begin{cases} 0 & \text{for } t < 0 \\ t & \text{for } 0 \leqq t \leqq 1 \\ 1 & \text{for } 1 < t \end{cases}$ $\quad x_2(t) = \begin{cases} 0 & \text{for } t < 0 \\ 2t & \text{for } 0 \leqq t \leqq 1 \\ 2 & \text{for } 2 \leqq t \end{cases}$

in the intervals $[-1, 0]$, $[-1, +1]$, $[-1, +2]$.

4. Assume that for some given vector differential equation the vector functions

$$\mathbf{x}_1(t) = \begin{pmatrix} t \\ t^2 - 1 \end{pmatrix}, \quad \mathbf{x}_2(t) = \begin{pmatrix} 2t \\ t^2 \end{pmatrix}.$$

are linearly independent solutions.
Show that the vector functions

$$\mathbf{y}_1(t) = \begin{pmatrix} t \\ 1 \end{pmatrix}, \quad \mathbf{y}_2(t) = \begin{pmatrix} 0 \\ t^2 - 2 \end{pmatrix}$$

are also linearly independent solutions.

2.5 Fundamental system

Any set of n linearly independent solutions of a homogeneous linear differential equation is called a *fundamental system* of this equation. Furthermore, if a fundamental system is known, any other fundamental system can be obtained as a set of n linearly independent combinations of the solutions of the first system.

It should be noted that all this is true so long as the matrix function $A(t)$ is integrable in $[a, b]$. For those values of t where this condition is violated, all kinds of strange behavior may result. For example, *Bessel's equation*, written as a system, is

$$x' = y, \qquad y' = -x - t^{-1}y;$$

hence there is a singularity at the point $t = 0$. Indeed, infinitely many solutions of this equation become infinite as $t \to 0$, and for all those which remain finite, $y \to 0$ as $t \to 0$. [Hence $y(0)$ cannot be given arbitrarily.]

3 Associated Matrix Equation

3.1 Fundamental matrix

Given equation (3.2.1), we may write the following equation

$$X' = A(t)X, \tag{3.3.1}$$

called the *associated matrix equation* of (3.2.1), where $X = X(t)$ is now a matrix of dimension $(n \times n)$.

Since the derivative of a matrix is obtained by element-wise differentiation, equation (3.3.1) is equivalent to

$$x_j^{i'} = \sum_k a_k^i x_j^k \quad i, j = 1, 2, \ldots, n. \tag{3.3.2}$$

Hence every column of the matrix (x_j^i) is a vector which satisfies the vector differential equation (3.2.1).

Conversely, if we write as columns of $X(t)$ any n vectors $\mathbf{x}(t)$ satisfying the vector equation (3.2.1), the matrix we then obtain is a solution of (3.3.1).

A matrix $X(t)$ is called a *fundamental matrix solution* of (3.2.1) if its columns are n linearly independent solutions of this equation. Thus $X(t)$ also satisfies equation (3.3.1).

If $X(t)$ is a fundamental matrix of (3.2.1) and if, for $t = t_0$,

$$X(t_0) = E \tag{3.3.3}$$

(E = unit matrix), then $X(t)$ is called the *principal fundamental matrix.*

3.2. Let $X(t)$ be a fundamental matrix of (3.2.1). By definition, its columns are n linearly independent solutions of (3.2.1); hence by Theorem 3.2.4 every solution $\mathbf{x}(t)$ of this equation is a linear combination of these n columns

$$x^i(t) = \sum_j x^i_j(t)c^j \quad i = 1, 2, \ldots, n. \tag{3.3.4}$$

[Here c^j are arbitrary constants; we write j as a superindex because $\mathbf{c}$ is considered a column vector.]

Conversely, every such linear combination is a solution $\mathbf{x}(t)$ of equation (3.2.1). Hence (3.3.4) represents the general solution of this equation.

In matrix notation, (3.3.4) is

$$\mathbf{x}(t) = X(t)\mathbf{c}, \tag{3.3.5}$$

where $\mathbf{c}$ is an arbitrary constant vector. This is the importance of a fundamental matrix: it gives us all the solutions of the vector differential equation.

It should be noted that the product $\mathbf{c}X(t)$, where $\mathbf{c}$ is now a row vector, *does not*, in general, satisfy equation (3.2.1).

A direct proof that (3.3.5) satisfies equation (3.2.1) is found in

$$\mathbf{x}'(t) = X'(t)\mathbf{c} = A(t)X(t)\mathbf{c} = A(t)\mathbf{x}(t).$$

The shortness of this proof shows the advantages of using matrix notation.

Now let $X(t)$ be the principal fundamental matrix. Then the solution of (3.2.1), satisfying the initial condition $\mathbf{x}(t_0) = \mathbf{x}_0$, is

$$\mathbf{x}(t) = X(t)\mathbf{x}_0 .$$

Indeed, $\mathbf{x}_0$ is a constant vector; so $\mathbf{x}(t)$ satisfies the differential equation. For $t = t_0$, on the other hand, $\mathbf{x}(t_0) = X(t_0)\mathbf{x}_0 = E\mathbf{x}_0 = \mathbf{x}_0$.

3.3 *Theorem*

A matrix $X(t)$, *satisfying equation* (3.3.1), *is a fundamental solution if and only if* $\det X(t) \neq 0$ *for all* $t \in [a, b]$.

Proof

If $\det X(t) \neq 0$, the columns of $X(t)$ are linearly independent, which proves the sufficiency of this condition. In addition, it is sufficient that $\det X(t) \neq 0$ for any single value of t, because by the uniqueness of the solutions, any linear relationship between the columns existing for some value of t must also hold throughout the entire interval of existence. [Compare this with the proof of Theorem 3.2.4.]

Conversely, if $X(t)$ is a fundamental matrix, every vector solution $\mathbf{y}(t)$ can be represented as a linear combination of the columns of $X(t)$. Thus

$$y^i(t) = \sum_j x^i_j(t)c^j.$$

For each t_0, the initial condition

$$y^i(t_0) = \sum_j x^i_j(t_0)c^j \tag{3.3.6}$$

may be given arbitrarily, thereby obtaining unique values for the c^j. But this implies $\det (x^i_j(t_0)) \neq 0$ for each t_0.

3.4 *Remark*

Not every matrix $X(t)$, whose columns are linearly independent vector functions in $[a, b]$, has a nonzero determinant. It is easily seen that the columns of

$$\begin{pmatrix} t & 1+t \\ 0 & 0 \end{pmatrix}$$

are linearly independent, while the determinant of this matrix is identically zero. As a consequence of Theorem 3.3.3, a situation such as this cannot arise if the columns of the matrix are solutions of a linear system of differential equations. Note that for each value of t, the columns of this matrix, when considered as numerical vectors, are linearly dependent in the sense that they belong to some proper subspace, but this condition is not sufficient to make them linearly dependent in terms of functions, as defined in 3.2.3.

3.5 Theorem

If $X(t)$ is any fundamental matrix of (3.2.1) and C is any constant $(n \times n)$ matrix, then $X(t)C$ is a solution matrix of (3.3.1). If, furthermore, C is nonsingular, then $X(t)C$ is another fundamental matrix of (3.2.1).

Proof

Part I:

$$\frac{d}{dt}[X(t)C] = X'(t)C = [A(t)X(t)]C = A(t)[X(t)C].$$

Part II:

$$\det [X(t)C] = \det X(t) \det C \neq 0.$$

3.6 Theorem

If the matrix $X(t) = (x^i_j(t))$ is continuously differentiable in $[a, b]$ and if det $X(t)$ *is never zero in $[a, b]$, then $X(t)$ is the fundamental matrix of a certain unique differential equation of the type* (3.2.1).

Proof

Assume that such a differential system (having $X(t)$ as a fundamental matrix)

$$x^{i'}_k(t) = \sum_j a^i_j(t)x^j_k(t) \quad i, k = 1, 2, \ldots, n,$$

exists; then let us determine its coefficients $a^i_j(t)$. Note that $x^j_k(t)$ and $x^{i'}_k(t)$ are known functions. For fixed i $(1 \leqq i \leqq n)$, the differential system constitutes an algebraic system of n linear equations for the n unknowns $a^i_1, a^i_2, \ldots. a^i_n$. The determinant of the coefficients of this linear algebraic system is det $X(t)$, and since it is not zero, there will be a unique solution for each value of t. We have thus obtained the ith column of $A(t)$. In the same way, we can determine all columns, constructing the matrix $A(t)$ whose existence is therefore proved. It also follows that $A(t)$ is uniquely determined. Finally, it is easy to see that the functions $a^i_j(t)$ are continuous in $[a, b]$.

3.7 *Liouville's Theorem* [*also known as Abel's Theorem*]

If $X(t)$ *satisfies* (3.3.1), *then*

$$\det X(t) = \det X(t_0) \exp \int_{t_0}^{t} \operatorname{tr} A(s)\, ds. \tag{3.3.7}$$

Here, $\exp u = e^u$ *and* $\operatorname{tr} A = $ *trace of* $A = a_1^1 + a_2^2 + \cdots + a_n^n$.

Proof

By the rule of differentiation of a determinant,

$$\frac{d}{dt} \det X(t) = \begin{vmatrix} x_1^{1'} & x_2^{1'} & \cdots & x_n^{1'} \\ x_1^2 & x_2^2 & \cdots & x_n^2 \\ \cdots & \cdots & \cdots & \cdots \\ x_1^n & x_2^n & & x_n^n \end{vmatrix} + \begin{vmatrix} x_1^1 & x_2^1 & \cdots & x_n^1 \\ x_1^{2'} & x_2^{2'} & \cdots & x_n^{2'} \\ \cdots & \cdots & \cdots & \cdots \\ x_1^n & x_2^n & & x_n^n \end{vmatrix} + \cdots$$
$$+ \begin{vmatrix} x_1^1 & x_2^1 & \cdots & x_n^1 \\ x_1^2 & x_2^2 & \cdots & x_n^2 \\ \cdots & \cdots & \cdots & \cdots \\ x_1^{n'} & x_2^{n'} & \cdots & x_n^{n'} \end{vmatrix}.$$

In the first determinant in the sum on the right side, we may replace $x_j^{1'}$ by $\sum_k a_k^1 x_j^k$. Then

$$\begin{vmatrix} \sum a_k^1 x_1^k & \sum a_k^1 x_2^k & \cdots & \sum a_k^1 x_n^k \\ x_1^2 & x_2^2 & \cdots & x_n^2 \\ \cdots & \cdots & \cdots & \cdots \\ x_1^n & x_2^n & \cdots & x_n^n \end{vmatrix} = \sum a_k^1 \begin{vmatrix} x_1^k & x_2^k & \cdots & x_n^k \\ x_1^2 & x_2^2 & \cdots & x_n^2 \\ \cdots & \cdots & \cdots & \cdots \\ x_1^n & x_2^n & \cdots & x_n^n \end{vmatrix} = a_1^1 \det X(t),$$

since all determinants in the sum on the right are zero, except when $k = 1$.

Proceeding in this way with the other determinants in the sum, we obtain

$$\frac{d}{dt} \det X(t) = (a_1^1 + a_2^2 + \cdots + a_n^n) \det X(t).$$

This is an equation of the type $y' = b(t)y$, which is integrated by separation of variables. Thus (3.3.7) is obtained.

As a corollary of this theorem, if $\det X(t)$ is different from zero for some value of t, then it is different from zero for every t within the interval $[a, b]$.

3.8 Reduction of order of a system

If one solution

$$\mathbf{u}(t) = \begin{pmatrix} u^1(t) \\ u^2(t) \\ \vdots \\ u^n(t) \end{pmatrix}$$

of equation (3.2.1) *is known, then the problem of finding the general solution of this equation is reduced to finding the general solution of some differential system of order* $n - 1$, *plus an integration.*

To prove this statement, let us assume that some component of the vector $\mathbf{u}(t)$ ($u^1(t)$, for example) is different from zero throughout $[a, b]$. If this were not the case, we could subdivide $[a, b]$ into subintervals, on each of which such a statement is correct, and then carry out the reduction of order on each subinterval. Note that, by the uniqueness of the trivial solution, not all components of $\mathbf{u}(t)$ may vanish simultaneously at any point.

Define the matrices

$$U(t) = \begin{pmatrix} u^1 & 0 & 0 & \cdots & 0 \\ u^2 & 1 & 0 & \cdots & 0 \\ u^3 & 0 & 1 & \cdots & 0 \\ \cdots & \cdots & \cdots & \cdots & \cdots \\ u^n & 0 & 0 & \cdots & 1 \end{pmatrix}, \quad U^{-1}(t) = \begin{pmatrix} \dfrac{1}{u^1} & 0 & 0 & \cdots & 0 \\ \dfrac{-u^2}{u^1} & 1 & 0 & \cdots & 0 \\ \dfrac{-u^3}{u^1} & 0 & 1 & \cdots & 0 \\ \cdots & \cdots & \cdots & \cdots & \cdots \\ \dfrac{-u^n}{u^1} & 0 & 0 & \cdots & 1 \end{pmatrix}$$

and make the change of variables

$$\mathbf{x} = U\mathbf{y}, \qquad \mathbf{y} = U^{-1}\mathbf{x}.$$

Then

$$\mathbf{x}' = A\mathbf{x} = AU\mathbf{y}$$

and

$$\mathbf{x}' = (U\mathbf{y})' = U'\mathbf{y} + U\mathbf{y}'.$$

Hence

$$\mathbf{y}' = U^{-1}(AU - U')\mathbf{y} = B\mathbf{y}, \tag{3.3.8}$$

where B is defined by this equation. Carrying out the computation,

$$B = \begin{pmatrix} 0 & a_2^1 & \cdots & a_n^1 \\ 0 & -\dfrac{u^2}{u^1}a_2^1 + a_2^2 & \cdots & -\dfrac{u^2}{u^1}a_n^1 + a_n^2 \\ \cdots & \cdots & \cdots & \cdots \\ 0 & -\dfrac{u^n}{u^1}a_2^1 + a_2^n & \cdots & -\dfrac{u^n}{u^1}a_n^1 + a_n^n \end{pmatrix}$$

is obtained. Since this matrix has only zeros in its first column, the system (3.3.8) is of the form

$$\begin{cases} y^{1'} = f^1(y^2, y^3, \ldots, y^n), \\ y^{i'} = f^i(y^2, y^3, \ldots, y^n) \quad i = 2, 3, \ldots, n. \end{cases}$$

Hence we can first solve the system for the $n - 1$ unknowns $y^2, y^3, \ldots, y^n$ and then obtain y^1 by integration.

Similarly, one can prove that if for the system (3.2.1), m linearly independent solutions are known, the remaining problem reduces to the solution of a differential system of order $(n - m)$, plus m integrations.

Problems

1. In the following examples, prove that $X(t)$ satisfies the differential equation $X' = AX$. Indicate in which cases $X(t)$ is a fundamental matrix.

(a) $A = \begin{pmatrix} 1 & 0 \\ 0 & 2 \end{pmatrix}, \quad X = \begin{pmatrix} e^t & e^t \\ e^{2t} & -e^{2t} \end{pmatrix}.$

(b) $A = \begin{pmatrix} 1 & 0 \\ 0 & -1 \end{pmatrix}, \quad X = \begin{pmatrix} e^t & e^t \\ 0 & 0 \end{pmatrix}.$

(c) $A = \begin{pmatrix} 0 & 1 \\ -1 & 0 \end{pmatrix}, \quad X = \begin{pmatrix} \sin t & \cos t \\ \cos t & -\sin t \end{pmatrix}.$

(d) $A = \begin{pmatrix} 0 & 1 \\ 1 & 0 \end{pmatrix}, \quad X = \begin{pmatrix} \operatorname{Sinh} t & \operatorname{Cosh} t \\ \operatorname{Cosh} t & \operatorname{Sinh} t \end{pmatrix}.$

(e) $A = \begin{pmatrix} 0 & 1 \\ 1 & 0 \end{pmatrix}, \quad X = \begin{pmatrix} e^{-t} & -e^{-t} \\ -e^{-t} & e^{-t} \end{pmatrix}.$

2. Find the differential equations $\mathbf{x}' = A(t)\mathbf{x}$ having the following matrices as fundamental solutions

(a) $\begin{pmatrix} 2 & t \\ t^2 & t^3 \end{pmatrix}$, (b) $\begin{pmatrix} t & 0 \\ 1 & t^{-1} \end{pmatrix}$,

(c) $\begin{pmatrix} \cos t & -\cos t \\ \sin t & \sin t \end{pmatrix}$, (d) $\begin{pmatrix} e^t & -e^{2t} \\ e^{2t} & e^{3t} \end{pmatrix}$.

4 Adjoint Equation

4.1 Definition

Given equation (3.2.1), *where* $\mathbf{x}$ *is a column vector of dimension* n, *the adjoint equation is defined to be*

$$\mathbf{y}' = -\mathbf{y}A(t), \tag{3.4.1}$$

where now $\mathbf{y} = \mathbf{y}(t)$ *is a row vector of the same dimension* n.

Letting $\mathbf{z} = \mathbf{y}^T$, the transpose of $\mathbf{y}$, equation (3.4.1) can be written in the form

$$\mathbf{z}' = -A^T(t)\mathbf{z}, \tag{3.4.1$'$}$$

where $\mathbf{z}$ is now a column vector. In the form (3.4.1), it is easy to forget that, once developed componentwise, this system has as a coefficient matrix the transpose of A. For this reason, the form (3.4.1′) is preferred, but both are obviously equivalent.

4.2 Properties

If $\mathbf{x}(t)$ and $\mathbf{y}(t)$ satisfy, respectively, equations (3.2.1) and (3.4.1), then the scalar product $\mathbf{y}(t)\mathbf{x}(t)$ is constant. The proof is

$$\frac{d}{dt}(\mathbf{yx}) = \mathbf{y}'\mathbf{x} + \mathbf{y}\mathbf{x}' = -\mathbf{y}A\mathbf{x} + \mathbf{y}A\mathbf{x} = 0.$$

The same holds if the matrices $X(t)$, $Y(t)$ satisfy, respectively, the equations $X' = AX$, $y' = -YA$. The proof is the same.

Conversely, if $X(t)$ is nonsingular and satisfies $X' = AX$, while $Y(t)$ is such that $YX = \text{constant}$, then

$$0 = \frac{d}{dt}(YX) = Y'X + YX' = Y'X + YAX;$$

hence

$$Y' = (-YAX)X^{-1} = -YA.$$

This is the case, for example, if $Y(t) = X^{-1}(t)$.

4.3 Remarks

The last proof *cannot* be applied to vectors having a constant scalar product. Indeed, given $\mathbf{x}(t)$, the condition $\mathbf{y}(t)\mathbf{x}(t) = \text{constant}$ does not determine $\mathbf{y}(t)$. On the other hand, the last proof can be applied to the case when $\mathbf{y}(t)$ is a row vector and $X(t)$ is a nonsingular matrix. Then $\mathbf{y}(t)X(t) = \text{constant}$ implies $\mathbf{y}' = -\mathbf{y}A(t)$ (provided $X' = A(t)X$). In other words, if the scalar product of $\mathbf{y}(t)$, with *all solutions* of (3.2.1), is constant, then $\mathbf{y}(t)$ satisfies the adjoint equation.

If $X(t)$ is a fundamental matrix of (3.2.1), then a necessary and sufficient condition for $Y(t)$ to be a fundamental matrix of (3.4.1) is that there exists a nonsingular constant matrix C such that $YX = C$.

Problems

1. Given the C^1 matrix $A(t)$, let $\mathbf{x}(t)$ be a column vector and $\mathbf{y}(t)$ a row vector. Define

$$L[\mathbf{x}] = A\mathbf{x} + \mathbf{x}',$$
$$M[\mathbf{y}] = \mathbf{y}A - \mathbf{y}'.$$

Show that

$$\mathbf{y}L[x] - M[\mathbf{y}]\mathbf{x} = (\mathbf{y}\mathbf{x})'.$$

2. Let $x(t)$ and $y(t)$ be scalar functions. The linear homogeneous differential operators $L[\cdot]$ and $M[\cdot]$ are defined to be *adjoint* to each other, if for arbitrary (sufficiently differentiable) $x(t)$, $y(t)$, $yL[x] - xM[y]$ is the derivative of some expression in x and y and their derivatives. [This is important because then the $\int_a^b \{yL[x] - xM[y]\}\,dt$ can be expressed in terms of the values at a and b.] Show that if

$$L[x] = a_0(t)x + a_1(t)x' + \cdots + a_n x^{(n)},$$

then

$$M[y] = a_0(t)y - (a_1(t)y)' + \cdots \pm (a_n y)^{(n)}$$

is an adjoint of $L[x]$.

5 Nonhomogeneous Linear Differential Equation

5.1 Definition: relation with the homogeneous equation

The equation

$$\mathbf{x}' = A(t)\mathbf{x} + \mathbf{b}(t), \tag{3.5.1}$$

where $\mathbf{x}(t)$ *is an n vector and the matrix* $A(t)$ *and the vector* $\mathbf{b}(t)$ *are given, is called linear nonhomogeneous if* $\mathbf{b}(t)$ *is not identically zero. We always assume that* $A(t)$ *and* $\mathbf{b}(t)$ *satisfy the conditions for existence and uniqueness of solutions.*

If a particular solution $\mathbf{u}(t)$ is known, then $\mathbf{y}(t) = \mathbf{x}(t) - \mathbf{u}(t)$ satisfies the *homogeneous* equation

$$\mathbf{y}' = \mathbf{x}' - \mathbf{u}' = A(t)\mathbf{x} + \mathbf{b}(t) - A(t)\mathbf{u} - \mathbf{b}(t) = A(t)\mathbf{y}.$$

Hence:

The general solution of the nonhomogeneous equation (3.5.1) *is obtained by adding a particular solution of this equation to the general solution of the corresponding homogeneous equation* (3.2.1).

Any solution $\mathbf{x}(t)$ is obtained in this way, and every $\mathbf{x}(t)$ so obtained satisfies the nonhomogeneous equation.

In many cases, a particular solution of the nonhomogeneous equation (3.5.1) can be found by a special device (a good guess or the method of undetermined coefficients, for example). In the following sections a formula for the general solution will be presented in terms of a fundamental matrix solution of the homogeneous equation (3.2.1).

5.2 *Variation of constants*

The general solution of the *homogeneous* equation (3.2.1) is given by

$$\mathbf{x}(t) = X(t)\mathbf{c}, \tag{3.5.2}$$

where $X(t)$ is a fundamental matrix and $\mathbf{c}$ an arbitrary constant vector. We shall try to satisfy the *nonhomogeneous* equation (3.5.1) with the same expression but allow $\mathbf{c}$ to be a function of t. [Hence the name of the method is *variation of constants*.]

Differentiating (3.5.2), we get

$$\mathbf{x}' = X'\mathbf{c} + X\mathbf{c}' = AX\mathbf{c} + X\mathbf{c}' = A\mathbf{x} + X\mathbf{c}'.$$

In order to satisfy

$$\mathbf{x}' = A\mathbf{x} + \mathbf{b},$$

we must have

$$X\mathbf{c}' = \mathbf{b} \quad \text{or} \quad \mathbf{c}' = X^{-1}\mathbf{b}.$$

Thus we obtain as a solution of (3.5.1)

$$\mathbf{x}(t) = X(t) \int X^{-1}(s)\mathbf{b}(s)\, ds.$$

If, in this formula, we integrate from t_0 to t, then the solution $\mathbf{x}(t)$ obtained will satisfy the initial condition $\mathbf{x}(t_0) = 0$. If, instead, the initial condition is $\mathbf{x}(t_0) = \mathbf{x}_0$, we have to add that particular solution of the homogeneous equation which satisfies this initial condition. This solution of the homogeneous equation is $X(t)\mathbf{c}$, where $\mathbf{c}$ is a suitable constant. The solution $\mathbf{x}(t)$ of the nonhomogeneous equation, for which $\mathbf{x}(t_0) = \mathbf{x}_0$, is therefore given by

$$\mathbf{x}(t) = X(t)\left\{\mathbf{c} + \int_{t_0}^{t} X^{-1}(s)\mathbf{b}(s)\, ds\right\}.$$

If $X(t)$ is the principal fundamental matrix, then the initial condition $\mathbf{x}(t_0) = \mathbf{x}_0$ is satisfied by taking $\mathbf{c} = \mathbf{x}_0$ as an arbitrary constant, and the expression of the solution is

$$\mathbf{x}(t) = X(t)\left\{\mathbf{x}_0 + \int_{t_0}^{t} X^{-1}(s)\mathbf{b}(s)\, ds\right\}. \tag{3.5.3}$$

5.3 *Remark*

Formula (3.5.3) is a vector equation. In order to get the scalar expression for the components of the solution, we must develop this formula by components. For example, for a *second-order system*, let

$$\mathbf{x}(t) = (x^1(t), x^2(t))^T$$

be the solution. Let

$$\mathbf{b}(t) = (b^1(t), b^2(t))^T, \qquad \mathbf{x}_0 = (x_0^1, x_0^2)^T.$$

Let $x_j^i(t)$ be the elements of the fundamental matrix

$$X(t)$$

and

$$y_j^i(t)$$

be the elements of

$$X^{-1}(t), \quad (i, j = 1, 2).$$

Then

$$x^1(t) = x_1^1(t)\left\{x_0^1 + \int_{t_0}^t [y_1^1(s)b^1(s) + y_2^1(s)b^2(s)]\, ds\right\}$$

$$+ x_2^1(t)\left\{x_0^2 + \int_{t_0}^t [y_1^2(s)b^1(s) + y_2^2(s)b^2(s)]\, ds\right\},$$

$$x^2(t) = x_1^2(t)\left\{x_0^1 + \int_{t_0}^t [y_1^1(s)b^1(s) + y_2^1(s)b^2(s)]\, ds\right\}$$

$$+ x_2^2(t)\left\{x_0^2 + \int_{t_0}^t [y_1^2(s)b^1(s) + y_2^2(s)b^2(s)]\, ds\right\}.$$

[The complexity of this formula clearly shows the advantage of using (and remembering) the formulas in the vector-matrix notation.]

Problems

1. The Euler equation

$$t^2x'' - 2tx' + 2x = 0$$

has the general solution

$$x(t) = At + Bt^2.$$

Verify this.

Using the formula of variations of constants, solve the nonhomogeneous equation

$$t^2x'' - 2tx' + 2x = t, \qquad x(1) = 0, \quad x'(1) = 0.$$

Note that the equation is singular at $t = 0$. [*Hint*: write the equation as a system of order two and apply the formula given in Section 3.5.3.]

2. The equation

$$x'' + k^2x = 0$$

has the general solution

$$x(t) = A \cos kt + B \sin kt.$$

Verify this.

Using the formula of variation of constants, prove that the nonhomogeneous equation

$$x'' + k^2x = \sin \omega t$$

has one periodic solution, except when $\omega = k$. [*Hint*: write the equation as a system, apply the formula given in section 3.5.3, and consider the periodicity of the integral and the other factor appearing in the solution.]

---

6 *Linear Equation of Order n*

6.1 *Equivalent system*

The scalar, linear homogeneous differential equation of order n

$$x^{(n)} = a_0(t)x + a_1(t)x' + \cdots + a_{n-1}(t)x^{(n-1)} \tag{3.6.1}$$

is equivalent to the system

$$\left.\begin{array}{l} x_1' = x_2, \\ x_2' = x_3, \\ \cdots\cdots\cdots \\ x_n' = a_0 x_1 + a_1 x_2 + \cdots + a_{n-1}x_n. \end{array}\right\} \tag{3.6.2}$$

The correspondent matrix

$$A = \begin{pmatrix} 0 & 1 & 0 & \cdots & 0 \\ 0 & 0 & 1 & \cdots & 0 \\ \cdots & \cdots & \cdots & \cdots & \cdots \\ 0 & 0 & 0 & \cdots & 1 \\ a_0 & a_1 & a_2 & \cdots & a_{n-1} \end{pmatrix} \tag{3.6.3}$$

is called the *companion matrix* of the polynomial

$$\lambda^n - a_{n-1}\lambda^{n-1} - \cdots - a_1\lambda - a_0 = 0. \tag{3.6.4}$$

This equation is precisely the characteristic equation of the matrix A, which is easily obtained if we develop the determinant

$$\det(\lambda E - A) = \begin{vmatrix} \lambda & -1 & 0 & \cdots & 0 \\ 0 & \lambda & -1 & \cdots & 0 \\ 0 & 0 & \lambda & \cdots & 0 \\ \cdots & \cdots & \cdots & \cdots & \cdots \\ 0 & 0 & 0 & \cdots & -1 \\ -a_0 & -a_1 & -a_2 & \cdots & \lambda - a_{n-1} \end{vmatrix} \tag{3.6.5}$$

by its last row.

It should be noted particularly that equation (3.6.4) can be obtained from (3.6.1) by substituting λ^i for $x^{(i)}$.

6.2 Existence and uniqueness of solutions

All properties of linear systems of differential equations are applicable to the linear equation of order n. So, if the coefficients $a_i(t)$ are continuous (or at least integrable) in $[a, b]$, then for any $t_0 \in [a, b]$, the initial value problem

$$x(t_0) = x_0, \qquad x'(t_0) = x'_0, \ldots, \qquad x^{(n-1)}(t_0) = x_0^{(n-1)} \tag{3.6.6}$$

has a unique solution in $[a, b]$.

Sometimes the nth order equation is written as

$$a_n x^{(n)} + a_{n-1}x^{(n-1)} + \cdots + a_1 x' + a_0 x = 0.$$

To guarantee the existence and uniqueness of solutions in this case, the leading coefficient $a_n(t)$ should not be allowed to vanish for any value of $t \in [a, b]$, so that this equation can be solved for $x^{(n)}$.

In what follows we will *always* assume that the coefficients $a_i(t)$ satisfy the conditions for existence and uniqueness of solutions. To simplify the writing, we will also assume that $t_0 = 0$, which obviously does not restrict generality.

6.3 Wronskian

If the scalar function $u_1(t)$ satisfies the nth order equation (3.6.1), then the vector

$$\begin{pmatrix} u_1^1 \\ u_1^2 \\ u_1^3 \\ \vdots \\ u_1^n \end{pmatrix} = \begin{pmatrix} u_1 \\ u_1' \\ u_1'' \\ \vdots \\ u_1^{(n-1)} \end{pmatrix} \tag{3.6.7}$$

satisfies the system (3.6.2). Conversely, every solution of (3.6.2) gives us a solution of (3.6.1), namely, the first component of the vector solution.

Let the matrix

$$U = \begin{pmatrix} u_1 & u_2 & \cdots & u_n \\ u_1' & u_2' & \cdots & u_n' \\ \cdots & \cdots & \cdots & \cdots \\ u_1^{(n-1)} & u_2^{(n-1)} & \cdots & u_n^{(n-1)} \end{pmatrix} \tag{3.6.8}$$

be a fundamental matrix of (3.6.2). Then the functions $u_1(t)$, $u_2(t)$, $\ldots$, $u_n(t)$ are linearly independent solutions of (3.6.1). We prove this by assuming

$$\sum_i c_i u_i(t) \equiv 0$$

for some constants $c_1, c_2, \ldots, c_n$. Then

$$\sum_i c_i u_i'(t) \equiv 0.$$

The same relationship holds between their derivatives of any order and also between the columns of (3.6.8). But because this is a fundamental solution matrix, its columns are linearly independent; hence

$$c_1 = c_2 = \cdots = c_n = 0.$$

Conversely, if $u_1(t)$, $\ldots$, $u_n(t)$ are linearly independent solutions of (3.6.1), the columns of (3.6.8) are also linearly independent; hence (3.6.8) is a fundamental solution for (3.6.2). Furthermore, any other solution $u(t)$ of (3.6.1) gives a solution vector of (3.6.2) that must be a linear combination of the columns of (3.6.8); hence $u(t)$ itself is a linear combination of $u_1(t)$, $\ldots$, $u_n(t)$. This shows that:

> *The linear homogeneous differential equation of nth order* (3.6.1) *has n linearly independent solutions, and every other solution is a linear combination of these. Such a system of n linearly independent solutions is called a fundamental system of* (3.6.1).

The determinant of the matrix $U(t)$ in (3.6.8) is called the wronskian of the functions $u_1(t), \ldots, u_n(t)$. If this determinant vanishes for a single value of t, then it must vanish identically, and the functions $u_i(t)$ are linearly dependent. This follows from Theorem 3.3.7 and its corollary. Or, more precisely, it can be stated as:

6.4 Theorem

If $u_i(t)$, $i = 1, 2, \ldots, n$, are solutions of equation (3.6.1) *and if $W(u_1, u_2, \ldots, u_n)$ denotes the wronskian of these functions, then*

$$W(u_1(t), \ldots, u_n(t)) = W(u_1(0), \ldots, u_n(0)) \exp \int_0^t a_{n-1}(s)\, ds. \quad (3.6.9)$$

This follows from Theorem 3.3.7.

6.5 Theorem

Let $u_1(t), \ldots, u_n(t)$ be n times continuously differentiable on $[a, b]$ and assume that the wronskian $W(u_1, \ldots, u_n)$ does not vanish at any point of $[a, b]$. Then there is a uniquely defined nth order linear homogeneous differential equation such that $\{u_1(t), \ldots, u_n(t)\}$ is a fundamental system of solutions.

This follows from Theorem 3.3.6.

6.6 Nonhomogeneous nth order equation

All the results given for nth order systems are applicable to the linear nonhomogeneous equation of order n. The method of *variation of constants* is valid, in particular. The corresponding formula, written for the scalar nth order equation (that is, for the first component of the system), is no simpler than the general formula given in 3.5.2. Hence we will not write it out.

6.7 Remark

A necessary condition for the functions $u_1(t), \ldots, u_n(t)$ (which may be assumed continuously differentiable at least n times) to be linearly dependent in $[a, b]$ is that their wronskian $W(u_1, \ldots, u_n)$ be identically zero in $[a, b]$. Obviously, if there is some linear relationship between the functions, the same relationship will hold between the columns of the wronskian matrix; hence the determinant is zero.

It is remarkable that, in general, the converse does not hold: the wronskian $W(u_1, \ldots, u_n)$ may be identically zero in $[a, b]$; nevertheless, the functions may be linearly independent. An example of such a case is

$$u(t) = \begin{cases} t^4 & \text{for } t \geqq 0 \\ 0 & \text{for } t \leqq 0 \end{cases}, \qquad v(t) = \begin{cases} 0 & \text{for } t \geqq 0 \\ t^4 & \text{for } t \leqq 0 \end{cases}.$$

These functions are linearly independent in the interval $[-1, +1]$, but the wronskian $W(u, v)$ is identically zero. In this example the functions are continuously differentiable three times. With the aid of the function e^{-t^2}, all of whose derivatives vanish at $t = 0$, it is even possible to construct examples where the functions are continuously differentiable an infinite number of times.

Comparing this result with the above theorems, we can state that such functions cannot be solutions of a linear homogeneous differential equation whose coefficients satisfy the conditions of the existence theorem.

Problems

1. Find out which of the following sets of functions are linearly independent on R.
   (a) $x = e^t, \quad x = e^{2t}, \quad x = e^{3t}$,
   (b) $x = 1, \quad x = e^t, \quad x = \text{Sinh } t$,
   (c) $x = e^{-t}, \quad x = \text{Sinh } t, \quad x = \text{Cosh } t$,
   (d) $x = 2t + 3, \quad x = t - 1, \quad x = -t + 2$.
2. Find three linearly independent solutions of the equation $x''' = x''$. Write the wronskian of these solutions and then directly verify Theorem 3.6.4.
3. Find the second-order differential equation which has the solutions $x = t$ and $x = t^2$. Note that the equation is singular for $t = 0$.

7 *Differential Equations in the Complex Domain*

7.1 Functions of a complex variable

The differential equation

$$x' = f(t, x) \tag{3.7.1}$$

may be considered in various ways when complex variables are involved. If we consider both variables, t and x, as complex valued (i.e., $f(t, x)$ a complex function of two complex variables), then the requirement that $x'(t)$ exist means that $x(t)$ must be an analytic function. This is very restrictive because the knowledge of $x(t)$ in any neighborhood of any point t_0, by analytic continuation, completely determines the whole function $x(t)$. In this case, the best representation of $x(t)$ is by Taylor series (nonnegative integer powers of $t - t_0$), or Laurent series (integer, possibly including negative, powers), or Puiseux series (fractional powers). The basic point of such a theory, which has been developed in much detail, is to study the singularities of $x(t)$. [In this book we will not concern ourselves with this problem.]

7.2 Complex functions of a real variable

If in equation (3.7.1) we assume t to be a real variable but allow $x(t)$ to be complex, we may write

$$x = u + iv, \qquad f = g + ih;$$

hence

$$u' = g(t, u, v), \qquad v' = h(t, u, v).$$

We have, therefore, a system of two real differential equations with two real unknown functions $u(t)$ and $v(t)$.

The most important case is when the equation is linear. Consider the vector case when $\mathbf{x}(t)$ is a complex vector, that is, a vector with complex components. Assume that t is real and also that the matrix $A(t)$ is real. Then the linear homogeneous equation $\mathbf{x}' = A(t)\mathbf{x}$ decomposes into two real equations $\mathbf{u}' = A(t)\mathbf{u}$, $\mathbf{v}' = A(t)\mathbf{v}$. Therefore: *both the real and the imaginary parts of any solution* $\mathbf{x}(t)$ *satisfy the same differential equation.*

In problems dealing with real quantities, it is often convenient to consider complex solutions of linear differential equations. The most common example is the equation of the harmonic oscillator $x'' = k^2x$. If to the real solution

$$u(t) = A \cos kt + B \sin kt,$$

which may represent the true physical description of the behavior of a system, we add an imaginary solution

$$iv(t) = Ai \sin kt - Bi \cos kt,$$

with no physical meaning, we get

$$\begin{aligned} x(t) &= u(t) + iv(t) \\ &= A(\cos kt + i \sin kt) - iB(\cos kt + i \sin kt) \\ &= Ae^{ikt} - iBe^{ikt} = (A - iB)e^{ikt}. \end{aligned}$$

In this solution, only the real part will have a physical significance; nevertheless, the notation is so simple that it is usual to say that the factor e^{ikt} represents a complex oscillation, while $A - iB$ is the complex amplitude of oscillation. Because operating with the exponential function is easier than employing trigonometric functions, this rather strange interpretation is widely used in technical applications. We will use it in Chapters 4 and 5.

References

Our use of matrix analysis in presenting the theory of linear differential equations generally follows the structural forms of Coddington and Levinson. [See References, Chapter 2.] For a more detailed exposition of matrix algebra and analysis we refer the reader to the following books:

C. C. MacDuffee. *The Theory of Matrices.* New York: Chelsea, 1946 (reprint).

F. R. Gantmacher. *The Theory of Matrices*, Volumes I and II (Russian transl.). New York: Chelsea, 1960.

The proof of the Jordan canonical form can be found in

S. Lefschetz. *Differential Equations: Geometric Theory*, Appendix 1. New York: Wiley (Interscience), 1957.

For the study of differential equations in the complex domain, we refer the reader to

E. Hille. *Lectures on Ordinary Differential Equations.* Reading, Mass.: Addison-Wesley, 1969.

L. Bieberbach. *Theorie der Gewöhnlichen Diffenrentialgleichungen auf Funktionentheoretischer Grundlage dargestellt*, in the series *Grundlehren der Mathematischen Wissenschaften.* Berlin: Springer-Verlag, 1965 (2nd ed.).

4

Linear Systems with Constant Coefficients

There are two primary reasons for a detailed study of linear differential systems with constant coefficients. In the first place, an enormous variety of phenomena from the physical and natural sciences, which have a wide range of technical applications, can be aptly described by these systems.

Secondly, because these systems can be explicitly solved, they give a greater insight into the possible behavior of the more general systems with variable coefficients.

1 *Direct Method of Solution*

1.1 Exponential substitution

Given the homogeneous linear differential system

$$x^{i'} = \sum_j a^i_j x^j \quad i = 1, 2, \ldots, n, \tag{4.1.1}$$

where the coefficients a^i_j are constant, we try the solution

$$x^i = c^i e^{\lambda t} \quad i = 1, 2, \ldots, n. \tag{4.1.2}$$

Substituting (4.1.2) in (4.1.1), we may cancel the exponential factor (which is never zero) and obtain

$$\lambda c^i = \sum_j a^i_j c^j \quad i = 1, 2, \ldots, n. \tag{4.1.3}$$

Therefore: *λ must be an eigenvalue of the matrix (a^i_j) and (c^i) a corresponding eigenvector.*

Equation (4.1.3) is equivalent to

$$\sum_j (a^i_j - \lambda\, \delta^i_j)c^j = 0 \quad i = 1, 2, \ldots, n. \tag{4.1.4}$$

Here δ^i_j (*Kronecker's delta*) is equal to 1 if $i = j$, or it is equal to 0 if $i \neq j$, as defined in formula (3.1.4).

A necessary and sufficient condition for the algebraic linear system (4.1.4) to have a nontrivial solution, is that the determinant of the coefficients

$$\det(\lambda E - A) = \begin{vmatrix} \lambda - a^1_1 & -a^2_1 & \cdots & -a^n_1 \\ -a^1_2 & \lambda - a^2_2 & \cdots & -a^n_2 \\ \cdots & \cdots & \cdots & \cdots \\ -a^1_n & -a^2_n & \cdots & \lambda - a^n_n \end{vmatrix} = 0. \tag{4.1.5}$$

Expanding this determinant, an algebraic equation of degree n is obtained. Note that its degree is precisely n [it cannot be smaller in particular cases] because the coefficient of λ^n turns out to be 1. This equation is called the *characteristic equation* of the matrix A, and its solution is the first step in solving the differential system (4.1.1).

Let $\lambda_1, \lambda_2, \ldots, \lambda_n$ be the roots of the characteristic equation. They may be real or complex; they may all be distinct or some may be repeated (of multiplicity greater than one). For every *characteristic root* λ_r, there is at least one *eigenvector* $c_r = (c^i_r)$ satisfying (4.1.4) for $\lambda = \lambda_r$. Hence, by (4.1.2),

$$x^i_r(t) = c^i_r\, e^{\lambda_r t} \quad (i = 1, 2, \ldots, n) \tag{4.1.6}$$

is a solution of the differential system (4.1.1).

Given λ_r satisfying the characteristic equation (4.1.5): in order to obtain the numerical value of the constants c^i_r, we have to solve the algebraic linear homogeneous system (4.1.4), which is compatible by the choice of λ_r. If (c^i_r) is a solution, then (kc^i_r) is also a solution for every constant k. If (c^i_r) and (d^i_r) are two solutions, then every linear combination $(k_1 c^i_r + k_2 d^i_r)$ is also a solution for k_1, k_2 constants. We will see in Section 4.1.4 that two such linearly independent solutions (c^i_r), (d^i_r) can only exist if the corresponding eigenvalue λ_r is a multiple root of the characteristic equation.

1.2 Case of real distinct characteristic roots

If all the roots of (4.1.5) happen to be real and distinct, then the general solution of the differential system (4.1.1) can be obtained immediately. Indeed, for every root λ_r we get an eigenvector (c_r^i), as explained above, and therefore a particular solution of (4.1.1). Furthermore, we will prove that solutions of the type (4.1.6) corresponding to different values of λ_r are linearly independent. Hence we have obtained n linearly independent solutions, and we know from the theory of linear differential systems that a general linear combination of these solutions constitutes the general solution of the system (4.1.1).

In order to prove the linear independence of the nontrivial solutions (c_r^i), $r = 1, 2, \ldots, n$, assume that some linear combination

$$\sum_r p^r \mathbf{c}_r = 0. \tag{4.1.7}$$

[r is a superindex, not an exponent!] Then, multiplying by the powers of the matrix A, we get

$$A \sum_r p^r \mathbf{c}_r = \sum_r p^r A \mathbf{c}_r = \sum_r p^r \lambda_r \mathbf{c}_r = 0,$$

$$A^2 \sum_r p^r \mathbf{c}_r = \sum_r p^r A^2 \mathbf{c}_r = \sum_r p^r \lambda_r^2 \mathbf{c}_r = 0.$$

$$\cdots\cdots\cdots\cdots\cdots\cdots\cdots\cdots\cdots\cdots$$

$$A^{n-1} \sum_r p^r \mathbf{c}_r = \sum_r p^r A_r^{n-1} \mathbf{c}_r = \sum_r p^r \lambda_r^{n-1} \mathbf{c}_r = 0.$$

Writing these vector equations componentwise, we obtain for the ith component,

$$\begin{aligned}
&p^1 c_1^i + p^2 c_2^i + \cdots + p^n c_n^i = 0\\
&p^1 c_1^i \lambda_1 + p^2 c_2^i \lambda_2 + \cdots + p^n c_n^i \lambda_n = 0\\
&p^1 c_1^i \lambda_1^2 + p^2 c_2^i \lambda_2^2 + \cdots + p^n c_n^i \lambda_n^2 = 0\\
&\cdots\cdots\cdots\cdots\cdots\cdots\cdots\cdots\cdots\cdots\\
&p^1 c_1^i \lambda_1^{n-1} + p^2 c_2^i \lambda_2^{n-1} + \cdots + p^n c_n^i \lambda_n^{n-1} = 0.
\end{aligned}$$

This is an algebraic linear homogeneous system for the values $p^j c_j^i$, $j = 1, 2, \ldots, n$, whose coefficients form the determinant

$$\begin{vmatrix}
1 & 1 & 1 & \cdots & 1\\
\lambda_1 & \lambda_2 & \lambda_3 & \cdots & \lambda_n\\
\lambda_1^2 & \lambda_2^2 & \lambda_3^2 & \cdots & \lambda_n^2\\
\cdots & \cdots & \cdots & \cdots & \cdots\\
\lambda_1^{n-1} & \lambda_2^{n-1} & \lambda_3^{n-1} & \cdots & \lambda_n^{n-1}
\end{vmatrix}.$$

This is called a *Vandermonde determinant* and can be shown to factor into the expression $\sum_{i>j}(\lambda_i - \lambda_j)$. Hence it cannot be zero if all the λ_i's are distinct. According to (4.1.7), all the coefficients p^r must therefore be zero and the linear independence of the vectors c_r is proved.

Note: In order to factor the Vandermonde determinant, subtract from the last row the second-to-the-last row, multiplied by λ_1, then subtract from the second-to-the-last row the one before, multiplied by λ_1, and so on. This way the first column becomes $(1, 0, 0, \ldots, 0)$; hence we may delete the first row and first column. But in the new determinant of order $(n-1)$, the first column has the common factor $(\lambda_2 - \lambda_1)$, the second column the common factor $(\lambda_3 - \lambda_1)$, and so on. Taking out these factors, a Vandermonde determinant of order $(n-1)$ remains, which can be treated in the same way. Finally, only the product of all the factors taken out remains, which is the result indicated above.

Example 1

Consider the third-order differential system

$$\begin{cases} x^{1\prime} = 2x^1 \\ x^{2\prime} = -2x^1 + x^2 - 2x^3 \\ x^{3\prime} = x^1 + 3x^3. \end{cases}$$

Substituting $x^i = c^i e^{\lambda t}$, $i = 1, 2, 3$, we obtain the algebraic equations for c^1, c^2, c^3,

$$\begin{cases} (\lambda - 2)c^1 = 0 \\ 2c^1 + (\lambda - 1)c^2 + 2c^3 = 0 \\ -c^1 + (\lambda - 3)c^3 = 0. \end{cases}$$

The characteristic equation is

$$\begin{vmatrix} \lambda - 2 & 0 & 0 \\ 2 & \lambda - 1 & 2 \\ -1 & 0 & \lambda - 3 \end{vmatrix} = (\lambda - 2)(\lambda - 1)(\lambda - 3),$$

and the characteristic roots are 2, 1, 3.

For $\lambda = 2$, we may take $c^1 = 1$, $c^2 = 0$, $c^3 = -1$ in the above algebraic system. Hence the first solution is

$$\begin{cases} x^1 = e^{2t} \\ x^2 = 0 \\ x^3 = -e^{2t}. \end{cases}$$

For $\lambda = 1$, we obtain $c^1 = c^3 = 0$, $c^2 = 1$. So, the second solution is

$$\begin{cases} x^1 = 0 \\ x^2 = e^t \\ x^3 = 0. \end{cases}$$

For $\lambda = 3$ we may take $c^1 = 0$, $c^2 = 1$, $c^3 = -1$. This gives

$$\begin{cases} x^1 = \quad 0 \\ x^2 = \quad e^{3t} \\ x^3 = -e^{3t}. \end{cases}$$

The general solution is

$$\begin{cases} x^1 = \quad k^1 e^{2t} \\ x^2 = \quad k^2 e^t + k^3 e^{3t}. \\ x^3 = -k^1 e^{2t} - k^3 e^{3t}. \end{cases}$$

1.3 Complex roots

If some of the characteristic roots are complex, then all results given above are still valid. The correspondent solutions, however, are complex-valued, and this is inconvenient.

In this case, assuming that the coefficients of the differential system are real, if $\lambda = \sigma + i\omega$ is a characteristic root, its conjugate $\bar{\lambda} = \sigma - i\omega$ will also be a root. Also, the correspondent eigenvectors will be complex conjugate: $\mathbf{c}$ and $\bar{\mathbf{c}}$. In this way, we get a pair of solutions

$$\begin{aligned} \mathbf{c}e^{\lambda t} &= \mathbf{c}e^{\sigma t}(\cos \omega t + i \sin \omega t), \\ \bar{\mathbf{c}}e^{\bar{\lambda} t} &= \bar{\mathbf{c}}e^{\sigma t}(\cos \omega t - i \sin \omega t). \end{aligned}$$

According to 3.7.2, instead of these solutions we may take their real and imaginary parts, which we know also satisfy the differential equation (4.1.1). Writing $\mathbf{c} = \mathbf{a} - i\mathbf{b}$, $\bar{\mathbf{c}} = \mathbf{a} + i\mathbf{b}$, we obtain

$$\begin{aligned} &e^{\sigma t}(\mathbf{a} \cos \mathbf{b}\omega t + \sin \omega t) \\ &e^{\sigma t}(\mathbf{a} \sin \omega t - \mathbf{b} \cos \omega t). \end{aligned} \tag{4.1.8}$$

The calculation of the constant vectors $\mathbf{a} = (a^i)$ and $\mathbf{b} = (b^i)$ can be made either by calculating $\mathbf{c} = \mathbf{a} - i\mathbf{b}$, as before, or by substituting the expressions (4.1.8) directly into the differential system and solving the resulting linear algebraic system, which will be compatible according to the choice of σ, ω. The latter procedure, the method of undetermined coefficients, is usually the preferred one.

Example 2

Consider the following system:

$$\begin{cases} x^{1\prime} = x^2 \\ x^{2\prime} = x^3 \\ x^{3\prime} = -3x^2 - 2x^3. \end{cases}$$

The characteristic equation is

$$\begin{vmatrix} \lambda & -1 & 0 \\ 0 & \lambda & -1 \\ 0 & 3 & \lambda+2 \end{vmatrix} = \lambda(\lambda^2 + 2\lambda + 3) = 0,$$

with the roots $\lambda_1 = 0$, $\lambda_{2,3} = -1 \pm i\sqrt{2}$.

For $\lambda = 0$, the solution is constant: $x^i(t) = c_i$, $i = 1, 2, 3$. The equations for the values c_i are

$$\begin{aligned} -c_2 &= 0 \\ c_3 &= 0 \\ -3c_2 - 2c_3 &= 0. \end{aligned}$$

This gives a solution $c_1 = 1$, $c_2 = c_3 = 0$, or

$$\begin{cases} x^1 = 1 \\ x^2 = 0 \\ x^3 = 0. \end{cases}$$

For $\lambda = \sigma \pm i\omega$, with $\sigma = -1$, $\omega = \sqrt{2}$, we write the solution in the form

$$\begin{aligned} x^1 &= e^{\sigma t}\{a_1 \cos \omega t + b_1 \sin \omega t\} \\ x^2 &= e^{\sigma t}\{a_2 \cos \omega t + b_2 \sin \omega t\} \\ x^3 &= e^{\sigma t}\{a_3 \cos \omega t + b_3 \sin \omega t\}. \end{aligned}$$

The derivatives are

$$x^{i\prime} = e^{\sigma t}\{(\sigma a_i + \omega b_i) \cos \omega t + (\sigma b_i - \omega a_i) \sin \omega t\}.$$

Substituting in the differential system, we obtain

$$\begin{aligned} (\sigma a_1 + \omega b_1 - a_2) \cos \omega t + (-\omega a_1 + \sigma b_1 - b_2) \sin \omega t &= 0 \\ (\sigma a_2 + \omega b_2 - a_3) \cos \omega t + (-\omega a_2 + \sigma b_2 - b_3) \sin \omega t &= 0 \\ (3a_2 + (\sigma + 2)a_3 + \omega b_3) \cos \omega t + (3b_2 - \omega a_3 + (\sigma + 2)b_3) \sin \omega t &= 0. \end{aligned}$$

Equating separately the coefficients of $\cos \omega t$ and of $\sin \omega t$ to zero, we get a system of six linear homogeneous equations in the unknown $a_1, b_1, a_2, b_2, a_3, b_3$. This system is compatible because both σ, ω were chosen to satisfy the characteristic equation. The solution of this system contains two arbitrary constants. This can be seen "physically" by noting that we may shift the time origin in such a way that some preassigned $\cos \omega t$ term vanishes. [It is sufficient to put the time origin at a point where $a_1 \cos \omega t + b_1 \sin \omega t = 0$.] Afterward, we still have an arbitrary factor in the amplitude of oscillation because the system is linear homogeneous.

Introducing the values $\sigma = -1$, $\omega = \sqrt{2}$ and solving, we obtain the two linearly independent solutions

$$\begin{array}{llll} a_1 = -1/3, & b_1 = \sqrt{2}/3, & a_2 = 1, & b_2 = 0, \\ & & a_3 = -1, & b_3 = -\sqrt{2}; \\ a_1 = -\sqrt{2}/3, & b_1 = -1/3, & a_2 = 0, & b_2 = 1, \\ & & a_3 = \sqrt{2}, & b_3 = -1. \end{array}$$

Putting together the three solutions, each one multiplied by an arbitrary constant, the expression for the general solution is

$$\begin{aligned} x^1 &= k_1 + e^{-t}\left\{\left(-\frac{1}{3}k_2 - \frac{\sqrt{2}}{3}k_3\right)\cos\sqrt{2}\,t + \left(\frac{\sqrt{2}}{3}k_2 - \frac{1}{3}k_3\right)\sin\sqrt{2}\,t\right\} \\ x^2 &= e^{-t}\{k_2 \cos\sqrt{2}\,t + k_3 \sin\sqrt{2}\,t\} \\ x^3 &= e^{-t}\{(-k_2 + \sqrt{2}k_3)\cos\sqrt{2}\,t + (-\sqrt{2}k_2 - k_3)\sin\sqrt{2}\,t\}. \end{aligned}$$

1.4 Multiple roots

If all characteristic roots are distinct, n linearly independent solutions are obtained by the procedure just indicated. If some multiple roots appear, this method may give less than n linearly independent solutions; hence we still have to look for additional solutions in order to obtain them all.

Let λ be a characteristic root of multiplicity m, with $1 < m \leqq n$. Corresponding to this value of λ, there can exist (according to each particular case) any number from 1 to m of linearly independent solutions of the algebraic system (4.1.4) and therefore of the differential system (4.1.1). How many such linearly independent solutions actually exist can be determined by algebraically analyzing system (4.1.4), from which we can calculate them explicitly.

If, corresponding to the root λ of multiplicity m, there exist m linearly independent solutions (eigenvectors), then we are in exactly the same position as with distinct roots; that is, we get as many solutions as the degree of the characteristic equation indicates.

If, instead, there are $m' < m$ linearly independent solutions of the form $\mathbf{x} = \mathbf{c}e^{\lambda t}$, then we must also look for solutions of the form

$$\mathbf{x} = (\mathbf{c} + \mathbf{d}t)e^{\lambda t}, \tag{4.1.9}$$

where $\mathbf{c}$, $\mathbf{d}$ are constant vectors. Written in components, this solution is

$$x^i = (c^i + d^i t)e^{\lambda t} \quad i = 1, 2, \ldots, n. \tag{4.1.10}$$

If the latter still does not give us m linearly independent solutions, we must try solutions of the form

$$\mathbf{x} = (\mathbf{c} + \mathbf{d}t + \mathbf{k}t^2)e^{\lambda t}$$

and, in general, polynomials in t times the exponential. In the next section we will prove that if λ has multiplicity m, we may at most have polynomials of degree $m - 1$, thereby obtaining all the solutions we need.

As before, the practical way to get the numerical value of the coefficients of solutions is through the method of undetermined coefficients. Substituting the corresponding expression in the differential system, we get a linear algebraic system for all the coefficients. This system is compatible if λ is a root of the corresponding characteristic equation.

Example 3

Consider the system

$$\begin{cases} x^{1'} = 3x^1 - x^3 \\ x^{2'} = 2x^2 \\ x^{3'} = x^1 + x^3. \end{cases}$$

The characteristic equation is

$$\begin{vmatrix} \lambda - 3 & 0 & 1 \\ 0 & \lambda - 2 & 0 \\ -1 & 0 & \lambda - 1 \end{vmatrix} = \lambda^3 - 6\lambda^2 + 12\lambda - 8 = (\lambda - 2)^3 = 0.$$

Hence we have a triple root, $\lambda_1 = \lambda_2 = \lambda_3 = 2$.

Substituting

$$\lambda = 2 \quad \text{and} \quad x^1 = c_1 e^{2t}, \quad x^2 = c_2 e^{2t}, \quad x^3 = c_3 e^{2t},$$

we obtain the following system for c_1, c_2, c_3:

$$\begin{cases} -c_1 + 0 + c_3 = 0 \\ \ \ 0 \ + 0 + 0 \ = 0 \\ -c_1 + 0 + c_3 = 0. \end{cases}$$

Here we have two arbitrary constants: c_2 and $c_1 = c_3$. We therefore obtain only two linearly independent solutions. In order to obtain three linearly independent solutions, we must try solutions of the form

$$\begin{cases} x^1 = c_1 e^{2t} + d_1 te^{2t} \\ x^2 = c_2 e^{2t} + d_2 te^{2t} \\ x^3 = c_3 e^{2t} + d_3 te^{2t}. \end{cases}$$

The derivatives are

$$x^{i'} = e^{2t}\{2c_i + d_i + 2td_i\} \quad i = 1, 2, 3.$$

Introducing these values in the differential system and solving for the undetermined coefficients, we obtain the relations

$$d_2 = 0, \qquad d_3 = d_1, \qquad c_1 - c_3 = d_1.$$

For c_2 we do not obtain any relation; hence c_2 is arbitrary. We may also put $c_1 = 1$, $d_1 = 0$, or $c_1 = 1$, $d_1 = 1$. In this way, we obtain three linearly independent solutions

$$\begin{aligned} c_1 &= 1 \quad 0 \quad 1 \\ c_2 &= 0 \quad 1 \quad 0 \\ c_3 &= 1 \quad 0 \quad 0 \\ d_1 &= 0 \quad 0 \quad 1 \\ d_2 &= 0 \quad 0 \quad 0 \\ d_3 &= 0 \quad 0 \quad 1. \end{aligned}$$

Multiplying them by three arbitrary constants and adding, we obtain the general solution

$$\begin{cases} x^1 = k_1 e^{2t} + k_3(e^{2t} + te^{2t}) \\ x^2 = k_2 e^{2t} \\ x^3 = k_1 e^{2t} + k_3 te^{2t}. \end{cases}$$

1.5 Linear equation of order n

The nth order linear homogeneous equation with constant coefficients

$$x^{(n)} + a_{n-1}x^{(n-1)} + \cdots + a_1 x' + a_0 x = 0 \tag{4.1.11}$$

is solved by first trying the solution $x = e^{\lambda t}$. Substituting this expression in the differential equation, we obtain the *characteristic equation*

$$\lambda^n + a_{n-1}\lambda^{n-1} + \cdots + a_1\lambda + a_0 = 0. \tag{4.1.12}$$

As in the case of a system, if the n characteristic roots are distinct, we obtain n linearly independent solutions; hence the general solution is the general linear combination of them

$$x = k_1 e^{\lambda_1 t} + k_2 e^{\lambda_2 t} + \cdots + k_n e^{\lambda_n t}.$$

If λ is a root of multiplicity m, then functions

$$e^{\lambda t}, \quad te^{\lambda t}, \quad t^2 e^{\lambda t}, \quad \ldots, \quad t^{m-1}e^{\lambda t},$$

will also satisfy the equation. The proof, although somewhat lengthy, is basically a straightforward, direct substitution and then application of the algebraic condition that λ is an m-fold root.

If $\lambda = \sigma + i\omega$ is a root, then $x = e^{\sigma t}\cos\omega t$ and $x = e^{\sigma t}\sin\omega t$ are solutions. If $\lambda = \sigma + i\omega$ is, for example, a root of multiplicity three, then the functions

$$e^{\sigma t}\cos\omega t, \quad e^{\sigma t}\sin\omega t, \quad te^{\sigma t}\cos\omega t, \quad te^{\sigma t}\sin\omega t, \quad t^2 e^{\sigma t}\cos\omega t, \quad t^2 e^{\sigma t}\sin\omega t$$

are solutions of the differential equation. [The proof will be given in Section 2.]

1.6 Nonhomogeneous linear equation

In Chapter 3, Section 5, we dealt with the nonhomogeneous equation. We saw then that if we are able to find one particular solution of the nonhomogeneous differential equation, it is sufficient to add the general solution of the correspondent homogeneous equation in order to obtain the general solution of the nonhomogeneous equation.

The formula of variation of constants (3.5.3) gives the explicit solution of the nonhomogeneous equation, but for practical purposes, a good guess of the solution is sometimes quicker and easier. The method of *undetermined coefficients* is helpful in finding such a good guess.

Let us consider, for example, the equation

$$x'' + x = 2t^2 - 1.$$

It seems reasonable to try to satisfy this equation by a polynomial in t, and we see that we need (at least) a polynomial of the second degree. Therefore we try

$$\begin{aligned} x &= at^2 + bt + c, \\ x' &= 2at + b, \\ x'' &= 2a. \end{aligned}$$

Replacing these values in the equation, we obtain

$$x'' + x = at^2 + bt + c + 2a = 2t^2 - 1.$$

Hence, by equating coefficients $a = 2$, $b = 0$, $c = -5$, the particular solution is $x = 2t^2 - 5$ and the general solution is

$$x = A \cos t + B \sin t + 2t^2 - 5.$$

The method works similarly for systems. In general, if the right member is a polynomial, we will try to satisfy the equation with x equal a polynomial; if the right side contains exponential or trigonometric functions, we will try exponential or combinations of trigonometric functions.

Problems

1. Solve $\mathbf{x}' = A\mathbf{x}$, $\mathbf{x}(0) = \mathbf{x}_0$ for

(a) $A = \begin{pmatrix} 2 & 1 \\ 1 & 2 \end{pmatrix}$, $\mathbf{x}_0 = \begin{pmatrix} 2 \\ -8 \end{pmatrix}$; (b) $A = \begin{pmatrix} 1 & 2 \\ 3 & 2 \end{pmatrix}$, $\mathbf{x}_0 = \begin{pmatrix} 2 \\ -2 \end{pmatrix}$;

(c) $A = \begin{pmatrix} 1 & -3 \\ 3 & -5 \end{pmatrix}$, $\mathbf{x}_0 = \begin{pmatrix} 1 \\ -1 \end{pmatrix}$; (d) $A = \begin{pmatrix} -3 & 4 \\ 1 & 1 \end{pmatrix}$, $\mathbf{x}_0 = \begin{pmatrix} 1 \\ 0 \end{pmatrix}$;

(e) $A = \begin{pmatrix} 1 & 5 \\ -2 & -1 \end{pmatrix}$, $\mathbf{x}_0 = \begin{pmatrix} 5 \\ 4 \end{pmatrix}$; (f) $A = \begin{pmatrix} 0 & 1 \\ -5 & -4 \end{pmatrix}$, $\mathbf{x}_0 = \begin{pmatrix} -1 \\ -2 \end{pmatrix}$.

2. Solve $\mathbf{x}' = A\mathbf{x}$, $\mathbf{x}(0) = \mathbf{x}_0$, for

(a) $A = \begin{pmatrix} -1 & 1 & 1 \\ 0 & 1/2 & 1/2 \\ 0 & 1/2 & 1/2 \end{pmatrix}$, $\mathbf{x}_0 = \begin{pmatrix} 0 \\ 2 \\ 0 \end{pmatrix}$;

(b) $A = \begin{pmatrix} 1 & 0 & 0 \\ 2 & 3 & 0 \\ 4 & 5 & 6 \end{pmatrix}$, $\mathbf{x}_0 = \begin{pmatrix} -5 \\ 11 \\ -3 \end{pmatrix}$;

(c) $A = \begin{pmatrix} 0 & 1 & -1 \\ 0 & 1 & 1 \\ 0 & 2 & 2 \end{pmatrix}$, $\quad \mathbf{x}_0 = \begin{pmatrix} 1 \\ 3 \\ -1 \end{pmatrix}$;

(d) $A = \begin{pmatrix} -1 & -1 & 1 \\ 1 & -1 & 0 \\ 0 & 0 & -1 \end{pmatrix}$, $\quad \mathbf{x}_0 = \begin{pmatrix} 2 \\ 0 \\ 1 \end{pmatrix}$;

(e) $A = \begin{pmatrix} 0 & 1 & 1 & 0 \\ -1 & 0 & 0 & 1 \\ 0 & 0 & 0 & 1 \\ 0 & 0 & -1 & 0 \end{pmatrix}$, $\quad \mathbf{x}_0 = \begin{pmatrix} 0 \\ 0 \\ 1 \\ 0 \end{pmatrix}$.

3. Solve
   (a) $x''' + 3x'' + 3x' + x = 0, \quad x(0) = 0, \quad x'(0) = 1, \quad x''(0) = 4;$
   (b) $x''' + x' = 0, \quad x(0) = x'(0) = 0, \quad x''(0) = -1;$
   (c) $x''' - x' = 0, \quad x(0) = x'(0) = 0, \quad x''(0) = -1;$
   (d) $x^{\text{IV}} - 16x = 0, \quad x(0) = 8, \quad x'(0) = 2, \quad x''(0) = 20, \quad x'''(0) = -24.$

4. Solve
   (a) $x''' + 3x'' + 3x' + x = \cos t, \quad x(0) = 1, \quad x'(0) = -1, \quad x''(0) = 2;$
   (b) $x''' + x' = 2, \quad x(0) = 1, \quad x'(0) = 2, \quad x''(0) = 0;$
   (c) $x''' - x' = 2t + 1, \quad x(0) = 2, \quad x'(0) = 0, \quad x''(0) = 3;$
   (d) $x'' + x = e^{-3t}, \quad x(0) = x'(0) = 1;$
   (e) $x'' + 2x' + 2x = 2e^{-t}, \quad x(0) = 4, \quad x'(0) = -4.$

5. Solve, by variations of constants,
   (a) $\mathbf{x}' = \begin{pmatrix} 0 & 1 \\ -1 & 0 \end{pmatrix}\mathbf{x} + \begin{pmatrix} t \\ 0 \end{pmatrix}, \quad \mathbf{x}(0) = \mathbf{x}_0;$
   (b) $\mathbf{x}' = \begin{pmatrix} 0 & 1 \\ 1 & 0 \end{pmatrix}\mathbf{x} + \begin{pmatrix} 1 \\ t \end{pmatrix}, \quad \mathbf{x}(0) = 0.$

6. Prove that if $\int_0^\infty |f(t)|\,dt = k$, then the solution $x(t)$ of

$$x'' + \omega^2 x = f(t), \qquad x(0) = x_0,\ x'(0) = y_0$$

satisfies for all $t > 0$:

$$|x(t)| \leqq \sqrt{x_0^2 + y_0^2} + k.$$

7. Find an upper bound of $\|\mathbf{x}(t)\|$ for all $t \geqq 0$, knowing that

$$\mathbf{x}' = \begin{pmatrix} -1 & 1 \\ -1 & -1 \end{pmatrix}\mathbf{x} + \begin{pmatrix} f(t) \\ g(t) \end{pmatrix}, \qquad \mathbf{x}(0) = 0,$$

and $|f(t)| \leqq 1$, $|g(t)| \leqq 1$.

8. Prove that if $|f(t)|$ is bounded for $t \geqq 0$, then all solutions of

$$x'' + 5x' + 6x = f(t)$$

are bounded for $t \geqq 0$. Assuming

$$|x(0)| < M,\ |x'(0)| < M,\ |f(t)| < M$$

for $t \geqq 0$, find an upper bound for $|x(t)|$ in $t \geqq 0$.

2 Solution in Matrix Form

2.1 Matrix solution

The explicit solution of linear differential systems with constant coefficients is obtained most elegantly and completely in the matrix form.

Let

$$\mathbf{x}' = A\mathbf{x} \tag{4.2.1}$$

be the differential system written in vector-matrix form. Then

$$X' = AX \tag{4.2.2}$$

will be the associated matrix equation [Section 3.3.1], whose solution with the initial condition

$$X(0) = E \tag{4.2.3}$$

yields the general solution of (4.2.1)

$$\mathbf{x}(t) = X(t)\mathbf{x}(0). \tag{4.2.4}$$

The solution of the matrix equation (4.2.2, 4.2.3) is given by the exponential function

$$X(t) = e^{At} = E + At + \frac{A^2t^2}{2} + \frac{A^3t^3}{3!} + \cdots. \tag{4.2.5}$$

Differentiating term by term, the series Ae^{At} is obtained. Term by term differentiation is allowed because, as we proved in Section 3.1.8, the series for e^{At} converges as fast as the numerical series $e^{\|A\|t}$.

It should be noted particularly that, in general, it is not true that

$$\frac{d}{dt}e^{B(t)} = e^{B(t)}B'(t).$$

In fact, if we try to differentiate the series for $e^{B(t)}$, we obtain terms of the form

$$\frac{d}{dt} B^n = B'BB \cdots + BB'B \cdots + \cdots,$$

and this expression can be set equal $nB'B^{n-1}$ only if B' commutes with B. This is the case, for example, when $B(t)$ is a diagonal matrix, or when $B(t) = At$.

2.2 Jordan canonical form

Given equation (4.2.1), let us make the coordinate transformation

$$\mathbf{x} = P\mathbf{y}, \qquad \mathbf{y} = P^{-1}\mathbf{x}, \tag{4.2.6}$$

where P is a nonsingular constant matrix. Substituting in the differential equation, we obtain

$$P\mathbf{y}' = A\mathbf{x} = AP\mathbf{y}$$

and finally

$$\mathbf{y}' = P^{-1} AP\mathbf{y}. \tag{4.2.7}$$

Hence:

In the new coordinate system, the differential equation takes the form $\mathbf{y}' = B\mathbf{y}$, *where* B *is a matrix which is similar to* A. *Therefore, by an appropriate choice of the transformation* P, *we can insure that the new matrix* B *is in the Jordan canonical form.* [*See Section* 3.1.6.]

In the following, we will solve the differential equation (4.2.1) assuming that the matrix A is already in the Jordan canonical form.

2.3 General form of the solution

Assume that in the matrix differential equation (4.2.2), the constant matrix A is in the Jordan canonical form

$$A = J = \begin{pmatrix} J_0 & & & \\ & J_1 & & \\ & & \ddots & \\ & & & J_p \end{pmatrix} \tag{4.2.8}$$

where

$$J_0 = \begin{pmatrix} \lambda_1 & & & & \\ & \lambda_2 & & & \\ & & \ddots & & \\ & & & \lambda_{r-1} & \\ & & & & \lambda_r \end{pmatrix}, \qquad J_i = \begin{pmatrix} \lambda_{r+i} & 1 & & & \\ & \lambda_{r+i} & 1 & & \\ & & \ddots & \ddots & \\ & & & \lambda_{r+i} & 1 \\ & & & & \lambda_{r+i} \end{pmatrix}$$

and where all nonwritten elements are zero. The solution is the exponential function e^{At}, as seen above.

In order to calculate $e^{At} = e^{Jt}$, we write the development of the exponential series, and remembering formula (3.1.19) in Section 3.1.9, we obtain

$$e^{Jt} = \sum_{n=0}^{\infty} \frac{J^n t^n}{n!} = \sum_{n=0}^{\infty} \frac{t^n}{n!} \begin{pmatrix} J_0^n & & & \\ & J_1^n & & \\ & & \ddots & \\ & & & J_p^n \end{pmatrix} = \begin{pmatrix} e^{J_0 t} & & & \\ & e^{J_1 t} & & \\ & & \ddots & \\ & & & e^{J_p t} \end{pmatrix}. \tag{4.2.9}$$

Therefore, once we know how to find the exponential of the Jordan submatrices $J_i t$, we can immediately write the exponential for the whole matrix Jt.

2.4 Distinct characteristic roots

If the characteristic roots of the original matrix A are all distinct, they will be the same after the transformation into the canonical form J, which must then take the form

$$Jt = J_0 t = \begin{pmatrix} \lambda_1 t & & & \\ & \lambda_2 t & & \\ & & \ddots & \\ & & & \lambda_n t \end{pmatrix}. \tag{4.2.10}$$

The exponential, which is the fundamental solution of the associated matrix equation, is then

$$e^{Jt} = e^{J_0 t} = \begin{pmatrix} e^{\lambda_1 t} & & & \\ & e^{\lambda_2 t} & & \\ & & \ddots & \\ & & & e^{\lambda_n t} \end{pmatrix}. \tag{4.2.11}$$

The general solution of the vector differential equation is

$$\begin{pmatrix} x^1 \\ x^2 \\ \vdots \\ x^n \end{pmatrix} = \begin{pmatrix} e^{\lambda_1 t} & & & \\ & e^{\lambda_2 t} & & \\ & & \ddots & \\ & & & e^{\lambda_n t} \end{pmatrix} \cdot \begin{pmatrix} c^1 \\ c^2 \\ \vdots \\ c^n \end{pmatrix}$$

or, componentwise,

$$x^i(t) = c^i e^{\lambda_i t} \tag{4.2.12}$$

with c^i arbitrary constants.

In case the matrix A in (4.2.1) is not in canonical form, we should first make the coordinate transformation (4.2.6), so that B is canonical. Then we can apply formula (4.2.12). Finally, we should use the inverse transformation defined by (4.2.6) to obtain the solution of the original system. Let us take a closer look at how this transformation (or any similar transformation) alters the solution.

If we know the solution $y^i(t) = c^i e^{\lambda_i t}$ in the y-coordinates and if we want to change to the x-coordinates, we have, from (4.2.6),

$$x^i = \sum_j p^i_j y^j = \sum_j p^i_j c^j e^{\lambda_j t} = \sum_j q^i_j e^{\lambda_j t}. \tag{4.2.13}$$

Hence:

> *The components x^i are linear combinations of the exponentials $e^{\lambda_j t}$. But not all the coefficients q^i_j are arbitrary; among the n^2 constants q^i_j there are only n arbitrary.* [*These correspond to the arbitrary choice of the n constants c^i.*]

For practical purposes, given the original system (2.2.1),which is not in canonical form, there is no need to find the transformation P which changes A into canonical form. Rather, we can start by finding the characteristic roots and then, if they are all distinct, apply directly the method of undetermined coefficients. We simply substitute

$$x^i = \sum_j q^i_j e^{\lambda_j t}$$

into equation (4.2.1). For each $i = 1, 2, \ldots, n$ we obtain n linear equations for the unknown coefficients, and therefore in the total n^2 equation. From the way the λ_j were chosen, in each system of n equations there is (at least) one equation which is linearly dependent on the others and which can therefore be disregarded. We therefore get $n^2 - n$ equations for the n^2 unknown q^i_j. Hence n of them can be taken arbitrarily, which means that we obtain n linearly independent solutions. The actual calculation, quite lengthy and tedious, contains no essential difficulty and is identical to the method described in Section 3.1.

2.5 *Multiple characteristic roots*

If some characteristic roots are of multiplicity greater than one, the canonical form might still be diagonal. [Compare Section 2.1.6.] In that case, everything is identical with the case of distinct roots discussed above. But, in general, a multiple characteristic root may appear in the canonical form in a submatrix of the type (3.1.14). By (4.2.9), we should therefore try to obtain $e^{J_i t}$ for such a Jordan submatrix J_i. This is what we shall do next.

Let

$$J_i = \begin{pmatrix} \lambda & 1 & & & & & \\ & \lambda & 1 & & & & \\ & & \lambda & 1 & & & \\ & & & \ddots & \ddots & & \\ & & & & \ddots & \ddots & \\ & & & & & \lambda & 1 \\ & & & & & & \lambda \end{pmatrix} = E + Z, \tag{4.2.14}$$

where E is the identity matrix and

$$Z = \begin{pmatrix} 0 & 1 & & & & & \\ & 0 & 1 & & & & \\ & & 0 & 1 & & & \\ & & & \ddots & \ddots & & \\ & & & & \ddots & \ddots & \\ & & & & & 0 & 1 \\ & & & & & & 0 \end{pmatrix}. \tag{4.2.15}$$

We note that

$$Z^2 = \begin{pmatrix} 0 & 0 & 1 & & & \\ & 0 & 0 & 1 & & \\ & & \cdots & \cdots & & \\ & & & 0 & 0 & 1 \\ & & & & 0 & 0 \\ & & & & & 0 \end{pmatrix}, \qquad Z^3 = \begin{pmatrix} 0 & 0 & 0 & 1 & & & \\ & \cdot & \cdot & \cdot & \cdot & \cdot & \\ & & \cdot & \cdot & \cdot & \cdot & \\ & & & \cdot & \cdot & \cdot & 1 \\ & & & & \cdot & \cdot & 0 \\ & & & & & \cdot & 0 \\ & & & & & & 0 \end{pmatrix}, \qquad \ldots,$$

$$Z^{n-1} = \begin{pmatrix} 0 & \cdots & 0 & 1 \\ 0 & \cdots\cdots & & 0 \\ \cdots & \cdots & \cdots & \cdots \\ 0 & \cdots\cdots & & 0 \end{pmatrix}, \qquad Z^n = 0. \tag{4.2.16}$$

The exponential series can therefore be written as

$$e^{J_i t} = E + \frac{\lambda E + Z}{1} t + \frac{\lambda^2 E + 2\lambda Z + Z^2}{2!} t^2 + \frac{\lambda^3 E + 3\lambda^2 Z + 3\lambda Z^2 + Z^3}{3!} t^3$$
$$+ \cdots = E\left(1 + \frac{\lambda t}{1} + \frac{\lambda^2 t^2}{2!} + \cdots\right) + Zt\left(1 + \frac{\lambda t}{1} + \frac{\lambda^2 t^2}{2!} + \cdots\right)$$

$$+\frac{Z^2t^2}{2!}\left(1+\frac{\lambda t}{1}+\cdots\right)+\cdots+\frac{Z^{n-1}t^{n-1}}{(n-1)!}\left(1+\frac{\lambda t}{1}+\cdots\right)$$
$$=e^{\lambda t}\left(E+Zt+Z^2\frac{t^2}{2!}+\cdots+Z^{n-1}\frac{t^{n-1}}{(n-1)!}\right)$$

or as

$$e^{J_1 t}=e^{\lambda t}\begin{pmatrix} 1 & t & \frac{t^2}{2!} & \cdots & \frac{t^{n-1}}{(n-1)!} \\ & 1 & t & \cdots & \frac{t^{n-2}}{(n-2)!} \\ & & \cdots & \cdots & \cdots \\ & & & & 1 \end{pmatrix}, \tag{4.2.17}$$

where all elements below the main diagonal are zero.

This formula, together with (4.2.9), solves the problem of obtaining the solution of the differential system, when the matrix is in canonical form.

In the general case when the matrix is not in canonical form, one may reduce it to that form by a linear coordinate transformation. After writing the solution and transforming back to the original coordinates, the components of the solution vector $\mathbf{x}(t)$ will then be linear combinations of the exponentials $e^{\lambda_i t}$ multiplied, in case λ_i is a root of multiplicity m, by a polynomial of degree (at most) $m-1$ in t. So, once the roots λ_i and their multiplicity are known, we will know what kind of terms might possibly appear. We can then determine the coefficients of all these terms by the method of undetermined coefficients and have thus returned to the *direct solution*, given in the previous section.

There are formulas giving the exponential e^{At} directly without reduction to canonical form, and in some particular cases these formulas are quite simple. [See, for example, T. M. Apostol, "Some explicit formulas for the exponential matrix e^{At}," *Amer. Math. Monthly* **76** (1969): 289–292.]

2.6 Complex characteristic roots

Until now all the expressions given for the solution of equation (4.2.1) are valid, even when some characteristic roots λ_i are complex (in which case we must naturally consider complex solutions). These expressions of the solution are indeed valid for differential systems with complex coefficients, which is the general case.

In most practical applications, the given differential system has real coefficients, and we are interested in real solutions. In this case, expressing the solutions as linear combinations of complex solutions is not very satisfactory, and we prefer to use a *real canonical form*, which will be defined next.

Consider directly the case of a complex multiple root $\lambda = \sigma + i\omega$. Since the matrix A of the differential system is assumed real, $\bar{\lambda} = \sigma - i\omega$ will also be a characteristic root of the same multiplicity. Writing only the submatrices corresponding to these two roots, we consider

$$J = \begin{pmatrix} J_1 & \\ & J_2 \end{pmatrix},$$

where

$$J_1 = \begin{pmatrix} \sigma + i\omega & 1 & & & \\ & \sigma + i\omega & 1 & & \\ & & \cdot & \cdot & \\ & & & \cdot & 1 \\ & & & & \sigma + i\omega \end{pmatrix}$$

and $J_2 = \bar{J}_1$. Here E is the identity matrix, and Z is defined in (4.2.15). Hence

$$J = \begin{pmatrix} \sigma E + Z + i\omega E & 0 \\ 0 & \sigma E + Z - i\omega E \end{pmatrix}.$$

In order to get the real canonical form, we apply a similarity transformation $R = P^{-1}JP$ with

$$P = \begin{pmatrix} E & -iE \\ -iE & E \end{pmatrix}, \qquad P^{-1} = \frac{1}{2}\begin{pmatrix} E & iE \\ iE & E \end{pmatrix}.$$

We obtain

$$R = \frac{1}{2}\begin{pmatrix} E & iE \\ iE & E \end{pmatrix} \cdot J \cdot \begin{pmatrix} E & -iE \\ -iE & E \end{pmatrix} = \begin{pmatrix} \sigma E + Z & \omega E \\ -\omega E & \sigma E + Z \end{pmatrix}. \tag{4.2.18}$$

This is the real canonical form.

Now we have to find the exponential of this matrix, which is the solution of the corresponding differential equation. To do this, we note that

$$P^{-1}e^{Jt}P = \exp\,[P^{-1}JPt] = e^{Rt}, \tag{4.2.19}$$

a result which follows from (3.1.15) if the exponential is considered to be the limit of the approximating polynomials. Hence

$$e^{Rt} = \frac{1}{2}\begin{pmatrix} E & iE \\ iE & E \end{pmatrix} \cdot \begin{pmatrix} e^{J_1 t} & \\ & e^{J_2 t} \end{pmatrix} \cdot \begin{pmatrix} E & -iE \\ -iE & E \end{pmatrix} = \begin{pmatrix} \operatorname{Re}\{e^{J_1 t}\} & \operatorname{Im}\{e^{J_1 t}\} \\ -\operatorname{Im}\{e^{J_1 t}\} & \operatorname{Re}\{e^{J_1 t}\} \end{pmatrix}. \tag{4.2.20}$$

Note that in the last step we multiplied submatrices as if they were elements and also took into account the fact that

$$\exp[J_2 t] = \exp[\overline{J_1 t}] = \overline{\exp[J_1 t]}.$$

Denoting

$$B = \begin{pmatrix} 1 & t & t^2/2 & \cdots \\ & 1 & t & \cdots \\ & & 1 & \cdots \\ & & & \ddots \end{pmatrix},$$

we obtain from (4.2.17)

$$\exp[J_1 t] = \exp[(\sigma + i\omega)t] \cdot B = e^{\sigma t}(\cos \omega t + i \sin \omega t) \cdot B$$

and

$$e^{Rt} = e^{\sigma t}\begin{pmatrix} B\cos \omega t & B \sin \omega t \\ -B \sin \omega t & B \cos \omega t \end{pmatrix}. \tag{4.2.21}$$

If the complex root is simple (of multiplicity one), $E = 1$, $Z = 0$, $B = 1$ should be taken in these formulas.

2.7 *Nonhomogeneous system*

For the nonhomogeneous vector differential equation

$$\mathbf{x}' - A\mathbf{x} = \mathbf{f}(t), \tag{4.2.22}$$

formula (3.5.3) gives the solution by variation of constants. In the case of constant coefficients, the fundamental solution is

$$X(t) = e^{A(t-t_0)}$$

with the inverse

$$X^{-1}(t) = e^{-A(t-t_0)}.$$

Hence formula (3.5.3) gives

$$\begin{aligned} \mathbf{x}(t) &= e^{A(t-t_0)}\mathbf{x}(t_0) + \int_{t_0}^{t} e^{A(t-t_0)}e^{-A(s-t_0)}\mathbf{f}(s)\,ds \\ &= e^{A(t-t_0)}\mathbf{x}(t_0) + \int_{t_0}^{t} e^{A(t-s)}\mathbf{f}(s)\,ds. \end{aligned} \tag{4.2.23}$$

Here the integration kernel $e^{A(t-s)}$ is a kind of "influence factor" of the excitation $\mathbf{f}(s)$, at time s, upon the response $\mathbf{x}(t)$, at time t.

Problems

1. Solve the matrix differential equations $X' = AX$, $X(0) = X_0$ for

(a) $A = \begin{pmatrix} 2 & 0 & 0 \\ 0 & 3 & 0 \\ 0 & 0 & 4 \end{pmatrix}$, $\quad X_0 = \begin{pmatrix} 1 & 0 & 0 \\ 0 & 1 & 0 \\ 0 & 0 & 1 \end{pmatrix}$;

(b) $A = \begin{pmatrix} 0 & 0 & 0 \\ 1 & 0 & 0 \\ 0 & 1 & 0 \end{pmatrix}$, $\quad X_0 = \begin{pmatrix} 1 & 1 & 1 \\ 1 & 1 & 1 \\ 1 & 1 & 1 \end{pmatrix}$;

(c) $A = \begin{pmatrix} 3 & 1 & 0 \\ 0 & 3 & 1 \\ 0 & 0 & 3 \end{pmatrix}$, $\quad X_0 = \begin{pmatrix} 0 & 1 & 0 \\ 0 & 0 & 1 \\ 1 & 0 & 0 \end{pmatrix}$.

2. Integrate the given equation $\mathbf{x}' = A\mathbf{x}$ and the adjoint equation $\mathbf{y}' = -A^T\mathbf{y}$. Check $\mathbf{y}^T\mathbf{x} = \text{constant}$.

(a) $A = \begin{pmatrix} 2 & 0 \\ 0 & 3 \end{pmatrix}$. $\qquad$ (b) $A = \begin{pmatrix} 0 & 1 \\ -1 & 0 \end{pmatrix}$.

(c) $A = \begin{pmatrix} 2 & 1 \\ 1 & 3 \end{pmatrix}$. $\qquad$ (d) $A = \begin{pmatrix} 2 & 1 \\ -1 & 2 \end{pmatrix}$.

(e) $A = \begin{pmatrix} 1 & 1 \\ 0 & 1 \end{pmatrix}$.

3 *Operational Method*

3.1 Idea of the method

Oliver Heaviside (1850–1925) gave a set of rules for solving linear differential equations with constant coefficients by introducing the *operator* $D = d/dt$ and by then using it in a formal way. The results proved to be true in the simple cases, which are important in many technical applications, but Heaviside never gave the mathematical proof. Hence a controversy arose about the mathematical value of this method.

With the aid of the *Laplace transform*, a rigorous treatment was given later, which justified the Heaviside method. Recently, Mikusinski gave a direct, purely algebraic foundation for this method with complete mathematical rigor. Here we will follow the traditional approach using the Laplace transform because of its wide use in applied fields.

3.2 The operator D

We define $D = d/dt$ as a linear operator. Addition and product with real numbers are defined in the natural way. For example,

$$(D + D)[x(t)] = (2D)[x(t)] = 2D[x(t)] = 2\frac{dx}{dt}.$$

Positive integer powers of D are defined by

$$D^2[x] = D[D[x]] = D[x'] = x'', \qquad D^3[x] = x''',$$

and so on.

The general expression appearing in the nth order linear differential equation with constant coefficients can be written as

$$a_n x^{(n)} + \cdots + a_1 x' + a_0 x = (a_n D^n + \cdots + a_1 D + a_0)x = L[x]. \tag{4.3.1}$$

Here $L[\cdot]$ is a linear differential operator with constant coefficients. Such operators obey the ordinary addition and multiplication rules of elementary algebra. So, for example,

$$\begin{aligned}(D - 1)(D^2 + D + 1)x(t)\\ &= (D - 1)(x'' + x' + x)\\ &= x''' + x'' + x' - x'' - x' - x\\ &= (D^3 - 1)x(t).\end{aligned}$$

The product of differential operators must always be regarded as the successive application of them. Hence it is easy to operate with such expressions. As an example, let us obtain the single differential equation equivalent to a given system. Let

$$\left.\begin{array}{l} Dx^1 = a_1^1 x^1 + \cdots + a_n^1 x^n - f^1(t) \\ \cdots\cdots\cdots\cdots\cdots\cdots\cdots\cdots \\ Dx^n = a_1^n x^1 + \cdots + a_n^n x^n - f^n(t) \end{array}\right\} \tag{4.3.2}$$

be the differential system, a_j^i constant coefficients, $f^i(t)$ known functions, and $x^i(t)$ the unknown functions. This system may be written as

$$\left.\begin{array}{l} (a_1^1 - D)x^1 + a_2^1 x^2 + \cdots + a_n^1 x^n = f^1(t) \\ \cdots\cdots\cdots\cdots\cdots\cdots\cdots\cdots\cdots\cdots \\ a_1^n x^1 + a_2^n x^2 + \cdots + (a_n^n - D)x^n = f^n(t). \end{array}\right\}$$

This system can be solved in the usual algebraic way, and using Cramer's rule,

$$\Delta(D)x^i = \Delta_1^i(D)f^1 + \cdots + \Delta_n^i(D)f^n. \tag{4.3.3}$$

Here $\Delta(D)$ is the main determinant and $\Delta_j^i(D)$ the adjoint determinants appearing in the linear algebraic system. D is treated as a scalar variable and x^i as scalar unknowns. But (4.3.3) is the single differential equation of order n for the function $x^i(t)$.

Continuing with the formal application of algebraic operations, we are tempted to solve the system (4.3.2) by

$$x^i = \frac{1}{\Delta(D)} \sum_j \Delta_j^i f^j(t). \tag{4.3.4}$$

But, until now, this expression makes no sense because the division by an operator $\Delta(D)$ has not been defined. Formally, such a quotient should be defined by the condition

$$u(t) = \frac{1}{\Delta(D)} \quad \text{is equivalent to} \quad \Delta(D)u(t) = 1. \tag{4.3.5}$$

This last equality is a differential equation, and we know quite well that its solution $u(t)$ is not unique. [We need the initial conditions to determine it uniquely.] Hence equation (4.3.4) does not mean anything more than (4.3.3).

The algebraic, operational procedure did not give us the solution to our problem. This solution will, however, appear in a natural way by applying the Laplace transform.

3.3 Laplace Transform

Let $x(t)$ be a real function defined for $t \geqq 0$. The Laplace transform of $x(t)$, denoted by $\mathscr{L}[x(t)]$, is defined by

$$X(p) = \mathscr{L}[x(t)] = \int_0^\infty x(t)e^{-pt}\,dt,$$

provided this integral exists. [Otherwise, $x(t)$ has no Laplace transform.]

The properties of the Laplace transform can be found in several texts on the subject; so we will not develop them here. Among these properties, we will use the following:

(a) *The operator $\mathscr{L}[\cdot]$ is linear, that is, for any constants a, b,*

$$\mathscr{L}[ax(t) + by(t)] = a\mathscr{L}[x(t)] + b\mathscr{L}[y(t)].$$

(b) *The Laplace transforms of the derivatives are given by*

$$\begin{aligned} \mathscr{L}[x'(t)] &= pX(p) - x(0), \\ \mathscr{L}[x''(t)] &= p^2X(p) - x'(0) - px(0), \\ \mathscr{L}[x^{(n)}(t)] &= p^nX(0) - x^{(n-1)}(0) - px^{(n-2)}(0) - \cdots - p^{n-1}x(0). \end{aligned} \tag{4.3.6}$$

Here $X(p) = \mathscr{L}[x]$ and $x(0), x'(0), \ldots, x^{(n)}(0)$ are the initial values of $x(t)$ and its derivatives.

(c) *The convolution of two functions $f(t)$, $g(t)$ is defined by*

$$h(t) = \int_0^t f(s)g(t-s)\,ds$$

and its Laplace transform is

$$\mathscr{L}[h(t)] = \mathscr{L}[f(t)]\mathscr{L}[g(t)]. \tag{4.3.7}$$

3.4 *Solution of the differential equation*

Here we will solve the differential equation of order n (4.3.3). The system (4.3.2) can be treated in a similar manner.

Let the nth order equation be

$$P(D)x = (a_n D^n + \cdots + a_1 D + a_0)x = f(t),$$

where $f(t)$ is known. Applying the Laplace transform to both members, we get

$$\begin{aligned} \mathscr{L}[P(D)x] \equiv (a_n p^n + \cdots &+ a_1 p + a_0)X(p) \\ &+ c_1 p^{n-1} + c_2 p^{n-2} + \cdots + c_n \\ &= \mathscr{L}[f(t)] = F(p). \end{aligned}$$

In the left member, the c_i are numerical coefficients depending on the values $x(0), x'(0), x''(0), \ldots,$ all of which are easy to calculate from (4.3.6). Since the right-hand side is known, being the transform of the known function $f(t)$, we can solve for the transform of the solution

$$X(p) = \frac{-c_1 p^{n-1} - \cdots - c_n}{a_n p^n + \cdots + a_1 p + a_0} + \frac{F(p)}{a_n p^n + \cdots + a_0} \tag{4.3.8}$$

Once $X(p) = \mathscr{L}[x(t)]$ is known, we have to find the *inverse transform* $x(t)$ of $X(p)$. In simple cases, this can be done by reference to a table of Laplace transforms. [See page 135.] More general techniques for obtaining the inverse Laplace transform are discussed in the References at the end of this chapter.

Formula (4.3.8) can be further analyzed. The first term in the right member depends only on the initial values of $x(t)$ and its derivatives. It corresponds to the response of the system when the input $f(t)$ is zero.

The second term in (4.3.8) depends only on $F(p)$ or its correspondent $f(t)$; it gives the response when all initial conditions are zero. By linearity, the complete solution is the superposition of these two partial solutions.

The last term in (4.3.8) may be written as the product

$$F(p) \cdot \frac{1}{a_n p^n + \cdots + a_0}.$$

If $g(t)$ is the inverse Laplace transform of $1/(a_n p^n + \cdots + a_0)$ and noting that $f(t)$ is the inverse transform of $F(p)$, we may use the convolution formula (4.3.7) to obtain the following inverse transform of the product:

$$\int_0^t f(s)g(t-s)\,ds = \int_0^t f(t-s)g(s)\,ds. \tag{4.3.9}$$

This expression therefore represents the response of the system to the input $f(t)$, when all initial conditions are zero. Furthermore, $g(t)$ can be imagined to be the response of the same system for the case $F(p) = 1$. It is interesting to note that the function $F(p) = 1$ is not the Laplace transform of any function $f(t)$ in the strict mathematical sense. It can nevertheless be thought of as being the limit of the Laplace transform of a function $f(t)$ defined by

$$f(t) = \delta(t) = \begin{cases} 1/\varepsilon & \text{for } 0 \leqq t < \varepsilon, \\ 0 & \text{for } \varepsilon \leqq t. \end{cases}$$

This is consistent with the formula defining the Laplace transform, as we will show in the next section. The function $\delta(t)$ is known as the *Dirac delta* function. Its physical meaning will be discussed next.

3.5 Impulse functions

In physics it is usual to consider such idealizations as

(a) *point mass*—a body of finite mass but infinitesimal size and therefore infinite density, as in lbs/cu in.;

(b) *point force*—a finite force distributed over an infinitesimal area or length and therefore having an infinite density, as in lbs/sq in. or lbs/in.;

(c) *instantaneous collision forces*—a force $f(t)$ that acts only during an infinitesimal time interval Δt but transmits a finite momentum, which is its integral $\int_t^{t+\Delta t} f(t)\,dt$. Hence $f(t)$ must be infinitely large.

Such abstractions are physically reasonable. As a mathematical counterpart, let us consider a function $f(t)$ such that

(a) $f(t) = 0$ for $t < 0$ and for $t > \varepsilon > 0$.

(b) $\int_0^\varepsilon f(t)\, dt = 1$.

Here ε should be considered "small." [Whatever this means—the idea is really to consider a limit process with $\lim \varepsilon = 0^+$.] Figure 4.1 shows different realizations of this function.

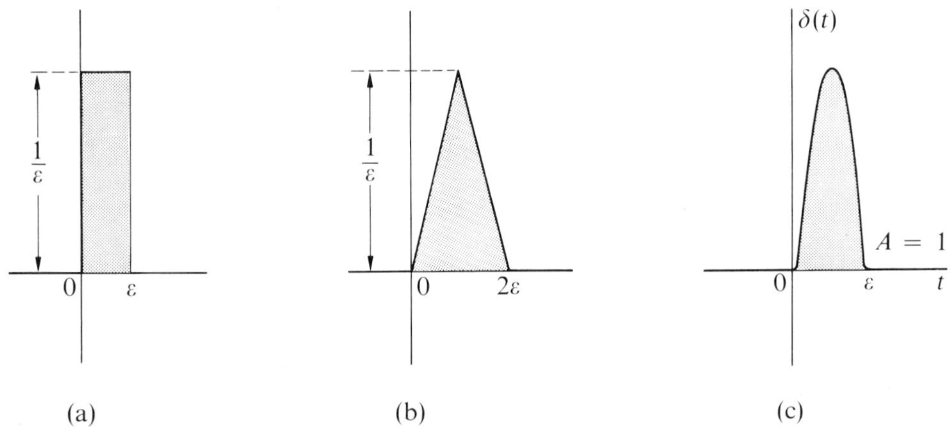

Figure 4.1

If we let ε tend to zero and call $\delta(t)$ the limit function, this function has the following impossible properties:

(a) $\delta(t) = 0$ for all values of t except $t = 0$.

(b) $\int_I \delta(t)\, dt = 1$ whenever the interval I contains the point $t = 0$.

Obviously, such a function does not exist, but it is still physically meaningful to consider an approximation where ε is not exactly zero but nevertheless small. Mathematically speaking, what happens is that for some results, which depend in some particular way on $f(t)$, if we let $f(t)$ tend to $\delta(t)$ (that is, let $\varepsilon \to 0$ above), the result tends to a well-determined value which may be interpreted as corresponding to the (nonexisting) function $\delta(t)$. In this sense, one can operate with this function, but much care must be taken in order to attain significant results. [The mathematical procedure does not automatically guarantee such meaningfulness or correctness.]

If we take as an approximation of the δ function the differentiable function given in Figure 4.1(c), we may differentiate it and obtain $\delta'(t)$. A function like the one represented in Figure 4.2 is thus obtained. If the first function was sufficiently smooth, higher derivatives could also be obtained.

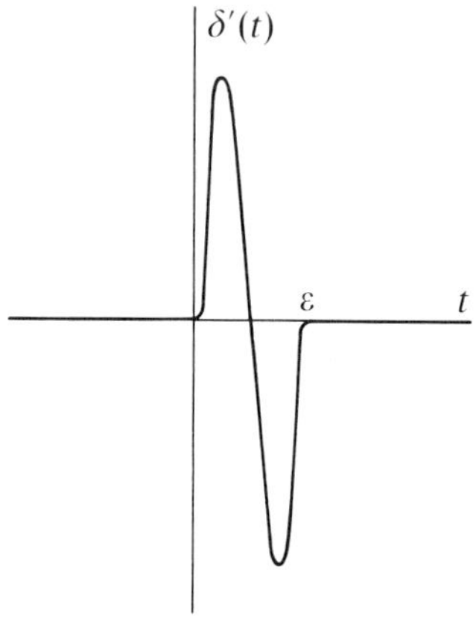

Figure 4.2

It is quite surprising that these functions, which are mathematical impossibilities, can be interpreted as having some well-determined Laplace transforms. So, if we take as an approximation of $\delta(t)$ the function

$$f(t) = 0 \quad \text{for } t < 0 \quad \text{or} \quad t > \varepsilon, \qquad f(t) = \frac{1}{\varepsilon} \quad \text{for } 0 \leqq t \leqq \varepsilon,$$

the Laplace transform will be

$$\mathscr{L}[\delta(t)] = \lim_{\varepsilon \to 0} \int_0^\varepsilon \frac{e^{-pt}}{\varepsilon}\, dt = \lim_{\varepsilon \to 0} \left[-\frac{e^{-pt}}{\varepsilon p} \right]_0^\varepsilon = \lim_{\varepsilon \to 0} \frac{1 - e^{-p\varepsilon}}{p\varepsilon} = 1.$$

It can be seen that if another approximation [such as (a) or (b) in Figure 4.1] of the δ-function is used, the same limit results. Hence

$$\mathscr{L}[\delta(t)] = 1. \tag{4.3.10}$$

We may also interpret the δ-function as the derivative of the *step function*

$$H(t) = \begin{cases} 0 & \text{for } t \leqq 0 \\ 1 & \text{for } t > 0. \end{cases}$$

If we take a smooth approximation of $H(t)$, as shown in Figure 4.3, the derivative will look like the approximation to $\delta(t)$ shown in Figure 4.1(c).

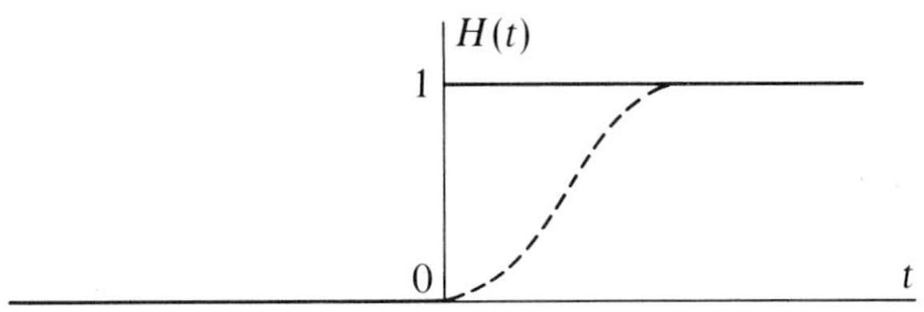

Figure 4.3

The Laplace transform of $H(t)$ is

$$\mathscr{L}[H] = \int_0^{\infty} e^{-pt}\, dt = 1/p.$$

Hence we can obtain $\mathscr{L}[\delta] = \mathscr{L}[H']$ applying formula (4.3.6):

$$\mathscr{L}[\delta] = \mathscr{L}[H'] = p\mathscr{L}[H] = 1,$$

in agreement with (4.3.10).

In the same way we can also obtain the Laplace transforms of the successive derivatives of the δ-function:

$$\mathscr{L}[\delta'] = p\mathscr{L}[\delta] = p$$
$$\mathscr{L}[\delta''] = p\mathscr{L}[\delta'] = p^2$$
$$\cdots\cdots\cdots\cdots\cdots\cdots$$

All these *impulse functions* do not exist as true functions in the mathematical sense. They refer to limit processes and were used as such in mathematical physics. In the late 1940s they became respectable mathematical objects in Laurent Schwartz's *theory of distributions*, where they appear as *continuous linear functionals over the space of infinitely differentiable functions*. In this theory, all operational rules with impulse functions are proved with mathematical rigor.

The impulse functions give meaningful results when they appear in such integrals as

$$\int_{-\infty}^{+\infty} f(t)\, \delta(t)\, dt = f(0), \tag{4.3.11}$$

provided $f(t)$ is continuous at $t = 0$;

$$\int_{-\infty}^{+\infty} f(t)\, \delta'(t)\, dt = -f'(0), \tag{4.3.12}$$

provided $f(t)$ is continuously differentiable at $t = 0$; and

$$\int_{-\infty}^{+\infty} f(t)\, \delta^{(n)}(t)\, dt = (-1)^n f^{(n)}(0), \tag{4.3.13}$$

provided $f(t)$ is continuously differentiable at least n times.

The first of these formulas may be obtained by a direct limit process [as we did in calculating the Laplace transform of $\delta(t)$]. The remaining formulas are easily obtained by integration by parts of a suitable approximation of the δ function [shown in the following example, where we will write $\delta(t)$ instead of the correspondent approximation].

$$\int_{-\varepsilon}^{+\varepsilon} f(t)\, \delta'(t)\, dt = [f(t)\, \delta(t)]_{-\varepsilon}^{+\varepsilon} - \int_{-\varepsilon}^{+\varepsilon} f'(t)\, \delta(t)\, dt = -f'(0)$$

because in the first term $\delta(-\varepsilon) = \delta(\varepsilon) = 0$; the value of the last term may be taken from (4.3.11) by replacing $f(t)$ with $f'(t)$.

Similarly,

$$\int_{-\varepsilon}^{+\varepsilon} f(t)\, \delta''(t)\, dt = [f(t)\, \delta'(t)]_{-\varepsilon}^{+\varepsilon} - \int_{-\varepsilon}^{+\varepsilon} f'(t)\, \delta'(t)\, dt = f''(0),$$

and so on.

As an application of these formulas, we will show how initial conditions for a linear differential equation can be converted into a nonhomogeneous term. Let us consider the differential equation

$$x'' + ax' + bx = 0$$

with

$$x(0) = x_0, \qquad x'(0) = y_0 .$$

Taking the Laplace transform, we obtain

$$p^2 X(p) - y_0 - px_0 + apX(p) - ax_0 + bX(p) = 0.$$

Hence

$$p^2 X(p) + apX(p) + bX(p) = px_0 + y_0 + ax_0 .$$

Taking the inverse Laplace transform, we get

$$x'' + ax' + bx = x_0\, \delta'(t) + (y_0 + ax_0)\, \delta(t),$$

now having the initial conditions

$$x'(0) = x(0) = 0.$$

Physically interpreted, the solution of this last equation starts at the origin with zero derivative, but in the infinitesimal time interval $[0, \varepsilon]$, it gets such a push from the acting forces (second member) that $x(t)$ jumps to x_0 and $x'(t)$ jumps to y_0.

Short Table of Laplace Transforms

$f(t)$	$F(p) = \mathscr{L}[f(t)]$
1	$\frac{1}{p} \quad (p > 0)$
$t^n \quad (n > 0 \text{ integer})$	$\frac{n!}{p^{n+1}} \quad (p > 0)$
$t^r \quad (r > 0 \text{ real})$	$\frac{\Gamma_{(r+1)}}{p^{r+1}} \quad (p > 0)$
e^{at}	$\frac{1}{p-a} \quad (p > a)$
$\sin at$	$\frac{a}{p^2 + a^2} \quad (p > 0)$
$\cos at$	$\frac{p}{p^2 + a^2} \quad (p > 0)$
$\sinh at$	$\frac{a}{p^2 - a^2} \quad (p > a)$
$\cosh at$	$\frac{p}{p^2 - a^2} \quad (p > a)$
$t^n e^{at} \quad (n > 0 \text{ integer})$	$\frac{n!}{(p-a)^{n+1}} \quad (p > a)$
$e^{at} \sin bt$	$\frac{b}{(p-a)^2 + b^2} \quad (p > a)$
$e^{at} \cos bt$	$\frac{p-a}{(p-a)^2 + b^2} \quad (p > a)$
$\delta(t)$	1
$\delta(t-c)$	e^{-cp}
$H(t-c)$	$\frac{e^{-cp}}{p} \quad (p > 0)$

Problems

1. Transform the following systems into a single differential equation for x. Do not solve it, but obtain the equations to get $y(t)$ and $z(t)$, once $x(t)$ is known. Indicate which are the initial conditions to be given for the system and how you get the initial conditions for the equation for $x(t)$. Use the operator notation $D = d/dt$.

(a) $\begin{cases} D^3x - z = 0 \\ D^2x + Dy - z = 0 \\ Dx + Dy - D^2z = 0. \end{cases}$ (b) $\begin{cases} D^2x + y + D^2z = 0 \\ D^2x + Dy - Dz = 0 \\ x - Dy - D^2z = 0. \end{cases}$

2. Using the operational method, solve the equations (a)–(f) in Problem 1, at the end of Section 1 of this chapter.

References

Matrix analysis has already been considered in Chapter 3. For operational methods and Laplace transforms, some good introductory textbooks are

R. V. Churchill. *Modern Operational Mathematics in Engineering*. New York: McGraw-Hill, 1944.

W. Kaplan. *Operational Methods for Linear Systems*. Reading, Mass.: Addison-Wesley, 1962.

J. G. Holbrook. *Laplace Transforms for Electronic Engineers*. London: Pergamon Press, 1966 (2nd ed.).

Probably the most encyclopedic and detailed book on the subject of the Laplace transform is

G. Doetsch. *Handbuch der Laplace Transformation*, Volumes I, II, and III. Basel: Birkhäuser Verlag, 1950, 1955, and 1956.

The purely algebraic approach to operational calculus is given in

J. Mikusinski. *Operational Calculus*. London: Pergamon Press, 1959.

A newer version of this algebraic approach is

G. Krabbe. *Operational Calculus*. New York–Berlin: Springer Verlag, 1970.

The classical reference for δ functions is

L. Schwartz. *Theorie des Distributions*, Volumes I and II. Paris: Herman, 1950 and 1951.

5

Applications

1 Systems

1.1 Mathematical models

In the early stages of its development, calculus was brilliantly utilized to solve problems in the physical sciences. The achievements of calculus associated with such names as Newton, Lagrange, Euler, and Fourier, enabled mathematics in general and the theory of differential equations in particular to become a primary tool in the construction of scientific solutions.

Natural laws can be simply and accurately expressed as differential equations which govern the behavior of the parameters describing a given system. In those cases in which all actions are well-known [in astronomy, for example, where there is little or no friction or similar perturbations], it is only a question of integrating the correspondent differential equations in order to predict the future behavior of the system.

To correctly apply mathematics to a physical system, one must naturally know the laws of physics governing its behavior. In most cases, only an *ideal system* [one which disregards many existing perturbations because of the complexity they bring to the problem] is considered. For example: the friction factor is not included in a study of moving bodies; the wind factor is deleted in the study of motion occurring in the air; and the rotation of the earth is ignored in dynamical experiments confined to a small region. In such cases one cannot expect absolutely precise results from the mathematical treatment; nevertheless the results are quite satisfactory.

Guided by an overwhelming success in physics, it was natural to try to apply mathematics, to such disciplines as biology, psychology, and the social sciences. Here, the situation is more difficult because many of the basic laws are unknown and because the number of factors influencing the evolution of a system is usually quite large. Such factors are also generally unknown or very difficult to determine. One is therefore led to consider highly idealized systems, whose behavior can be expected to only roughly approximate the actual system. Recently, however, many attempts at applying a mathematical description to problems of economics, biology, and even history have proved to be at least partially successful. But it is clear that we are only beginning to devise mathematical applications for other disciplines.

In order to distinguish carefully between the actual system and the mathematical problem, the latter is called the *mathematical model.* The basic problem is no longer to solve the equations of the mathematical model but to find that model which allows one to describe the behavior of the actual system with a satisfactory degree of exactness. This problem, finding the right mathematical model, is the key to further studies of a subject. Sometimes, when the intervening factors are impossible to take into account exactly, they are considered to be random disturbances, according to the laws of probability. In modern physics (quantum mechanics) the basic laws of nature themselves are categorized as probabilistic. [This certainly indicates the intrinsic impossibility of an exact knowledge of the world in the classical sense!]

As an example of a mathematical model, let us consider the growth of a population. Let $N(t)$ be the number of people (or a class of animals or bacteria) of an isolated community, at time t. It is reasonable to expect that, on the average, both birth and death rates are proportional to the size of the population. Hence in a small time interval dt the change in population will be

$$dN = kN(t)\,dt,$$

where k is a suitable constant. Integrating this differential equation, we obtain

$$N(t) = N(0)e^{kt},$$

an exponential increase.

Over relatively small time intervals, this agrees reasonably well with actual data, but it may be quite incorrect for long time intervals. If we want to try another mathematical model, we could suggest that the coefficient k itself be a function of $N(t)$. If we want to see some *saturation* effect, that is, a leveling off for increasing population, we may use the model

$$dN = (k_1 - k_2 N)N\,dt,$$

which can be integrated by separation of variables. In most cases, this model is a much better representation of the population growth than the first one.

1.2 State variable

Given a system, any description of it must contain some variables which determine what we want to know about the system. So, if our system is a single particle, the describing variable may be its position $\mathbf{x}(t)$. Then we must characterize the action on the system from the outside. This, in the case of the particle, will be the force $\mathbf{f}(t)$ acting upon it. But, in this example, the description of the system is not quite complete: The law governing the evolution of the system is Newton's law $\mathbf{x}'' = \mathbf{f}(t)/m$, where m is the mass. This is a second-order differential equation, and in order to determine the actual solution, we need the initial constants $\mathbf{x}(0)$ and $\mathbf{x}'(0)$, position and velocity, respectively. Hence the knowledge of the initial position, together with the acting force, is not enough to determine the future behavior. We can conveniently describe the instantaneous *state* of the system by giving both $\mathbf{x}(t)$ and $\mathbf{v}(t) = \mathbf{x}'(t)$. The vector $(\mathbf{x}, \mathbf{v})$ will then be called the state of the system. Its knowledge, as a function of time, aptly describes the evolution of the system, and its initial value, together with the exterior action upon the system, is all we need to be able to predict the future. Hence the following definition:

Definition

The state of a given system will be any set of variables which describe the system in a satisfactory way, such that their initial values, together with the exterior action, determine the future evolution of the system.

Usually these variables are imagined as components of a *state vector*, designated by $\mathbf{x}(t)$.

Sometimes, it is impossible to successfully characterize a system by a finite number of variables. In such cases we must consider a state variable $\mathbf{x}(t)$, which is an infinite-dimensional vector, or an element of an abstract space. A good part of the theory we will present can be generalized to infinite dimensional spaces, but it is beyond the scope of this book. We will always assume the state variable $\mathbf{x}(t)$ to be a finite dimensional vector.

1.3 Input and output

The exterior action upon a system will be expressed by a certain number of variables $u_1(t), u_2(t), \ldots$, considered to be components of a vector $\mathbf{u}(t)$. This vector is called the *input* of the system; other equivalent terms are *exterior force*, *forcing term*, *control*, and even *perturbation*. [In this last case, we imagine a "free" unperturbed system plus some perturbation acting on the system.]

The *output* of the system is a set of variables $y_1(t)$, $y_2(t)$, ..., considered to be components of the vector $\mathbf{y}(t)$, which give the data we want from the system. The output $\mathbf{y}(t)$ may be identical with the state $\mathbf{x}(t)$, but this is not necessarily so. Sometimes, only the vector $\mathbf{y}$ is suitable for observation, while the components of $\mathbf{x}$ are, so to speak, hidden.

In some rather abstract descriptions, a system is called a *black box*. Entering this box is the input $\mathbf{u}(t)$; coming out is the output $\mathbf{y}(t)$. The state vector $\mathbf{x}(t)$ is inside and hidden from observation.

Example 1

Consider the mechanical system represented in Figure 5.1. Two masses, free to move on parallel lines, are pushed equally by the input $u(t)$ through two springs. The positions of the two masses are $x_1(t)$, $x_2(t)$. The output $y(t)$ is a linear combination

$$y = ax_1 + bx_2, \qquad a + b = 1, \tag{5.1.1}$$

obtained, for example, by joining both masses with a lever, as shown in the figure.

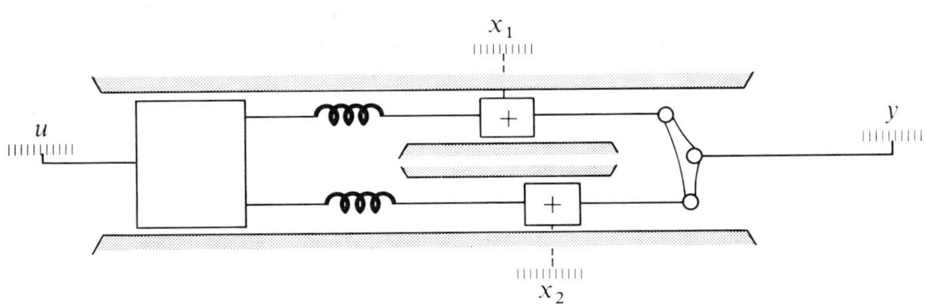

Figure 5.1

With k_1, k_2 suitable positive constants, the basic equations are

$$\left.\begin{aligned} x_1'' &= -k_1 x_1 + k_1 u \\ x_2'' &= -k_2 x_2 + k_2 u. \end{aligned}\right\} \tag{5.1.2}$$

From this we obtain

$$y'' = -ak_1x_1 - bk_2x_2 + (ak_1 + bk_2)u.$$

From (5.1.1)

$$x_2 = \frac{y}{b} - \frac{ax_1}{b},$$

so

$$y'' = -a(k_1 - k_2)x_1 - k_2 y + (ak_1 + bk_2)u. \tag{5.1.3}$$

If $k_1 \neq k_2$: we eliminate x_1 by differentiating twice, substituting x_1'', as given by (5.1.2), and substituting x_1, as obtained from (5.1.3), in terms of y and y''. In this way we get

$$\begin{aligned} y'''' = &-(k_1 + k_2)y'' - k_1 k_2 y + k_1 k_2 (a + b)u \\ &+ (ak_1 + bk_2)u''. \end{aligned} \tag{5.1.4}$$

We see that in this case ($k_1 \neq k_2$), $y(t)$ is governed by a fourth-order equation (equivalent to two second-order equations). We see also that the knowledge of $y(t)$ is equivalent to knowing $x_1(t)$ and $x_2(t)$. Indeed, we obtained, in the derivation of (5.1.4), x_1 in terms of y and y''; once y and x_1 are known, x_2 is obtained by (5.1.1).

If $k_1 = k_2 = k$, the situation is quite different. From (5.1.3), we obtain

$$y'' = -ky + ku, \tag{5.1.5}$$

a second-order equation which is evidently not equivalent to the original system. Here the determination of $y(t)$ is easier than in the first case [we need only $y(0)$ and $y'(0)$ as initial constants], but the knowledge of $y(t)$ is not sufficient to determine $x_1(t)$ and $x_2(t)$.

In this case it is not difficult to see that, with $u(t) \equiv 0$, it is possible for both masses to oscillate in opposite phase and with such amplitudes that $y(t) \equiv 0$. This motion, which cannot be observed from $y(t)$ alone, may be superposed to any other possible motion of the system. This is the reason why knowledge of $y(t)$ alone is not sufficient to obtain $x_1(t)$ and $x_2(t)$.

1.4 Linear systems

The basic solution of every homogeneous linear system is the *trivial, identically zero solution*. It corresponds to the *rest position* of the system.

The two causes which, without destroying linearity, can bring the system out of this rest position are nonzero initial data and the addition of a nonhomogeneous term. Hence such a nonhomogeneous term may be called a *forcing term*; it corresponds to the input.

The mathematical property of linearity, when phrased physically, becomes the

Principle of superposition

If a linear system is brought out of its rest position by several causes, the produced motion is equal to the superposition (understood as sum) of the motions corresponding to each individual cause.

Or, in particular:

If two causes act simultaneously, the motion produced is the superposition of the motions produced by each cause acting alone. And if the cause is multiplied by a constant factor, the motion will also be multiplied by the same factor.

In these statements, a *cause* may be either some nonzero initial condition or a nonzero forcing term.

Many physical systems having practical applications are linear. Furthermore, many systems which are not linear can be sufficiently well approximated by linear ones; that is, they can be replaced by linear systems without losing any essential feature of the original system. Consequently, the application of linear systems to physics and engineering is extremely important.

The simplest are those systems which remain, in some sense, invariant in time. More precisely: the reaction of the system to any input does not depend on the time at which the input is applied. In order to see the meaning of this condition we compare two solutions:

(a) $\mathbf{x}_1(t)$, solution of

$$\mathbf{x}_1'(t) = \mathbf{f}(t, \mathbf{x}_1(t)) + \mathbf{u}_1(t), \ \mathbf{x}_1(t_0) = \mathbf{x}_0;$$

(b) $\mathbf{x}_2(t)$, solution of the same equation

$$\mathbf{x}_2'(t) = \mathbf{f}(t, \mathbf{x}_2(t)) + \mathbf{u}_2(t),$$

but where the input is now applied a time h later,

$$\mathbf{u}_2(t) = \mathbf{u}_1(t - h),$$

and the initial condition is shifted h in time,

$$\mathbf{x}_2(t_0 + h) = \mathbf{x}_0 .$$

If the system is time-invariant, both solutions must agree, once the time is shifted; hence

$$\mathbf{x}_2(t) = \mathbf{x}_1(t - h).$$

Therefore

$$\mathbf{x}_2'(t) = \mathbf{x}_1'(t - h),$$

and finally,

$$\mathbf{f}(t, \mathbf{x}_1(t - h)) + \mathbf{u}_1(t - h) = \mathbf{f}(t - h, \mathbf{x}_1(t - h)) + \mathbf{u}_1(t - h).$$

Because this must hold for every solution (i.e., for every $\mathbf{x}$), we get

$$\mathbf{f}(t - h, \mathbf{x}) = \mathbf{f}(t, \mathbf{x}).$$

This means $\mathbf{f}(t, \mathbf{x})$ must be periodic in t. But this must also be true for every h. Hence:

> *If the differential system is time-invariant, it must be given by a differential equation of the type*
>
> $$\mathbf{x}' = \mathbf{f}(\mathbf{x}),$$
>
> *where the variable t does not appear in the function f.*

Systems which have this property are called *autonomous*.

If a system is linear, the function $\mathbf{f}(t, \mathbf{x})$ appearing in the differential equation is linear. If the system is autonomous, $\mathbf{f}(t, \mathbf{x})$ cannot depend on t; so it must be a linear system with *constant coefficients*. This shows why the linear systems with constant coefficients are so important—they represent the time-invariant linear systems.

1.5 Free and forced motions

If no exterior action is applied on a linear physical system, it may remain in its rest position. Taking this position as the origin of the coordinates, the system is represented by a homogeneous linear differential equation. If the system is not at its rest position but there is still no exterior action, we call this a *free* motion.

If there is an exterior action upon the system, this action will be represented by a nonhomogeneous term in the differential equation. The correspondent motion is called *forced*.

For nonlinear systems, the notion of free and forced motion is not precise anymore; nevertheless it is sometimes used. But it should be made clear in each case which terms are considered to represent an external action.

2 Rest Points and Critical Points

2.1 Rest points

For the differential equation

$$\mathbf{x}' = \mathbf{f}(t, \mathbf{x}), \tag{5.2.1}$$

$\mathbf{x}_0$ is called a *rest point* or an *equilibrium point*, if

$$\mathbf{x}(t) = \mathbf{x}_0 = \text{const}$$

is a solution. In order for this to happen,

$$\mathbf{f}(t, \mathbf{x}_0) = 0 \quad \text{for all } t \tag{5.2.2}$$

must be true.

The rest points can thus be determined by (5.2.2) without any need to integrate the equation (5.2.1).

2.2 Critical points

For the autonomous equation

$$\mathbf{x}' = \mathbf{f}(\mathbf{x}), \tag{5.2.3}$$

the rest points are usually called *critical points*. They are the roots of

$$\mathbf{f}(\mathbf{x}) = 0. \tag{5.2.4}$$

For example, in the plane (x_1, x_2), equation (5.2.3) is written as

$$\begin{cases} x_1' = f_1(x_1, x_2) \\ x_2' = f_2(x_1, x_2), \end{cases}$$

and the critical points are the intersections of the curves

$$f_1(x_1, x_2) = 0, \qquad f_2(x_1, x_2) = 0.$$

We see that these curves will usually intersect in a finite number of isolated points, but other configurations are also quite possible (for example, no intersections at all or infinitely many intersections, or even a continuous curve of critical points).

For linear homogeneous equations, the origin $\mathbf{x} = 0$ is always a rest point. Furthermore, if the equation has constant coefficients and if there is another critical point x_0, then all points kx_0 for $-\infty < k < \infty$ are also critical and constitute a *line of critical points.*

For example,

$$\begin{cases} x_1' = x_1 - x_2 \\ x_2' = -x_1 + x_2 \end{cases}$$

has the line of critical points $x_1 = x_2$.

2.3 Stability

If the linear homogeneous differential equation with constant coefficients

$$\mathbf{x}' = A\mathbf{x} \tag{5.2.5}$$

is such that all characteristic roots of the matrix A are negative or have negative real parts, we know that all solutions are linear combinations of solutions of the type

$$\mathbf{a}e^{\lambda t}P(t), \qquad e^{\sigma t}P(t)(\mathbf{a} \cos \omega t + \mathbf{b} \sin \omega t),$$

where λ and $\sigma + i\omega$ are characteristic roots of A, and $P(t)$ is a polynomial in t. Hence if all values of λ and σ are negative, all solutions, without exception, tend to zero as $t \to +\infty$.

Physically this means that for every initial state $\mathbf{x}_0$, the solution tends to the rest point $\mathbf{x} = 0$ in an exponential way. In practical approximation, the solution will appear to reach the origin in a finite time and remain there indefinitely. If, on the other hand, some characteristic root is positive or has positive real parts, there will be a solution of the type just given, which tends to infinity as $t \to +\infty$. This same solution will tend to zero as $t \to -\infty$.

Physically, this means that there are initial points as near to the origin as we want, such that the corresponding solution moves away from the origin without limit. In systems of this type, the origin is not a true state of rest for all practical purposes. Due to the inaccuracy of every measurement, the initial state may always be affected by some error; hence we can never be sure that the system, apparently at the rest point at the initial time, will not move away in the future.

For the first type of system, the opposite is true. Even if there is some error in positioning the system at the rest point, this error will decrease by itself with increasing time.

For the intermediate type of behavior, an initial positioning of the system near the rest point will not necessarily cause $\mathbf{x}(t)$ to tend to the rest point. However it will not move indefinitely away; rather $\mathbf{x}(t)$ will remain forever in some neighborhood of the rest point.

If the latter behavior is true, we say that the system is *stable*. In the first case the stability is of a stronger type, and we call it *asymptotically stable*. For the second type of behavior, we say that the system is *unstable*.

Hence we have established two criteria for linear equations with constant coefficients:

(a) *All characteristic roots have negative real parts implies that the system is asymptotically stable.*

(b) *Some characteristic roots have positive real parts implies that the system is unstable.*

In a case where all characteristic roots have nonpositive (negative *or* zero) real parts, various types of behavior are possible. The solutions of the type

$$\mathbf{a}e^{0t} = \mathbf{a} = \text{const} \quad \text{and} \quad \mathbf{a}\cos\omega t + \mathbf{b}\sin\omega t$$

are indeed stable, but if there are multiple roots, polynomials in t and therefore solutions of the type (which grow without limit and are therefore unstable),

$$e^{0t}(\mathbf{a} + \mathbf{b}t), \quad (\mathbf{a} + \mathbf{b}t)\cos\omega t, \quad \text{or} \quad (\mathbf{a} + \mathbf{b}t)\sin\omega t$$

may appear.

If we know that all characteristic roots have real parts $\leqq 0$ *and that those roots with zero real parts are simple, then we can assert that the system is stable.*

For linear systems with variable coefficients or for nonlinear systems, stability cannot be so simply characterized. It can nevertheless be defined in a similar way.

Definitions

Let $\mathbf{x}' = \mathbf{f}(t, x)$ *be the differential equation and* $\mathbf{x}_0$ *a rest point (that is,* $\mathbf{f}(t, \mathbf{x}_0) \equiv 0$*). Then we say that:*

(a) *The rest point* $\mathbf{x}_0$ *is stable if, for every* $\varepsilon > 0$, *there is some* $\delta > 0$ *such that every solution* $\mathbf{x}(t)$ *with* $\|\mathbf{x}(t_0) - \mathbf{x}_0\| < \delta$ *remains, for all* $t > t_0$, *in the neighborhood* $\|\mathbf{x}(t) - \mathbf{x}_0\| < \varepsilon$. *Obviously,* δ *will depend on* ε; *it may also depend on* t_0.

(b) *The rest point* $\mathbf{x}_0$ *is unstable if it is not stable.*

(c) *The rest point* $\mathbf{x}_0$ *is asymptotically stable if it is stable, and furthermore, there exists* $\eta > 0$ *such that all solutions with* $\|\mathbf{x}(t_0) - \mathbf{x}_0\| < \eta$ *tend to* $\mathbf{x}_0$ *as* $t \to +\infty$.

In later chapters we will present a more detailed look at the theory of stability.

Problem

1. Classify the following systems as unstable, stable, or asymptotically stable.
   (a) $x'' + 3x = 0$,
   (b) $x''' + x = 0$,
   (c) $x'' + 5x' + 6x = 0$,
   (d) $x'' + 2x' + 3x = 0$,
   (e) $x^{\text{IV}} + 2x'' + x = 0$,
   (f) $x''' + 5x'' + 6x' = 0$,
   (g) $x^{\text{IV}} + 5x''' + 6x'' = 0$.

3 Oscillations

3.1 Periodic and aperiodic solutions

Let us consider the simplest linear homogeneous differential equation capable of oscillations—the second-order linear homogeneous equation with constant coefficients

$$mx'' + cx' + kx = 0. \tag{5.3.1}$$

This is the equation of a material point of mass m, moving along a straight line, with x as its position coordinate, kx as a force pulling the point toward the origin [k is referred to as the *spring constant*], and c as a kind of *friction coefficient*, since cx' is a force proportional to the velocity and opposed to it. This is the mechanical model of (5.3.1). [See Figure 5.2.]

An electrical model for the same differential equation is represented in Figure 5.3. A resistor R, an inductance L, and a capacitor C are connected in series. The

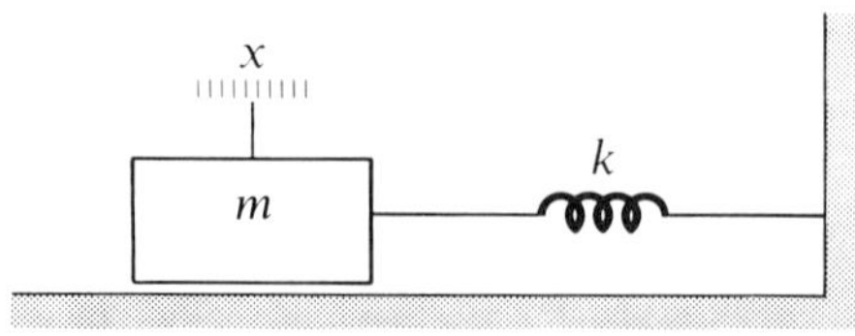

Figure 5.2

differential equation of this system, taking the current i as an unknown time function, is

$$Ri + L\frac{di}{dt} + \frac{1}{C}\int_0^t i(s)\,ds = 0,$$

where the integral term represents the voltage across the capacitor C and the lower limit of the integral has to be taken at any instant when this voltage is zero.

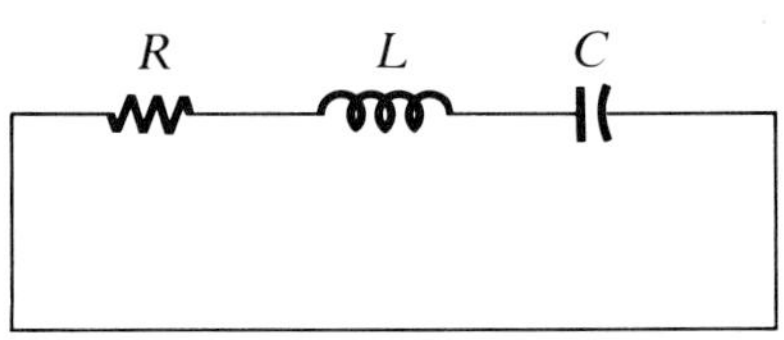

Figure 5.3

Differentiating, we obtain

$$Li'' + Ri' + \frac{1}{C}i = 0. \tag{5.3.2}$$

It is apparent that the inductance acts like a mass, bringing some inertia into the system; the resistance acts like a friction dissipating energy; and the capacitor acts like an elastic link, driving back the current.

If there is no mass in the mechanical system or no inductance in the electrical circuit, then the differential equation reduces to the first-order equation

$$x' = -\frac{k}{c}x,$$

and the solution is exponential. Hence

$$x = x_0 e^{-kt/c}.$$

The velocity is always proportional to the driving force.

If there is no spring in the mechanical system, the equation reduces to

$$mx'' + cx' = 0,$$

and every position x is a possible rest point. [In control terms, the system is *astatic*.] The same thing happens with the electric circuit equation (5.3.2) but not with the original equation, which then becomes

$$Li' + Ri = 0.$$

If there is no friction or no resistance, the middle term drops out and (5.3.1) becomes

$$x'' = -\frac{k}{m}x,$$

the equation of the *harmonic oscillator*. The solutions are

$$x = a \cos \omega t + b \sin \omega t$$

with $\omega^2 = k/m$.

In the general case, the characteristic equation for (5.3.1) will be

$$m\lambda^2 + c\lambda + k = 0$$

with characteristic roots

$$\lambda = -\frac{c \pm \sqrt{c^2 - 4mk}}{2m}.$$

If $c^2 < 4mk$ (that is, if the friction is sufficiently small), the solutions will be of the form

$$x = (a \cos \omega t + b \sin \omega t)e^{\sigma t}$$

with negative σ. This is the *oscillatory case*, and the solutions will tend asymptotically to zero, except when $c = 0$.

If $c^2 > 4mk$, the friction is large and the characteristic roots will be real and distinct. Both roots will be negative [the coefficients are positive]; hence the system will be stable. The solutions will be

$$x = ae^{\lambda_1 t} + be^{\lambda_2 t}.$$

This is the *aperiodic case*.

The change from one case to the other occurs when $c^2 = 4mk$. Then both roots coincide, $\lambda_1 = \lambda_2 = -c/2m$, and the solutions are

$$x = (a + bt)e^{\lambda t}.$$

Since λ is negative, the system is stable. This is called the *critical case*.

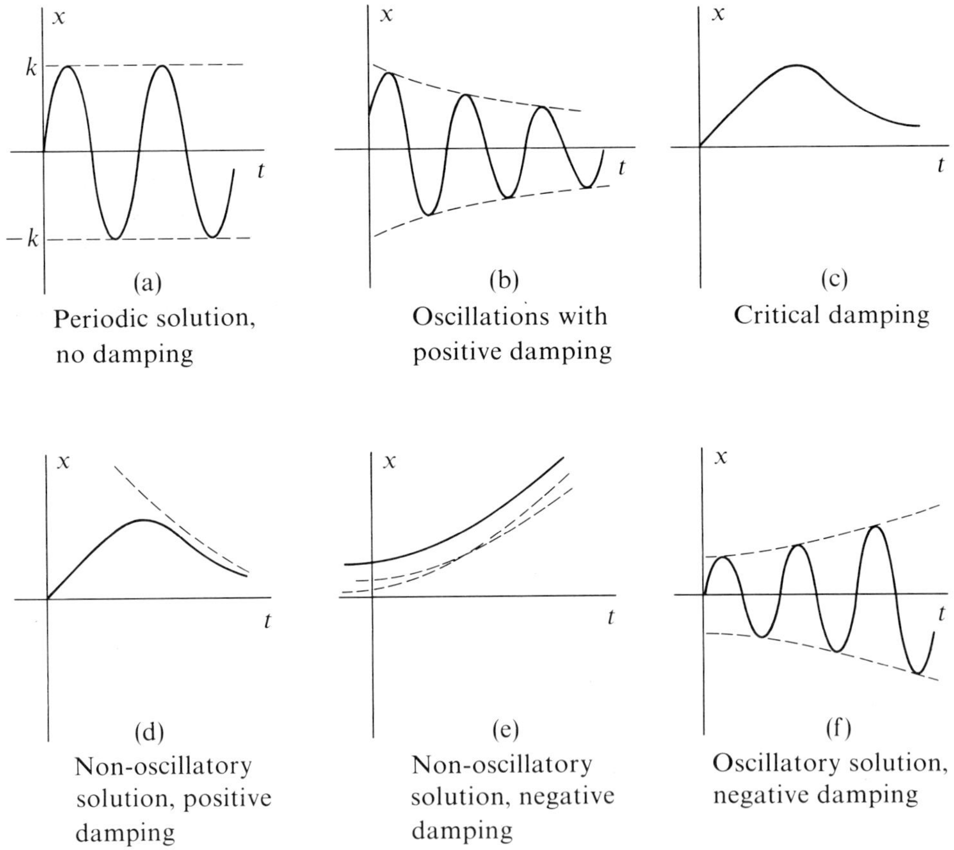

Figure 5.4

For the electrical circuit, the critical condition is

$$R^2 = \frac{4L}{C} \quad \text{or} \quad R = 2\sqrt{L/C}.$$

This value of the resistance gives the *critical damping*.

If there is no resistance, the electric circuit oscillates with the frequency $\omega/2\pi$, with $\omega = \sqrt{1/LC} = 1/\sqrt{LC}$. This is the *natural frequency*. In case there is some (small) resistance, the frequency of oscillation is changed slightly, as the formula for the characteristic roots indicates. If $\omega/2\pi$ is the frequency, ω is called the *angular frequency*.

The behavior of the solutions for both the aperiodic and oscillatory cases is represented in Figure 5.4, parts (a), (b), and (d).

Under some circumstances, it is possible to construct physical systems which obey the same differential equation (5.3.1) or (5.3.2), but with a negative middle term ($c < 0$ or $R < 0$). We speak, then, of a *negative friction* or a *negative resistance*. The formula for the solution is, of course, the same, but now the real parts are positive. Hence the solutions are unstable. In case both roots are real, the solutions look like those in Figure 5.4(e). In case the roots are complex, they look like those in Figure 5.4(f). [We leave to the reader the analysis of the cases where one characteristic root is zero.]

3.2 *Almost periodic solutions*

In higher order differential equations an interesting new phenomenon appears—the almost periodic solutions. Let us consider a simple example

$$x_1' = x_2, \qquad x_2' = -x_1, \qquad x_3' = x_4, \qquad x_4' = -k^2 x_3.$$

These are really two independent harmonic oscillators, and the solution is

$$x_1 = a\cos t + b\sin t, \qquad x_2 = b\cos t - a\sin t,$$
$$x_3 = c\cos kt + d\sin kt, \qquad x_4 = kd\cos kt - kc\sin kt.$$

Each group (x_1, x_2) and (x_3, x_4) makes a periodic motion of period 2π and $2\pi/k$, respectively. *If the constant k is rational (say, p/q with p, q integers), the vector*

$$\mathbf{x}(t) = (x_1(t), x_2(t), x_3(t), x_4(t))$$

describes a motion of period $2\pi q$. In this time period the first couple makes q full oscillations, while the second couple makes p full oscillations. [Its period is $2\pi q/p$.]

If the constant k is irrational, then the resulting motion $\mathbf{x}(t)$ is no longer periodic. If it were periodic, this would mean that the period must be a common multiple of both 2π and $2\pi/k$; hence integers p and q would exist such that $2\pi q = 2\pi p/k$ or $k = p/q$, contrary to the assumption that k is irrational.

The resulting type of function is called *almost periodic* because if we allow for a small error, it looks like a periodic motion.

We will give the definition of almost periodicity, according to Harald Bohr, but for the sake of completeness, we will first give the definition of periodicity.

Definition

The function $f(t)$, defined for $t \in (-\infty, \infty)$, is periodic of period T if

$$f(t+T) = f(t)$$

for all t.

Definition

The continuous function $f(t)$, defined in $(-\infty, \infty)$, is almost periodic if for any $\varepsilon > 0$, there is $L(\varepsilon)$ such that every real interval of length $L(\varepsilon)$ contains at least one number T so that for all $t \in (-\infty, \infty)$,

$$|f(t+T) - f(t)| < \varepsilon.$$

The reader should check to see whether this definition holds for functions of the form

$$f(t) = \sin at + \sin bt,$$

where a/b is irrational.

3.3 Coupled systems

Several physical systems are *coupled* when the state variables of those systems influence each other. Mathematically, this means that if we write the differential equations for these systems, we cannot solve one equation without taking the others into account; more precisely, the differential system as a whole cannot be broken into several independent systems.

Let us consider, for example, the system

$$\left.\begin{aligned} x_1'' + \omega_1^2 x_1 + k_1 x_2 &= 0 \\ x_2'' + \omega_2^2 x_2 + k_2 x_1 &= 0. \end{aligned}\right\} \tag{5.3.3}$$

These are two coupled harmonic oscillators. The absence of terms in x_1', x_2' indicates that there is no friction. ω_1 and ω_2 are the angular frequencies for the two systems, if we make $k_1 = k_2 = 0$ (no coupling). The quantities k_1 and k_2 are the coupling coefficients.

The substitution

$$x_1 = c_1 e^{\lambda t}, \qquad x_2 = c_2 e^{\lambda t}$$

gives

$$\begin{aligned} c_1\lambda^2 + c_1\omega_1^2 + c_2 k_1 &= 0 \\ c_1 k_2 + c_2 \lambda^2 + c_2 \omega_2^2 &= 0. \end{aligned}$$

Hence the characteristic equation is

$$\begin{vmatrix} \lambda^2 + \omega_1^2 & k_1 \\ k_2 & \lambda^2 + \omega_2^2 \end{vmatrix} = \lambda^4 + \lambda^2(\omega_1^2 + \omega_2^2) + \omega_1^2\omega_2^2 - k_1 k_2 = 0,$$

and its roots are determined by

$$\lambda^2 = -\tfrac{1}{2}[\omega_1^2 + \omega_2^2 \pm \sqrt{(\omega_1^2 - \omega_2^2)^2 + 4k_1k_2}]. \tag{5.3.4}$$

The square root appearing in this formula is never imaginary, as we assume $k_1, k_2 \geq 0$; therefore λ^2 is always real. If λ^2 is positive, the corresponding characteristic roots will be $\lambda_{1,2} = \pm\sqrt{\lambda^2}$; if $\lambda^2 = -\omega^2 < 0$, the roots will be $\lambda_{1,2} = \pm i\omega$.

Thus we have obtained four characteristic roots, at least two of which are complex conjugates.

3.4 *Normal modes of oscillation*

Let us expand our discussion of Example (5.3.3), concerning the two coupled oscillators. If one of the coupling coefficients k_1, k_2 is zero, formula (5.3.4) shows that the four characteristic roots are

$$\lambda_{1,2} = \pm i\omega, \qquad \lambda_{3,4} = \pm i\omega_2 .$$

This is the case for two uncoupled (independent) oscillators.

If the coupling coefficients are sufficiently small, both values of λ^2 will still be negative; so the roots will still be purely imaginary. If we denote the new characteristic roots by $\pm i\omega_1^*$, $\pm i\omega_2^*$, the general solution will be

$$\begin{cases} x_1 = A_1 \cos \omega_1^* t + B_1 \sin \omega_1^* t + C_1 \cos \omega_2^* t + D_1 \sin \omega_2^* t \\ x_2 = A_2 \cos \omega_1^* t + B_2 \sin \omega_1^* t + C_2 \cos \omega_2^* t + D_2 \sin \omega_2^* t, \end{cases}$$

where, of the eight constants, only four are arbitrary.

For each value of ω_1^*, ω_2^*, there is a family of solutions of the type

$$\begin{cases} x_1 = A \cos \omega_j^* t + B \sin \omega_j^* t \\ x_2 = C \cos \omega_j^* t + D \sin \omega_j^* t, \end{cases} \qquad j = 1, 2. \tag{5.3.5}$$

where now only two of the constants are arbitrary.

For each $j = 1, 2$, formula (5.3.5) gives a purely periodic oscillation of the whole system with a single frequency. These kinds of motions are called *normal modes of oscillation* of the system. In general, every possible motion of the system is the superposition of different normal modes of the system.

All of these considerations can be easily extended to an arbitrary finite number of coupled systems, each of the oscillatory type.

3.5 Complex notation for harmonic oscillations

Harmonic oscillations of the type

$$x(t) = a \cos \omega t + b \sin \omega t \tag{5.3.6}$$

appear frequently in physics and engineering; it is therefore worthwhile to present a more detailed study of the operations involving such functions.

Any two real numbers a, b can be expressed as

$$a = c \cos \varphi, \qquad b = -c \sin \varphi$$

by taking

$$c = +\sqrt{a^2 + b^2}, \qquad \sin \varphi = -\frac{b}{c}, \qquad \cos \varphi = \frac{a}{c}.$$

With this choice, the expression (5.3.6) becomes

$$x = c(\cos \omega t \cos \varphi - \sin \omega t \sin \varphi) = c \cos (\omega t + \varphi). \tag{5.3.7}$$

When an expression such as (5.3.6) appears as a solution of a differential equation, we may use the equivalent form (5.3.7), where the arbitrary constants are now c and φ instead of a and b.

Using the well-known Euler's formula

$$e^{i\alpha} = \cos \alpha + i \sin \alpha,$$

the expression (5.3.7) can be written as

$$x(t) = \text{Re}\,\{ce^{i(\omega t + \varphi)}\}, \tag{5.3.8}$$

where Re $\{z\}$ stands for the real part of z.

We can now establish the following convention:

Instead of writing $x(t)$, as in (5.3.8), we will write

$$x(t) = ce^{i(\omega t + \varphi)}, \tag{5.3.9}$$

with the understanding that whenever $x(t)$ is expressed as a complex quantity, only the real part is physically meaningful.

Because all operations which we will perform, including differentiation and integration, are linear, the result of these operations performed on the real parts will always coincide with the real part of the result obtained by performing the operations with the complex quantities. This convention is therefore justified since it gives the same results.

Using this convention,

$$x = c \cos(\omega t + \varphi)$$

can be written as

$$x = ce^{i(\omega t + \varphi)},$$

and

$$x = b \sin(\omega t + \varphi)$$

can be written as

$$x = e^{i[\omega t + \varphi - (\pi/2)]} = -ie^{i(\omega t + \varphi)}.$$

One of the advantages of the exponential form is the creation of more compact formulas [the rules of differentiation of products and quotients are easier to write with exponentials than with trigonometric functions]; another is the appealing geometric representation of complex numbers.

If we have several oscillatory magnitudes with the *same frequency* ω,

$$x_n = c_n e^{i(\omega t + \varphi_n)} = c_n e^{i\varphi_n} e^{i\omega t},$$

we may represent these vectors in the complex plane at the instant $t = 0$, thereby forming Figure 5.5. In drawing the graph we disregarded the factor $e^{i\omega t}$. For values of $t \neq 0$, the factor $e^{i\omega t}$ indicates that the whole graph is rotating with angular frequency ω, while the physically meaningful quantities are the real parts of each vector.

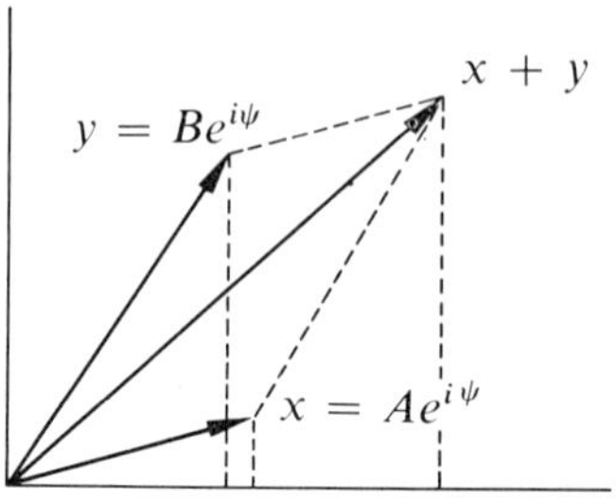

Figure 5.5

If we want to add these quantities, the graphical representation immediately shows us that we should add them as vectors. The whole figure so obtained will rotate with angular frequency ω, and at each instant t, the meaningful real parts will be the projections on the real axis of the vectors involved.

We may go one step further and consider the constants c_n as possibly complex numbers. If we do this, we may write

$$x = ce^{i\varphi}e^{i\omega t} = \gamma e^{i\omega t},$$

where the number γ is, in general, complex.

If we omit the factor $e^{i\omega t}$, we simply have $x = \gamma$; hence the quantity x itself may be considered to be a fixed vector. In order to obtain a physically meaningful quantity from x, we should multiply it by $e^{i\omega t}$ and then take the real part.

The advantage of this procedure is that we may perform numerous operations with these apparently constant vectors. We must remember, however, that for these vectors, differentiation is equivalent to multiplication by $i\omega$ and integration is equivalent to division by $i\omega$.

This method of operating with oscillatory quantities is used consistently throughout the theory of alternating currents, an important branch of electrical engineering, and also in acoustics and the general theory of vibrations.

Problems

1. Solve the system of equation (5.3.3) and find the normal modes of oscillation for

   (a) $\omega_1 = 3, \quad \omega_2 = \sqrt{5}, \quad k_1 = k_2 = \frac{3}{2},$ (b) $\omega_1 = 1, \quad \omega_2 = 2, \quad k_1 = k_2 = 2.$

2. Study the system of a harmonic oscillator coupled to a pure friction system:

$$x_1' = x_2, \quad x_2' = -x_1 + k(x_3 - x_1),$$
$$x_3' = x_4, \quad x_4' = -x_4 + k(x_1 - x_3).$$

Consider particularly the cases $k = 0$ (no coupling) and $k = \infty$. [The latter means $x_1 = x_3$, while $k(x_1 - x_3)$ is indeterminate but equal in the expressions for both x_2' and x_4', so that $x_2' + x_1 = -(x_4' + x_4)$.]

3. Express in complex notation $x(t) = Xe^{i\omega t}$:

(a) $x(t) = \sin \omega t - 3 \cos \omega t$,

(b) $x(t) = \cos \omega t + \cos\left(\omega t + \dfrac{\pi}{6}\right) + \cos\left(\omega t + \dfrac{\pi}{3}\right)$.

4. Prove that for every α,

$$\cos \alpha + \cos\left(\alpha + \frac{2\pi}{n}\right) + \cos\left(\alpha + 2\frac{2\pi}{n}\right) + \cdots + \cos\left(\alpha + (n-1)\frac{2\pi}{n}\right) = 0.$$

[*Hint*: Use $1 + r + r^2 + \cdots + r^{n-1} = (1 - r^n)/(1 - r)$.]

5. Explain, with the aid of the complex representation of harmonic motions, the *beats* phenomenon, which appears in motions like

$$x(t) = Ae^{i\omega_1 t} + Be^{i\omega_2 t},$$

when $\omega_1 - \omega_2$ is small, and which incorporates the fact that this kind of motion appears as a harmonic motion of either frequency ω_1 or ω_2, whose amplitude is changing periodically. [The basic harmonic motion is *modulated*, in electronic terminology. *Hint*: Write $Ae^{i\omega_1 t} + Be^{i\omega_2 t} = [A + Be^{i(\omega_2 - \omega_1)t}]e^{i\omega_1 t}$.]

---

4 Frequency Response

4.1 *Forced oscillations and resonance*

Let us consider the linear nonhomogeneous equation with constant coefficients

$$\mathbf{x}' - A\mathbf{x} = \mathbf{f}(t). \tag{5.4.1}$$

In the language of systems theory, $\mathbf{x}$ will be the *state vector*, which is also the *output*, while $\mathbf{f}$ will be the *input*. Accordingly, we can draw Figure 5.6, where in part (a) the vectors $\mathbf{x}$ and $\mathbf{f}$ are represented as single quantities but in part (b) the different components are shown explicitly.

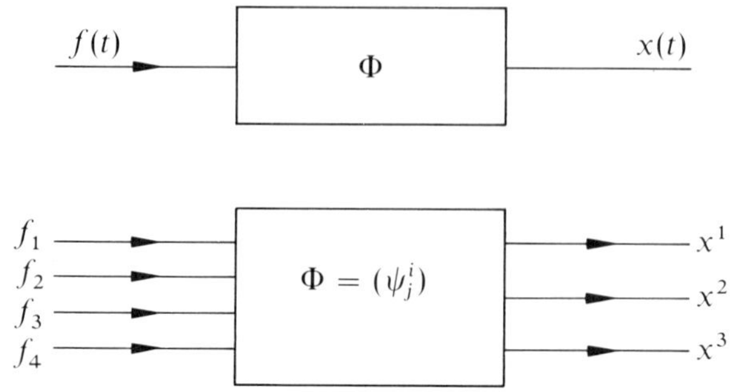

Figure 5.6

We know that in equation (5.4.1) the input $\mathbf{f}(t)$ does not, by itself, determine the output $\mathbf{x}(t)$, since the initial conditions must also be taken into account. The solution can always be written as a term depending only on the input $\mathbf{f}(t)$, plus a *complementary term* (the solution of the homogeneous equation), which adjusts the initial conditions. In almost all practical cases the system considered is asymptotically stable; so this complementary term dies out exponentially. If the first part of the solution is a periodic function, this part alone will remain after some time. This periodic solution is the most interesting one from a practical viewpoint. Such a periodic solution will generally exist if the input $\mathbf{f}(t)$ is periodic, which will be assumed.

Hence, from now on, we will assume that

$$\mathbf{f}(t) = \mathbf{F}e^{i\omega t}, \tag{5.4.2}$$

where $\mathbf{F}$ is now a constant vector (usually complex).

We will be interested in solutions of the type

$$\mathbf{x}(t) = \mathbf{X}e^{i\omega t}. \tag{5.4.3}$$

Substituting (5.4.2) and (5.4.3) in (5.4.1) and canceling the factor $e^{i\omega t}$, we obtain

$$i\omega\mathbf{X} - A\mathbf{X} = \mathbf{F}$$

or

$$(i\omega E - A)\mathbf{X} = \mathbf{F}, \tag{5.4.4}$$

from which we solve for

$$\mathbf{X} = (i\omega E - A)^{-1}\mathbf{F}. \tag{5.4.5}$$

[Here E is the identity matrix.] We may therefore state:

If $i\omega$ is not a characteristic root of the matrix A, then for every harmonic input of angular frequency ω, there is exactly one corresponding harmonic solution of the same frequency. If, on the other hand, $i\omega$ is a characteristic root of the matrix A [hence the inverse $(i\omega E - A)^{-1}$ does not exist], then there can be either no harmonic output at all or infinitely many output solutions to our problem.

In the latter case, the existence of infinitely many harmonic solutions, if only one such solution actually exists, results from the fact that if $i\omega$ is a characteristic root, the homogeneous system has "free" solutions of angular frequency ω. Hence if there is one solution to the nonhomogeneous equation, we may add any such solution to the homogeneous equation, thereby obtaining infinitely many solutions to the nonhomogeneous equation. In complex notation: if $\mathbf{X}$ satisfies (5.4.4) and there are solutions $\mathbf{Y}$ of $(i\omega E - A)\mathbf{Y} = 0$, then $\mathbf{X} + \mathbf{Y}$ also satisfies (5.4.4).

If $i\omega$ is a characteristic root of A [meaning that ω is a *natural frequency* of the system], the method of variation of constants shows that in general the solution corresponding to an input of frequency ω will diverge.

Indeed, in formula (3.5.3), which gives the solution of the nonhomogeneous system, $X(t)$ was the fundamental matrix of the free system; it will now contain terms of the type $e^{i\omega t}$, since $i\omega$ is a characteristic root. The inverse $X^{-1}(t)$ will then contain terms in $e^{-i\omega t}$. Hence, if in the integral

$$\int_{t_0}^{t} X^{-1}(s)b(s)\,ds$$

which appears in formula (3.5.3), the input $b(t)$ contains the factor $e^{i\omega t}$, the product $X^{-1}(s)b(s)$ will contain constant terms which, when integrated, will yield terms of the form kt that diverge as $t \to +\infty$.

For this reason: *if the input frequency ω coincides with any natural frequency of the free system, the solution will in general diverge, and one says that there is resonance.*

One also says that there is resonance when the characteristic root is $\lambda = -\sigma \pm i\omega_0$, with σ small, and the input frequency ω is equal (or nearly equal) to ω_0. In this case, in formula (5.4.5) the determinant of $i\omega E - A$ is nearly zero, in fact, of the order of σ (and possibly of $\omega - \omega_0$). Thus the solution does not diverge to infinity, but it does assume very large values.

For some special initial conditions, the diverging terms may have zero coefficients. Hence there may be, even in the case of resonance, solutions which do not diverge, but these solutions will be exceptional.

In Figure 5.7 typical resonance curves are represented; the amplitude of oscillation is shown as a function of the input frequency. For a system without damping ($\sigma = 0$), the curve goes to infinity; when there is damping, the curve reaches only a more or less sharp maximum.

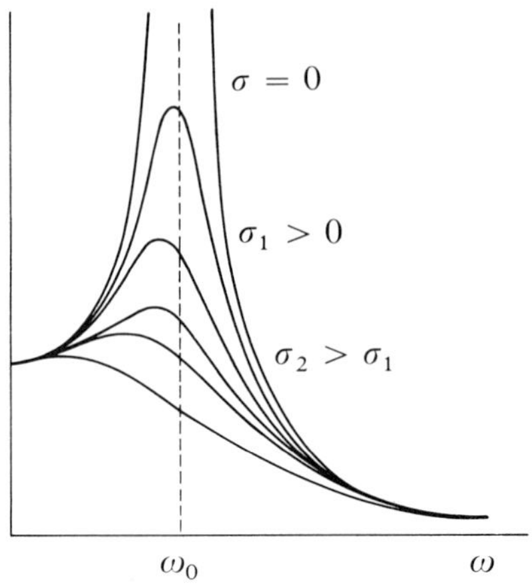

Figure 5.7

4.2 Impedance and admittance

Although the terminology has been borrowed from electrical engineering, where these problems were first considered, it is now widely used in other branches of oscillation theory. In electrical engineering an electric circuit is considered to be the system. A certain voltage is applied to this circuit, and electric current is produced. Thus the voltage may be considered to be the *input* and the current to be the *output*.

In some cases it is useful to reverse the order of a systems procedure. In other words, a current is sent through the system, as a consequence of which the voltage appears; hence the current becomes the input and the voltage is output.

In the theory of alternating currents, the input is a harmonic (sinusoidal) function, and interest is directed toward the harmonic output. We may therefore apply the analysis given in the previous section.

Denoting the voltage by $Ue^{i\omega t}$ and the current by $Ie^{i\omega t}$, where U and I are complex constants, formulas (5.4.4) and (5.4.5) may be rewritten as

$$U = ZI, \qquad I = YU, \tag{5.4.6}$$

where Z stands for $(i\omega E - A)$ and $Y = Z^{-1}$ (or conversely, depending upon whether we interpret U or I as input and output).

Z is called the *impedance* of the circuit, and Y is called the *admittance*. They are complex numbers if U and I are scalars. U and I may also be considered vectors of several components (n different input voltages and n output currents in an electric network, for example), in which case Z and Y will be matrices whose elements are complex numbers.

4.3 Series and parallel connection

If the same electric current $Ie^{i\omega t}$ passes through the impedances Z_1, Z_2, Z_3 [see Figure 5.8(a)], the voltage drops will be $IZ_1e^{i\omega t}$, $IZ_2 e^{i\omega t}$, and $IZ_3 e^{i\omega t}$, respectively. As these quantities are added, the total voltage drop will be $I(Z_1 + Z_2 + Z_3)e^{i\omega t}$. *Hence the total impedance will be the sum of the impedances. This is the series connection.*

If the same voltage $Ue^{i\omega t}$ is applied to the admittances Y_1, Y_2, Y_3 [see Figure 5.8(b)], the currents will be $UY_1e^{i\omega t}$, $UY_2 e^{i\omega t}$, and $UY_3 e^{i\omega t}$, respectively; the total current will therefore be $U(Y_1 + Y_2 + Y_3)e^{i\omega t}$. *Hence the total admittance will be the sum of the admittances. This is the parallel connection.*

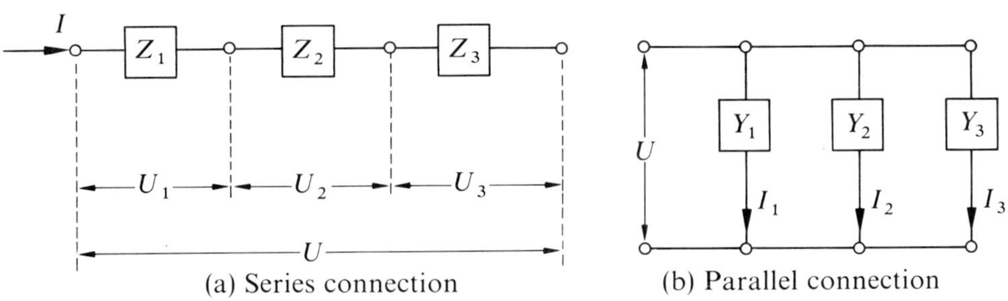

Figure 5.8

Usually, these formulas are written without the periodic factor $e^{i\omega t}$. We will do the same in some of the following formulas.

Taking into account the fact that $Y = Z^{-1}$, we may conclude:

(a) *For the series connection,*

$$Z = Z_1 + Z_2 + Z_3, \qquad Y^{-1} = Y_1^{-1} + Y_2^{-1} + Y_3^{-1};$$

(b) *For the parallel connection,*

$$Y = Y_1 + Y_2 + Y_3, \qquad Z^{-1} = Z_1^{-1} + Z_2^{-1} + Z_3^{-1}.$$

These formulas are written for just three elements, but they obviously apply to any (finite) number of them. Using these relations, one can obtain the impedance and admittance of complicated arrangements of circuit elements, providing the impedance of each element is known. The elementary elements of alternating current circuits include the following.

A *resistor* is characterized by its resistance R [which is a real and not a complex number], such that

$$Ue^{i\omega t} = RIe^{i\omega t},$$

or simply

$$U = RI.$$

An *inductor* is characterized by its (real) coefficient of self-induction L, such that

$$Ue^{i\omega t} = L\frac{d}{dt}[Ie^{i\omega t}] = i\omega LIe^{i\omega t},$$

or simply

$$U = i\omega LI.$$

A *capacitor* is characterized by its (real) capacity C, such that

$$Ie^{i\omega t} = C\frac{d}{dt}[Ue^{i\omega t}] = i\omega CUe^{i\omega t},$$

or simply

$$I = i\omega CU.$$

Hence

$$U = -\frac{i}{\omega C}I.$$

The impedance for these three circuit elements is:

(a) *for the resistor,*

$$Z = R \quad \text{the resistance.}$$

(b) *for the inductor,*

$$Z = i\omega L \quad \text{the reactance.} \tag{5.4.7}$$

(c) *for the capacitor,*

$$Z = -\frac{i}{\omega C} \quad \text{the capacitance.}$$

The admittance is always $Y = Z^{-1}$.

Example 2

As an example, let us consider the circuit of Figure 5.9(a).

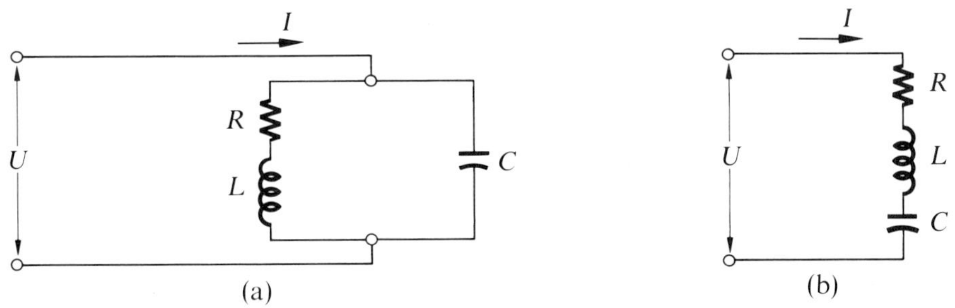

Figure 5.9

For the branch on the left, the resistance R is connected in series with the inductance $i\omega L$. Hence the impedance of this branch is

$$Z_1 = R + i\omega L.$$

The correspondent admittance is

$$Y_1 = \frac{1}{Z_1} = \frac{1}{R + i\omega L} = \frac{R - i\omega L}{R^2 + \omega^2 L^2}.$$

This admittance is connected in parallel with the branch on the right, where $Y_2 = i\omega C$. Hence the total admittance is

$$Y = Y_1 + Y_2 = \frac{R}{R^2 + \omega^2 L^2} + i\left\{\omega C - \frac{\omega L}{R^2 + \omega^2 L^2}\right\}.$$

In the particular case where the resistance R is zero [which can never be completely achieved], we obtain

$$Y = i\left\{\omega C - \frac{1}{\omega L}\right\}.$$

If, moreover, we make L and C satisfy the relation $LC = \omega^2$, we get the surprising result $Y = 0$; hence $Z = \infty$. This (somewhat ideal) circuit opposes an infinite resistance to any sinusoidal current of frequency ω.

For the circuit of Figure 5.9(b), the impedance is

$$Z = R + i\omega L - \frac{i}{\omega C}.$$

Here, if we make the ideal assumption that $R = 0$ and put the resonance condition $LC = \omega^2$, we will get $Z = 0$; hence $Y = \infty$. This circuit opposes no impedance to a current of angular frequency ω.

4.4 Transfer function

Let us consider the differential equation

$$P_n(D)x(t) = Q_m(D)f(t), \tag{5.4.8}$$

where P_n, Q_m are polynomials in the differential operator $D = d/dt$ of degree n and m, respectively, $m < n$. [$f(t)$ is the input, and $x(t)$ the output.] Such a differential equation may be obtained, for example, when a differential system of order n with a single scalar input $f(t)$ is reduced to a scalar differential equation of order n.

The coefficients of P_n, Q_m are assumed constant.

We apply the Laplace transform and write

$$\mathscr{L}[x(t)] = X(p), \qquad \mathscr{L}[f(t)] = F(p).$$

Then, for the case of zero initial values

$$\begin{aligned} x(0) &= x'(0) = \cdots = x^{(n-1)}(0) = 0, \\ f(0) &= f'(0) = \cdots = f^{(m-1)}(0) = 0, \end{aligned}$$

formulas (4.3.6) of Section 4.3.3 give

$$\mathscr{L}[x'] = pX(p), \ldots, \mathscr{L}[x^{(n)}] = p^n X(p),$$

and similarly for f. Hence (5.4.8) is transformed into

$$P_n(p)X(p) = Q_m(p)F(p). \tag{5.4.9}$$

[The operator D is simply replaced by p in the polynomials.]

Formally at least, we may write this equation as

$$X(p) = \frac{Q_m(p)}{P_n(p)} F(p). \tag{5.4.10}$$

The function

$$\Phi(p) = \frac{Q_m(p)}{P_n(p)},$$

which appears in (5.4.10), *is called the transfer function of the system described by equation* (5.4.8).

Giving the transfer function of a system is therefore equivalent to giving its differential equation.

Equation (5.4.10) or its equivalent

$$X(p) = \Phi(p)F(p) \tag{5.4.11}$$

is somewhat misleading because it assumes zero initial values. Nevertheless, we saw in Section 4.3.5 how initial data can be transformed into an input of impulse functions.

4.5 Harmonic response

Let us assume that input and output in (5.4.8) are harmonic of angular frequency ω. Thus

$$f(t) = Fe^{i\omega t}, \qquad x(t) = Xe^{i\omega t},$$

where F and X are complex constants. For harmonic functions, the operator D becomes

$$D[f(t)] = i\omega f(t);$$

so $D = i\omega$. Hence equation (5.4.8) becomes

$$P_n(i\omega)x(t) = Q_m(i\omega)f(t)$$

or

$$P_n(i\omega)Xe^{i\omega t} = Q_m(i\omega)Fe^{i\omega t}.$$

Canceling the factor $e^{i\omega t}$, the relation

$$X = \frac{Q_m(i\omega)}{P_n(i\omega)} F = Z(\omega)F \tag{5.4.12}$$

is finally obtained. Here $Q_m(i\omega)/P_n(i\omega)$ has the meaning of a complex impedance and is a function $\Phi(i\omega)$ of the frequency ω. The relation with the transfer function is very clear:

$$Z(\omega) = \Phi(i\omega).$$

This function $Z(\omega)$ is called the harmonic response. [Z is indeed equal X for $F = 1$.]

The harmonic response $\Phi(i\omega)$ is a complex function of the real variable ω. It may be expressed by its modulus A and its argument φ:

$$\Phi(i\omega) = A(\omega)e^{i\varphi(\omega)}.$$

In order to represent $\Phi(i\omega)$ graphically, graphs of both $A(\omega)$ and $\varphi(\omega)$ have to be given. If both curves are represented in the same graph, using a logarithmic scale for the frequencies ω, it is called a *Bode diagram*.

In the Bode diagram, instead of the values of A, usually the *gain*, $g = 20 \log_{10} A$ is indicated. This is convenient because in the series connection, the amplitude values of Z for the different components are multiplied; hence the corresponding values of the gain (essentially the logarithms of the amplitudes) are added.

Another way to represent the harmonic response Φ $(i\omega)$ is to plot its values in the complex plane, thereby obtaining a curve parametrized by ω. This is called the *Nyquist locus*.

Figure 5.10 represents both the Bode diagram and the Nyquist locus for a first-order system.

4.6 Relation with Fourier series and Fourier integrals

If the input $f(t)$ is not given as a single harmonic oscillation but as a finite sum of them,

$$f(t) = \sum_r F_r e^{i\omega_r t}, \tag{5.4.13}$$

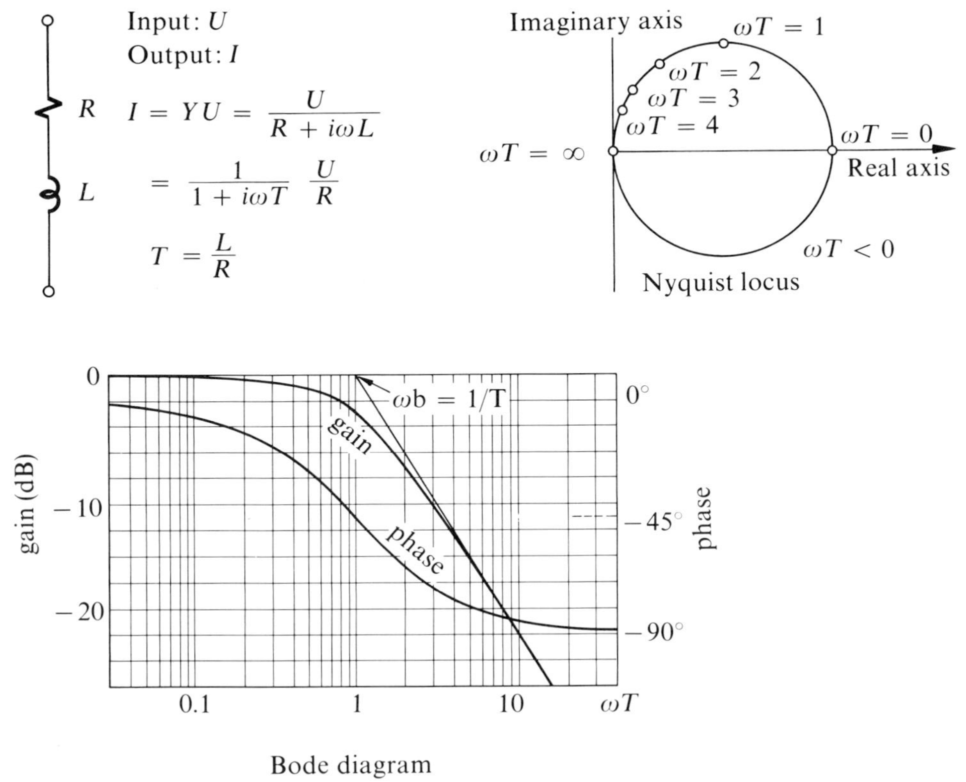

Figure 5.10

where F_r are coefficients (generally complex), then by the principle of superposition the output will be

$$x(t) = \sum_r \Phi(i\omega_r) F_r e^{i\omega_r t}. \tag{5.4.14}$$

In this formula we may admit an infinite number of terms, provided we impose conditions which guarantee appropriate (e.g., absolute and uniform) convergence of the series.

Hence:

> *If we develop the input in Fourier series* (5.4.13), *the output can then be obtained in the form of the Fourier series* (5.4.14), *provided the appropriate convergence is assured.*

We may go even further and consider input functions expressed as Fourier integrals;

$$f(t) = \int_{-\infty}^{+\infty} F(\omega)e^{i\omega t}\, d\omega. \tag{5.4.15}$$

Here the restrictive conditions assuring convergence are much more stringent and will not be considered in this book. But assuming convergence, we can also obtain the output in the form of the Fourier integral

$$x(t) = \int_{-\infty}^{+\infty} \Phi(i\omega)F(\omega)e^{i\omega t}\, d\omega. \tag{5.4.16}$$

Problems

1. Find the maximum amplitude of oscillation and the corresponding resonance frequency $\omega = \omega_r$ of

$$x'' + Rx' + 25x = \cos \omega t.$$

How does the resonance frequency ω_r compare with the natural frequency $\omega = b$ of the undamped system? Consider the values:
(a) $R = 0$, (b) $R = 4$, (c) $R = 10$, (d) $R = 20$.

2. Draw the graph of the harmonic response $\Phi(i\omega)$ in the complex plane (Nyquist locus) for
(a) $Dx = f$, (b) $(D + 1)x = f$,
(c) $(D^2 + 1)x = f$, (d) $(D^2 - 1)x = f$,
(e) $(D^2 + 5D + 6)x = f$, (f) $(D^2 + 2D + 2)x = f$,
(g) $(D^2 + 2D + 2)x = (D + 1)f$, (h) $(D^3 - D)x = f$.

References

A good, general text giving modern ideas and problems on systems and mathematical models is

R. E. Kalman, P. L. Falb, and M. A. Arbib. *Topics in Mathematical System Theory.* New York: McGraw-Hill, 1969.

On the special topic of almost periodic function, we refer the reader to

H. Bohr. *Fastperiodische Funktionen*, in the series *Ergebnisse der Mathematik und Ihren Grenzgebieten.* Berlin: Springer-Verlag, 1932; New York: Chelsea, 1947 (Eng. translation).

A. S. Besicovitch. *Almost Periodic Functions.* Cambridge, Eng.: Cambridge Univ. Press, 1932; New York: Dover, 1954 (reprint).

There is a vast literature concerning the applications of the frequency response method to electronics and system theory, from which we select

H. S. Tsien. *Engineering Cybernetics.* New York: McGraw-Hill, 1954.

G. J. Murphy. *Basic Automatic Control Theory.* Princeton, N.J.: Van Nostrand, 1957.

O. J. M. Smith. *Feedback Control Systems.* New York: McGraw-Hill, 1958.

K. H. Fasol. *Die Frequenzkennlinien.* Berlin: Springer-Verlag, 1968.

6

Geometric Theory and Phase Space

1 *Phase Space — Autonomous and Periodic Equations*

1.1 Phase or state space

We will consider the differential equation

$$\mathbf{x}' = \mathbf{f}(t, \mathbf{x}), \tag{6.1.1}$$

where t is a real variable and $\mathbf{x}$ (and therefore $\mathbf{f}$) is an n-dimensional vector. Many of our statements are still valid in a more general setting, namely when $\mathbf{x}$ is an element of a complete, locally compact metric space.

A *metric space* is a set of elements, the *points* of the space, on which a metric or distance function is defined. The set of vectors of dimension n forms such a space with the distance between the two elements $\mathbf{x}$ and $\mathbf{y}$ defined by $\|\mathbf{x} - \mathbf{y}\|$.

The space X of the unknown function $\mathbf{x}$ is called the *state space*, and a point $\mathbf{x} \in X$ is a *state* of the system. Historically, the term *phase space* was frequently used; it was derived from the first studies in statistical mechanics applied to thermodynamics. But we must clearly distinguish the following spaces:

(a) the space of the independent variable t, usually the real line $R = (-\infty, +\infty)$. [For many applications, it is more natural to take the real positive half-line $R^+ = [0, +\infty)$.]

(b) the space of the function $\mathbf{x}$. [We designate it by X, and in most cases, it will be the real n-dimensional space R^n. It is the *state space*.]

(c) the product space $R \times X$, whose points are the pairs $(t, \mathbf{x})$. [In physics, particularly in relativity theory, it is the four-dimensional *space time*. It may be called the *space of events*.]

Any solution of (6.1.1) may be interpreted as a *curve in the space of events* $R \times X$. It may also be interpreted as *a moving point in the state space* X. In the latter, the geometric locus in the state space of $\mathbf{x}(t)$ [i.e., the geometric curve described by $\mathbf{x}(t)$] will be called *trajectory* or *orbit*; it is the projection on the X space of the curve $\mathbf{x}(t)$ in the $R \times X$ space. When referring to this curve in $R \times X$ space, we will call it a *motion*. We may also visualize this curve (the motion) as a moving point which is now in the space of events; in this case its first coordinate (t) increases uniformly with time.

Regarding the function $\mathbf{f}(t, \mathbf{x})$ in (6.1.1), we will assume that it satisfies conditions which guarantee the existence and uniqueness of solutions of the initial value problem; for example, $\mathbf{f}(t, \mathbf{x})$ would be continuous in $(t, \mathbf{x})$ and Lipschitz in $\mathbf{x}$. Some consideration will also be valid, as the reader may see, in cases where we allow nonunique solutions.

1.2 Autonomous equations

If the function $\mathbf{f}(t, \mathbf{x})$ is independent of t, equation (6.1.1) takes the form

$$\mathbf{x}' = \mathbf{f}(\mathbf{x}), \tag{6.1.2}$$

and is called an *autonomous equation* or an *autonomous system*, if we think of it as a system of n scalar equations. Physically, this corresponds to a time-independent system.

The solution of an n-dimensional autonomous system can be reduced to the integration of a system of dimension $n - 1$. Indeed, writing (6.1.2) as a system for $\mathbf{x} = (x^1, x^2, \ldots, x^n)$,

$$\frac{dx^i}{dt} = f^i(x^1, x^2, \ldots, x^n) \qquad i = 1, 2, \ldots, n,$$

we can eliminate dt in the following way:

$$\frac{dx^1}{f^1} = \frac{dx^2}{f^2} = \cdots = \frac{dx^n}{f^n}. \tag{6.1.3}$$

If, for example, we assume $f^n(x) \neq 0$, we may write (6.1.2) as

$$\frac{dx^i}{dx^n} = \frac{f^i(x^1, x^2, \ldots, x^n)}{f^n(x^1, x^2, \ldots, x^n)} \qquad i = 1, 2, \ldots, n - 1 \tag{6.1.4}$$

A slight difference between (6.1.2) and (6.1.3) should be noted. While (6.1.2) defines a true *vector field*, (6.1.3) only defines a *direction field*; that is, at every point $\mathbf{x}$ only a direction (straight line) is determined, not the sense or the magnitude of the correspondent vector. We would not know, for example, in which sense the solution curve is to be described by the moving point $\mathbf{x}(t)$.

We assumed above that $f^n(x^1, \ldots, x^n) \neq 0$. At those points where this may not be true [this will happen precisely on the hypersurface defined by equation $f^n(x^1, \ldots, x^n) = 0$], we should take some other component x^i as the independent variable, such that $f^i(x^1, \ldots, x^n) \neq 0$.

The points where this method cannot be applied are those points $\mathbf{x} = (x^1, \ldots, x^n)$ for which all $f^i(x^1, \ldots, x^n) = 0$ $(i = 1, 2, \ldots, n)$. This means that at those points $\mathbf{x}$ the vector $\mathbf{f}(\mathbf{x}) = 0$. At such points, (6.1.3) does not determine any direction, and they are called *critical points*.

Given a nonautonomous system, it is possible to transform it into an autonomous one, increasing by one the dimension of the space. Let the system be

$$\frac{dx^i}{dt} = f^i(t, x^1, x^2, \ldots, x^n) \qquad i = 1, 2, \ldots, n. \tag{6.1.5}$$

Introducing the component x^{n+1}, which will be essentially coincident with t, and writing s for the new independent variable, we may write

$$\frac{dx^i}{ds} = f^i(x^{n+1}, x^1, x^2, \ldots, x^n) \qquad i = 1, 2, \ldots, n \tag{6.1.6}$$
$$\frac{dx^{n+1}}{ds} = 1.$$

[We may simply set $s = t$, of course, or leave s and set $x^{n+1} = t$.] The transformation of nonautonomous differential equations into autonomous ones is important because all properties valid for autonomous equations of dimension $n + 1$ are also valid for nonautonomous equations of dimension n.

It is important to note that the problems of integrating (6.1.2) and (6.1.3) are not really the same. If we first integrate (6.1.2), obtaining $x^1(t), \ldots, x^n(t)$, we only need to eliminate t, obtaining $n - 1$ algebraic equations in $x^1, \ldots, x^n$ which determines a trajectory curve that is a solution of (6.1.3).

Even if we know the trajectory, a solution of (6.1.3), however, we are still ignorant of the law of motion of the moving point along this trajectory. In order to determine the law of motion, we have to carry out an additional integration. If the trajectory is known, for example, in parametric form $x^i = \varphi^i(u)$, we need only to know the function $u(t)$ in order to have the complete motion. To get $u(t)$ we write

$$\frac{dx^1}{dt} = \frac{d\varphi^1}{du}\frac{du}{dt},$$

from which, assuming $d\varphi^1/du \neq 0$, we get

$$\frac{du}{dt} = \frac{dx^1/dt}{d\varphi^1/du} = \frac{f^1(x^1(u), \ldots, x^n(u))}{d\varphi^1/du},$$

using (6.1.2). Hence du/dt is obtainable by processes of differentiation. Letting $du/dt = g(u)$, we then obtain

$$t - t_0 = \int_{u(t_0)}^{u(t)} \frac{du}{g(u)}.$$

With respect to the transformation of the nonautonomous equation (6.1.5) into the autonomous equation (6.1.6), we should note that by integrating the latter, we obtain more solutions than from (6.1.5). Indeed, integrating the last equation of the system (6.1.6), we get $x^{n+1} = s + \text{constant}$; hence from any solution of (6.1.5), which is of the form

$$x^i = \varphi^i(t) \qquad i = 1, \ldots, n,$$

we get a one-parameter family of solutions of (6.1.6):

$$x^i = \varphi^i(s) \qquad i = 1, \ldots, n, \qquad x^{n+1} = s + \text{const.}$$

The most important characteristic of the autonomous equation is that its general solution

$$\mathbf{x} = \mathbf{F}(t, t_0, \mathbf{x}_0),$$

where t_0, $\mathbf{x}_0$ are the initial conditions, can be expressed with one less variable as

$$\mathbf{x} = \mathbf{G}(t - t_0, \mathbf{x}_0). \tag{6.1.7}$$

This means that since t appears only as dt in the differential equation (6.1.2), the origin of the t-coordinates is immaterial, and the solution therefore depends only on $t - t_0$. Analytically, if $\mathbf{F}(t, t_0, \mathbf{x}_0)$ is a solution, then $\mathbf{F}(t - t_0, 0, \mathbf{x}_0)$ is also a solution, and we may delete the zero and simply write $\mathbf{G}(t - t_0, \mathbf{x}_0)$.

Geometrically, this means that the curves representing the motion in $(t, \mathbf{x})$ space have, as projections on the $\mathbf{x}$ space, fixed trajectories, which do not depend on t_0 but only on the initial point $\mathbf{x}_0$. This is the reason that the study of the trajectories of a system is especially important in the autonomous case.

1.3 Periodic equations

The differential equation, or differential system, (6.1.1) is *periodic* in t, if there exists some *period* $T \neq 0$ such that

$$\mathbf{f}(t, \mathbf{x}) = \mathbf{f}(t + T, \mathbf{x}) \tag{6.1.8}$$

for every t and every $\mathbf{x}$.

Periodic equations constitute an intermediate case between general and autonomous differential equations. For the periodic equations, the solutions depend on the origin of time. Hence in order to determine a trajectory it is not enough to give any one of its points $\mathbf{x}_0$; the corresponding time t_0 must also be given. On the other hand, it is enough to know all solutions for any time interval of length T in order to obtain the solutions for every t. [Simply join together the pieces of length T.]

A rather trivial observation is that if (6.1.8) is true for a certain value T, it is also true for every multiple of T. When we speak of periodic functions of period T, we mean that T is the smallest positive number satisfying (6.1.8). Except in trivial cases (when $\mathbf{f}(t, \mathbf{x})$ is constant, for example), it can be shown that this smallest positive number does exist. [Note that in the autonomous case, $\mathbf{f}(\mathbf{x})$ has arbitrarily small periods.]

2 Critical Points and Periodic Solutions

2.1 Critical points

Let us consider autonomous differential equations of the type (6.1.2). If the point $\mathbf{x}_0$ satisfies

$$\mathbf{f}(\mathbf{x}_0) = 0, \tag{6.2.1}$$

then the function

$$\mathbf{x}(t) = \mathbf{x}_0 = \text{const} \tag{6.2.2}$$

satisfies the differential equation and $\mathbf{x}_0$ is called a *rest point* or *critical point*. Noncritical points are called *regular points*.

If $\mathbf{x}_0$ is a critical point, the solution (6.2.2) is, in the $(t, \mathbf{x})$ space, a straight-line parallel to the t-axis. Its projection on the X space reduces to the point $\mathbf{x}_0$, which is therefore a *degenerate trajectory*.

Written as (6.1.3), the system equations do not define any direction at the point $\mathbf{x}_0$, which is therefore a singular point for the direction field. Existence and uniqueness theorems cannot be applied at such a point. In fact, there may be a single solution, several solutions, or no solution at all through $\mathbf{x}_0$.

If, instead, we consider the system of equations of the type (6.1.2): there is no singularity at $\mathbf{x}_0$; existence and uniqueness theorems can be applied; and if these theorems hold, the single solution through $\mathbf{x}_0$ is (6.2.2). Considering the picture in the state space, we can therefore state:

Every critical point is a degenerated trajectory, and provided solutions are unique, no other solution can pass through the critical point.

This does not prevent, however, the existence of solutions $\mathbf{x}(t)$ which tend asymptotically to $\mathbf{x}_0$ as $t \to \pm\infty$.

Considering the condition for $\mathbf{x}_0$ to be critical in (6.1.2), the typical case will be that each equation

$$f^i(x^1, x^2, \ldots, x^n) = 0$$

defines a hypersurface in X space. The critical points will then become intersections of the n hypersurfaces $f^i = 0$ $(i = 1, 2, \ldots, n)$. Hence, in the typical case, the critical points will be isolated. In particular cases there may be no critical points at all, or there may be several curves or manifolds of various dimensions, all of whose points are critical.

Example 1

Consider the two-dimensional differential system

$$\begin{cases} x^{1'} = x^1 \\ x^{2'} = x^2. \end{cases}$$

The solution is

$$\begin{cases} x^1 = x_0^1\, e^t \\ x^2 = x_0^2\, e^t. \end{cases}$$

The trajectories are obtained by eliminating t. The differential equation is

$$\frac{dx^2}{dx^1} = \frac{x^2}{x^1},$$

with the solution

$$x^2 = \left(\frac{x_0^2}{x_0^1}\right) x^1.$$

Hence, in the state plane, the solutions are all the lines through the origin. But according to the original differential system, every line through the origin actually decomposes into three trajectories: two opposite half-lines, tending to the origin as $t \to -\infty$, and the origin itself, as the third, degenerate trajectory. It is customary to indicate the solutions in the state space by drawing the curves and indicating with arrows the sense in which the curves are described. This has been done in Figure 6.1 for Example 2.1. All solutions of the last differential equation pass through the origin, which is a critical point.

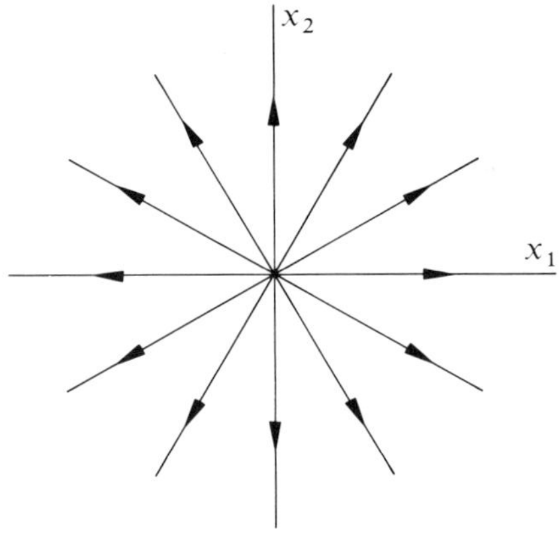

Figure 6.1

Example 2

Consider the differential system

$$\begin{cases} x^{1'} = x^2 \\ x^{2'} = -x^1, \end{cases}$$

whose solution is

$$\begin{cases} x^1 = x_0^1 \cos t + x_0^2 \sin t \\ x^2 = -x_0^1 \sin t + x_0^2 \cos t. \end{cases}$$

Eliminating t we obtain

$$\frac{dx^2}{dx^1} = -\frac{x^1}{x^2}$$

with solutions

$$(x^1)^2 + (x^2)^2 = (x_0^1)^2 + (x_0^2)^2 = \text{const.}$$

The trajectories are circles with the center at the origin, as shown in Figure 6.2. No trajectory approaches the origin, which is a critical point.

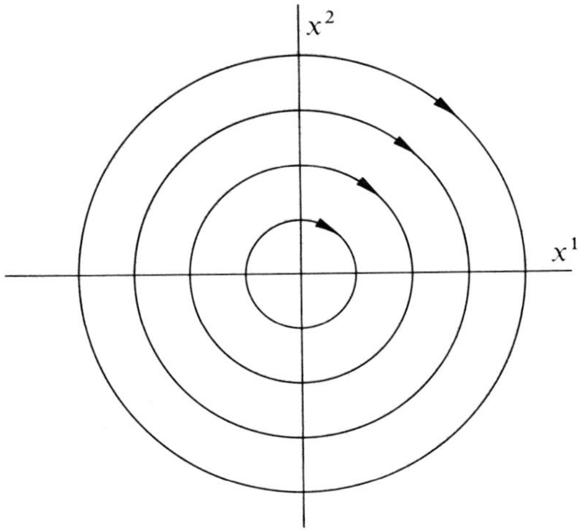

Figure 6.2

Example 3

Consider the differential system

$$\begin{cases} x^{1\prime} = (x^1)^2 + (x^2)^2 \\ x^{2\prime} = 0. \end{cases}$$

In order to eliminate t we may write

$$\frac{dx^2}{dx^1} = 0;$$

hence

$$x^2 = x_0^2 = \text{const.}$$

Exactly one solution of this last differential equation passes through the origin, which is a critical point. This solution decomposes into three trajectories: two half-lines and the critical point itself. [See Figure 6.3.]

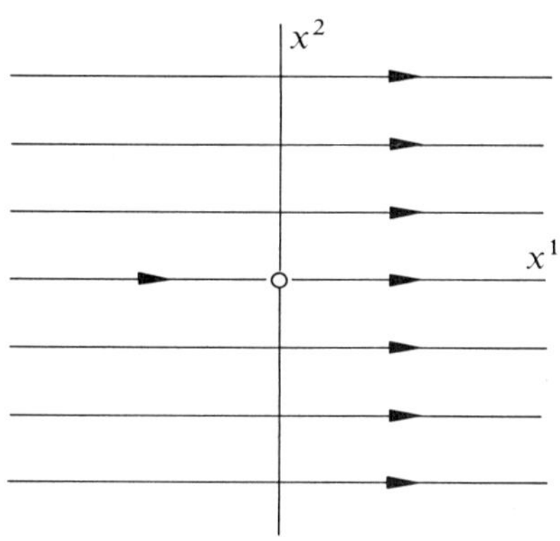

Figure 6.3

Example 4

Consider the differential system

$$\begin{cases} x^{1\prime} = -x^1 \\ x^{2\prime} = 0. \end{cases}$$

Eliminating t, we again obtain

$$\frac{dx^2}{dx^1} = 0,$$

but now every line $x^2 = \text{constant}$ decomposes into three trajectories and all points with $x^1 = 0$ are critical. [See Figure 6.4.]

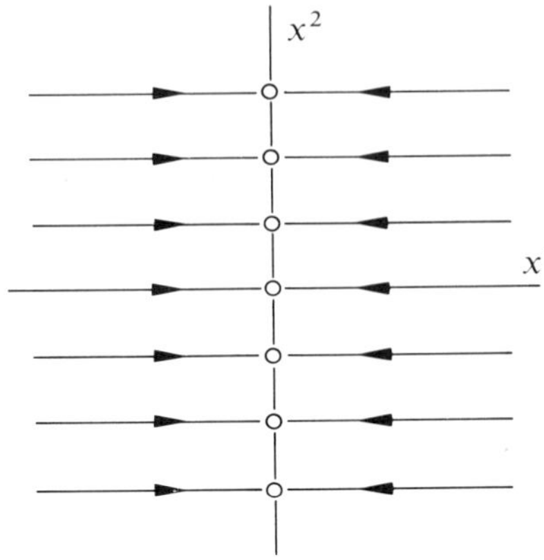

Figure 6.4

Example 5

Consider the differential system

$$\begin{cases} x^{1\prime} = x^2 \\ x^{2\prime} = 0. \end{cases}$$

Here again

$$\frac{dx^2}{dx^1} = 0,$$

but in this case every line $x^2 =$ constant is a single trajectory, with the exception of the x^1 axis ($x^2 = 0$), which is a line of critical points. [See Figure 6.5.]

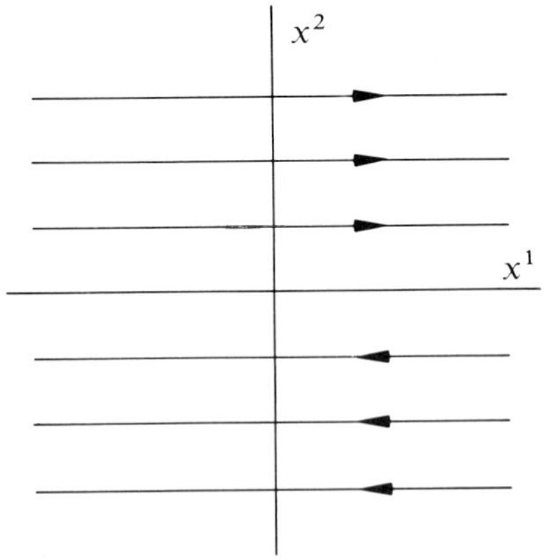

Figure 6.5

2.2 Periodic solutions

Let $\mathbf{F}(t - t_0, \mathbf{x}_0)$ be the general solution of the autonomous differential equation (6.1.2). If, for a given value of $\mathbf{x}_0$, $\mathbf{F}(t - t_0, \mathbf{x}_0)$ is periodic of period T; that is, if

$$\mathbf{F}(t + T - t_0, \mathbf{x}_0) = \mathbf{F}(t - t_0, \mathbf{x}_0) \tag{6.2.3}$$

for every $t - t_0$, then the corresponding trajectory is a closed curve in X space. In order for this to happen, equation (6.2.3) need only be satisfied for one value of $t - t_0$ because the uniqueness of this solution through each point and its invariance with respect to the initial time insure that $\mathbf{x}(t)$ will repeatedly describe the same closed trajectory. Hence the solution will be periodic.

Periodic solutions correspond physically to stationary oscillations; hence it is important to find periodic solutions of a given differential equation. If the equation is autonomous, this will correspond to finding the closed trajectories. If the equation is not autonomous, however, the situation becomes more complex. For example, if the solution $\mathbf{x}(t)$ starting at $(t_0, \mathbf{x}_0)$ is periodic, this does not imply that the solution starting at $(t_1, \mathbf{x}_0)$ must also be periodic too, since the vector field may have changed in the time interval $[t_0, t_1]$.

If the differential equation is periodic of period T, that is, if

$$\mathbf{f}(t, \mathbf{x}) = \mathbf{f}(t + T, \mathbf{x}),$$

then finding the solutions which are periodic with the same period T will correspond to finding those points $\mathbf{x}$ such that

$$\mathbf{x} = \mathbf{F}(T, 0, \mathbf{x}),$$

where $\mathbf{F}(t, t_0, \mathbf{x}_0)$ is the general solution. This insures that the trajectory will be closed.

In the latter case we can study the transformation of the X space, defined by

$$\tau[\mathbf{x}] = \mathbf{F}(T, 0, \mathbf{x}),$$

into itself. This transformation will be continuous and one-to-one, by virtue of the existence and uniqueness of solutions. Hence it is a *topological transformation* or *homeomorphism*. The problem of finding the periodic solutions of period T is then equivalent to finding the *fixed points* of the transformation τ. In many cases well-known fixed-point theorems from the theory of transformations of abstract spaces may be used to establish the existence of a periodic solution.

If $\mathbf{x}(t)$ is a periodic solution of period T/n with $n =$ integer > 1, then $\mathbf{x}(t)$ is also periodic of period T. On the other hand, if $\mathbf{x}(t)$ has period nT with $n =$ integer > 1, then it may not appear among the periodic solutions of period T. This latter type of solution is called *subharmonic*, while the former type (of period T/n) is called *harmonic*.

Problem

1. Find the critical points of the following differential systems:

(a) $\begin{cases} x' = y \\ y' = 0, \end{cases}$ (b) $\begin{cases} x' = x^2 - y^2 \\ y' = xy - 1, \end{cases}$

(c) $\begin{cases} x' = (x - y)(x + y + 1) \\ y' = (x - y)(x + 1), \end{cases}$ (d) $\begin{cases} x' = x - y + z - 2 \\ y' = 2x - y - z \\ z' = x + y - 3z + 2, \end{cases}$

(e) $\begin{cases} x' = (x^2 - y^2 + z^2)(x - 1) \\ y' = (x^2 - y^2 + z^2)(y - 1) \\ z' = (x^2 - y^2 + z^2)(z - 1). \end{cases}$

3 Topological Properties of Trajectories

3.1 Properties of critical points and regular points

In this section we will consider autonomous differential equations of type (6.1.2), which satisfy conditions insuring the existence and uniqueness of solutions, as well as their continuous dependence on the initial conditions.

We have already seen that for such equations, the trajectories are fixed curves in the state space X (independent of t). Since nonautonomous equations can be converted into autonomous ones by increasing the dimension of the space, all results obtained here will also be applicable (correctly interpreted) to nonautonomous equations considered in dimension $n + 1$.

For the sake of simplicity in the formulation of results, we will also assume that the differential equation (6.1.2) is defined on the whole space $X = R^n$. [For many results, however, this assumption will be immaterial.]

Under these assumptions we may state:

The set of critical points of (6.1.2) *is a closed set.*

This is equivalent to:

The set of regular points of (6.1.2) *is an open set, or, if* $\mathbf{x}_0$ *is a regular point of* (6.1.2), *then there exists a neighborhood of* $\mathbf{x}_0$ *all of whose points are regular.*

These statements are corollaries of the more general statement:

If $\mathbf{x}_0$ *is a regular point of* (6.1.2), *then given any* $\alpha > 0$, *there is a neighborhood of* $\mathbf{x}_0$ *such that for every* $\mathbf{x}$ *belonging to it,* $\mathbf{f}(\mathbf{x}) \neq 0$ *and the angle between* $\mathbf{f}(\mathbf{x}_0)$ *and* $\mathbf{f}(\mathbf{x})$ *is less than* α.

The proof follows from the fact that $\mathbf{f}(\mathbf{x}_0)$ is constant, $\mathbf{f}(\mathbf{x})$ is a continuous function of $\mathbf{x}$, and the angle between $\mathbf{f}(\mathbf{x}_0)$ and $\mathbf{f}(\mathbf{x})$ is a continuous function of these vectors, as long as neither is zero.

3.2 Transversal cells and arcs

A differentiable $n - 1$ *cell, through the point* $\mathbf{x}_0 \in R^n$, *is given by a differentiable mapping*

$$x^i = \varphi^i(u^1, u^2, \ldots, u^{n-1}) \qquad i = 1, 2, \ldots, n \tag{6.3.1}$$

from a neighborhood of the origin in R^{n-1} *into* R^n, *such that the Jacobian matrix*

$$\frac{\partial \varphi^i}{\partial u^j} \qquad i = 1, 2, \ldots, n \qquad j = 1, 2, \ldots, n - 1 \tag{6.3.2}$$

has rank $n - 1$ *and* $\varphi(0) = \mathbf{x}_0$.

Geometrically speaking, an $n - 1$ cell is a piece of an $n - 1$ dimensional manifold (the differentiable image of a neighborhood of the origin in R^{n-1}), such that this origin is mapped into the point $\mathbf{x}_0$ and such that (6.3.2) is of maximum rank.

We say that this cell is transversal to the vector field $\mathbf{f}(\mathbf{x})$, *if the vector* $\mathbf{f}(\mathbf{x}_0)$ *is not a linear combination of the vectors*

$$\frac{\partial \mathbf{x}}{\partial u^1}, \qquad \frac{\partial \mathbf{x}}{\partial u^2}, \qquad \ldots, \qquad \frac{\partial \mathbf{x}}{\partial u^{n-1}}.$$

Note: These vectors

$$\frac{\partial \mathbf{x}}{\partial u^i} \qquad i = 1, 2, \ldots, n-1$$

are tangent to the curves

$$u^2 = u^3 = \cdots = u^{n-1} = 0,$$
$$u^1 = u^3 = \cdots = u^{n-1} = 0, \qquad \ldots, \qquad u^1 = u^2 = \cdots = u^{n-2} = 0.$$

[See Figure 6.6.] Hence the cell is transversal if the vector $\mathbf{f}(\mathbf{x}_0)$ is not contained in the plane tangent to the cell at the point $\mathbf{x}_0$.

The tangent vectors mentioned above have the analytic representation

$$\begin{pmatrix} \dfrac{\partial \varphi^1}{\partial u^i} \\ \dfrac{\partial \varphi^2}{\partial u^i} \\ \dfrac{\partial \varphi^3}{\partial u^i} \\ \cdots \\ \dfrac{\partial \varphi^n}{\partial u^i} \end{pmatrix} \qquad i = 1, 2, \ldots, n-1,$$

where the derivatives are taken at the point $\mathbf{x}_0$ $(u^i = 0)$. The condition of transversality is therefore

$$\begin{vmatrix} \dfrac{\partial \varphi^1}{\partial u^1} & \dfrac{\partial \varphi^1}{\partial u^2} & \cdots & \dfrac{\partial \varphi^1}{\partial u^{n-1}} & f^1(\mathbf{x}_0) \\ \dfrac{\partial \varphi^2}{\partial u^1} & \dfrac{\partial \varphi^2}{\partial u^2} & \cdots & \dfrac{\partial \varphi^2}{\partial u^{n-1}} & f^2(\mathbf{x}_0) \\ \cdots & \cdots & \cdots & \cdots & \cdots \\ \dfrac{\partial \varphi^n}{\partial u^1} & \dfrac{\partial \varphi^n}{\partial u^2} & \cdots & \dfrac{\partial \varphi^n}{\partial u^{n-1}} & f^n(\mathbf{x}_0) \end{vmatrix} \neq 0, \tag{6.3.3}$$

where the derivatives are also taken at $u^i = 0$ $(i = 1, 2, \ldots, n-1)$.

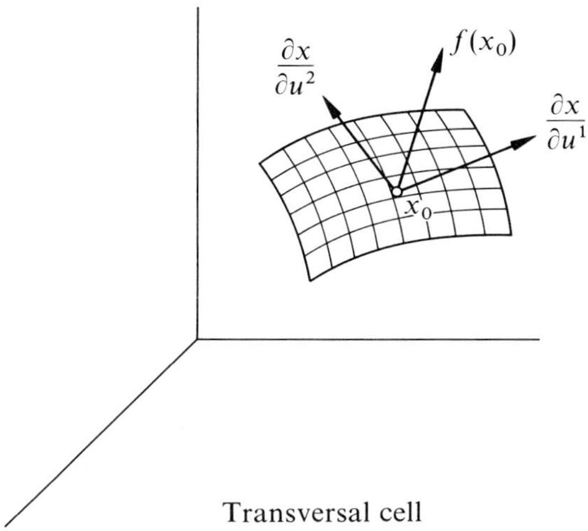

Transversal cell

Figure 6.6

Let us call this cell E_{n-1}. If we take it sufficiently small (that is, if we take the neighborhood of the origin in u space sufficiently small), the continuity of both $\mathbf{f}(\mathbf{x})$ and the partial derivatives of φ^i will insure that the condition of transversality (6.3.3) is satisfied at all points of E_{n-1}. [This will be assumed from now on.]

We will then proceed to make the restrictive assumption that the general solution of the differential equation (6.1.2) depends not only in a continuous but a continuously differentiable manner on the initial conditions. We can then prove the following lemma.

3.3 Lefschetz Lemma

Given the differential equation (6.1.2) *with the above-mentioned general assumptions and given a transversal cell E_{n-1} at $\mathbf{x}_0$, there exists a neighborhood U of $\mathbf{x}_0$ such that the trajectory through each point $\mathbf{y}$ of U intersects E_{n-1} in a point $\mathbf{x}$ which depends continuously on $\mathbf{y}$.*

Proof

Let $\mathbf{x}$ be a point of E_{n-1},

$$x^i = \varphi^i(u^1, u^2, \ldots, u^{n-1}) \qquad i = 1, 2, \ldots, n,$$

and let $\mathbf{y}$ be a general point on the solution through $(0, \mathbf{x})$,

$$\mathbf{y} = \mathbf{F}(t, \mathbf{x}) \quad \text{or} \quad y^i = F^i(t, x^1, x^2, \ldots, x^n).$$

Substituting the second-to-last value in the last expression, we get

$$\begin{aligned} y^i &= F^i[t, \varphi^1(u^1, \ldots, u^{n-1}), \ldots, \varphi^n(u^1, \ldots, u^{n-1})] \\ &= G^i(t, u^1, u^2, \ldots, u^{n-1}). \end{aligned} \tag{6.3.4}$$

From the assumptions, it follows that the functions G^i will be continuously differentiable.

We shall apply the implicit function theorem to the system

$$\Phi^i = y^i - G^i(t, u^1, \ldots, u^{n-1}) \qquad i = 1, 2, \ldots, n, \tag{6.3.5}$$

where $\mathbf{y}$ is a point near $\mathbf{x}_0$. We want to solve this system for the variables $t, u^1, \ldots, u^{n-1}$, as functions of $y^1, \ldots, y^n$. This will be possible because these functions are continuously differentiable and because the Jacobian

$$J = \frac{\partial(\Phi^1, \ldots, \Phi^n)}{\partial(u^1, \ldots, u^{n-1}, t)}$$

$$= \begin{vmatrix} -\sum_h \frac{\partial F^1}{\partial x^h}\frac{\partial x^h}{\partial u^1} \cdots -\sum_h \frac{\partial F^1}{\partial x^h}\frac{\partial x^h}{\partial u^{n-1}} \; f^1(y) \\ \cdots\cdots\cdots\cdots\cdots\cdots\cdots\cdots\cdots\cdots \\ -\sum_h \frac{\partial F^n}{\partial x^h}\frac{\partial x^h}{\partial u^1} \cdots -\sum_h \frac{\partial F^n}{\partial x^h}\frac{\partial x^h}{\partial u^{n-1}} \; f^n(y) \end{vmatrix}$$

becomes, for $t = u^1 = u^2 = \cdots u^{n-1} = 0$,

$$J(0, \ldots, 0) = \begin{vmatrix} -\frac{\partial x^1}{\partial u^1} \cdots -\frac{\partial x^1}{\partial u^{n-1}} \; f^1(x_0) \\ \cdots\cdots\cdots\cdots\cdots\cdots \\ -\frac{\partial x^n}{\partial u^i} \cdots -\frac{\partial x^n}{\partial u^{n-1}} \; f^n(x_0) \end{vmatrix}, \tag{6.3.6}$$

which is different from zero by the condition of transversality. It should be noted that, at $\mathbf{x}_0$, $\partial F^i/\partial x^j = \delta^i_j$ (Kronecker's delta) because $\mathbf{x}$ always belongs to E_{n-1} and therefore corresponds to $t = 0$. Hence $\mathbf{F}(0, \mathbf{x}) = \mathbf{x}$.

Since, for $t = u^1 = \cdots = u^{n-1} = 0$, equations (6.3.4) are satisfied by $y^i = x^i_0$ and since the Jacobian determinant does not vanish at this point, we may apply the implicit function theorem in the following form:

Given $n + m$ variables related by the system of equations

$$\Phi^i(z^1, \ldots, z^n, v^1, \ldots, v^m) = 0 \qquad i = 1, 2, \ldots, m$$

such that

(1) $\Phi^i(a^1, \ldots, a^n, b^1, \ldots, b^m) = 0 \quad i = 1, 2, \ldots, m$;

(2) the functions $\Phi^i(z^1, \ldots, z^n, v^1, \ldots, v^m)$ are continuous and continuously differentiable at all points of some neighborhood of $(a^1, \ldots, a^n, b^1, \ldots, b^m)$ in R^{n+m};

(3) the Jacobian $J = \partial\Phi/\partial v$ exists and is different from zero at the point $(a^1, \ldots, a^n, b^1, \ldots, b^m)$;

(4) the partial derivatives $\partial\Phi^i/\partial v^j$ $(i, j = 1, 2, \ldots, m)$ exist in the above mentioned neighborhood and are continuous at the point $(a^1, \ldots, a^n, b^1, \ldots, b^m)$.

Then, in some neighborhood of the point $(a^1, \ldots, a^n)$ in the space R^n of the variables $z^1, \ldots, z^n$, there is a unique system of single-valued functions

$$v^i = f^i(z^1, \ldots, z^n) \qquad i = 1, 2, \ldots, m,$$

which is continuously differentiable in that neighborhood, such that

$$b^i = f^i(a^1, \ldots, a^n) \qquad i = 1, 2, \ldots, m$$

and the relations

$$\Phi^i[z^1, \ldots, z^n, f^1(z^1, \ldots, z^n), \ldots, f^m(z^1, \ldots, z^n)] = 0$$

are satisfied identically in that neighborhood for $i = 1, 2, \ldots, m$. The total differentials of $v^i = f^i(z^1, \ldots, z^n)$ are obtained by algebraically solving the expressions for the total differentials of the functions Φ^i equated to zero.

This is the implicit function theorem. We can immediately verify that the functions Φ^i in (6.3.5) satisfy all the conditions of this theorem. Hence there exists a neighborhood U_0 of $\mathbf{x}_0$ in the state space X such that for every $\mathbf{y}$ belonging to U_0, the variables t, $u^1, \ldots, u^{n-1}$ are continuous and differentiable functions of $y^1, \ldots, y^n$. In other words, given any point $\mathbf{y}$ in the neighborhood U_0 of $\mathbf{x}_0$, values of t and $u^1, \ldots, u^{n-1}$ are determined such that

$$\mathbf{y} = \mathbf{F}(t, \mathbf{x}(\mathbf{u})),$$

where $\mathbf{x}(\mathbf{u})$ is the point of the cell E_{n-1} corresponding to the parameters $u^1, \ldots, u^{n-1}$.

Geometrically speaking, the trajectory through $\mathbf{y}$ meets E_{n-1} at a point $\mathbf{x}$ which varies continuously with $\mathbf{y}$. The lemma is therefore proved.

Note: According to this lemma, the trajectory through $\mathbf{y} \in U_0$ meets E_{n-1} in a well-determined point $\mathbf{x}$. This does not prevent the possibility that the same trajectory may, after some time, return to U_0 and intersect E_{n-1} again, or even several additional times. [Later, in the Poincaré-Bendixson theory, we will see examples of such situations.]

3.4 Regular arcs and trajectory tubes

The arc $\gamma = \widehat{\mathbf{x}, \mathbf{x}^*}$ is a *regular arc of trajectory*, if it is an arc of a trajectory and if all its points, including the end points $\mathbf{x}$, $\mathbf{x}^*$, are regular.

3.5 Lemma

Given the regular arc of trajectory γ with end points $\mathbf{x}_0$, $\mathbf{x}_0^*$, where $\mathbf{x}_0^* = \mathbf{F}(t_0^*, \mathbf{x}_0)$, and given two $n - 1$ cells E_{n-1}, E_{n-1}^*, transversal to the vector field at the points $\mathbf{x}_0$, $\mathbf{x}_0^*$, respectively, any sufficiently small $n - 1$ dimensional neighborhood U_{n-1} of $\mathbf{x}_0$ in E_{n-1} corresponds to an $n - 1$ dimensional neighborhood U_{n-1}^* of x_0^* in E_{n-1}^*, which is the image of U_{n-1} under the transformation

$$\mathbf{x}^* = \mathbf{F}(t^*, \mathbf{x}),$$

where $\mathbf{F}(t_0^*, \mathbf{x}_0) = \mathbf{x}_0^*$ and t^* depends continuously on $\mathbf{x}_0$. In other words, $\mathbf{x}^*$ is the intersection of the trajectory through $\mathbf{x}$ with the cell E_{n-1}^*, and if there is more than one such intersection, it is the one which, by continuity, tends to $\mathbf{x}_0^*$ as $\mathbf{x} \to \mathbf{x}_0$.

Proof

In order to prove this lemma, we apply the Lefschetz lemma and choose a neighborhood V^* of $\mathbf{x}_0^*$ in the space $X = R^n$ which is sufficiently small so that any trajectory intersecting V^* also intersects E_{n-1}^*.

Then, by the continuous dependence on initial conditions of the solution of the differential equation, there is a neighborhood V of $\mathbf{x}_0$ in $X = R^n$ such that every trajectory through V also passes through V^* and therefore meets E_{n-1}^*. Taking the intersection of V and E_{n-1} as U_{n-1}, we see that every trajectory through U_{n-1} also intersects E_{n-1}^*.

Let us denote, by $\mathbf{x}$ and $\mathbf{x}^*$ any two points in E_{n-1} and E_{n-1}^*, such that $\mathbf{x}^* = \mathbf{F}(t^*, \mathbf{x})$. From existence, uniqueness, and continuous dependence on initial conditions of the solutions of the differential equation, it follows that the relation between $\mathbf{x}$ and $\mathbf{x}^*$ is one-to-one and continuous both ways. Hence it is a homeomorphism, and it transforms the open set U_{n-1} in E_{n-1} into an open subset of E_{n-1}^*, which is precisely the neighborhood U_{n-1}^* of the statement of the lemma; thus the lemma is proved.

The point set union of all the arcs determined by U_{n-1} and U_{n-1}^* [each such arc is homeomorphic to the arc $\widehat{x_0, x_0^*}$] is called a *trajectory tube* or *path tube*. All points of such a trajectory tube are regular points, and with the exception of the end points on U_{n-1} and U_{n-1}^*, they are interior points of the tube. [See Figure 6.7.]

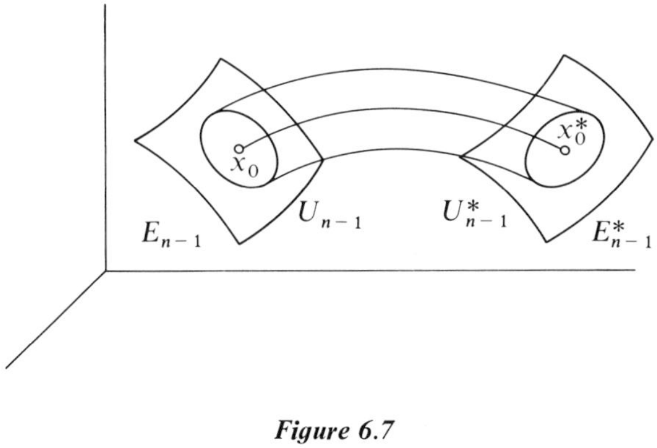

Figure 6.7

If the state space is of dimension two, the $n - 1$ cells become *transversal arcs* (a segment of a normal line to the trajectory, for example). The trajectory tube then becomes a *trajectory rectangle* or *path rectangle*.

If the differential equation, that is, the function $\mathbf{f}(\mathbf{x})$, is analytic, then the solution $\mathbf{F}(t, \mathbf{x})$ will also be analytic. Note this implies that the dependence of the solution on the initial conditions is analytic. Hence if we take the hypersurfaces E_{n-1}, E_{n-1}^* as analytic, then the correspondence $\mathbf{x} \to \mathbf{x}^*$ will also be analytic.

4 Dynamical Systems

4.1 In many problems concerning solutions of differential equations (stability, asymptotic behavior, periodicity, and so forth), the differential equation itself does not enter the picture except to ensure existence and uniqueness of the trajectory through each point. We may therefore ignore the differential equation and begin with the assumptions that a family of curves, the trajectories, has been given and that these curves have topological properties like those observed in the previous section. The rest of the theory follows from the assumptions made about the curves.

This point of view is appealing insofar as we assume only what is really necessary to the development of the theory and no more. For example, the usual assumptions on continuity and the Lipschitzian property of the function defining the differential equation, which guarantees existence and uniqueness of the solution, can be relaxed [as we saw in Chapter 2]. There is no real need to assume any such general conditions, as long as we have existence and uniqueness; hence the latter properties are the only ones we will assume.

Another advantage of this point of view is that the theory developed along this line is more suitable for generalization to universal abstract spaces, where the mere definition of a solution of a differential equation is already difficult to establish.

4.2 Definition of a dynamical system

Let $X = \{\mathbf{x}\}$ be a complete metric space* and t a real variable, $-\infty < t < +\infty$. Let $\mathbf{F}: R \times X \to X$ be a function such that

(1) $\mathbf{F}(t, \mathbf{x})$ is defined and continuous at every $(t, \mathbf{x})$, $t \in R$, $\mathbf{x} \in X$;

(2) $\mathbf{F}(0, \mathbf{x}) = \mathbf{x}$ for every $\mathbf{x} \in X$ (*initial condition*);

(3) $\mathbf{F}(t_1, \mathbf{x}_0) = \mathbf{x}_1$ and $\mathbf{F}(t_2, \mathbf{x}_1) = \mathbf{x}_2$ imply $\mathbf{F}(t_1 + t_2, \mathbf{x}_0) = \mathbf{x}_2$ (*group property*).

This property can also be written as

$$\mathbf{F}(t_2, \mathbf{F}(t_1, \mathbf{x}_0)) = \mathbf{F}(t_1 + t_2, \mathbf{x}_0).$$

Such a function $\mathbf{F}$ defines a *dynamical system* or *flow* on the space X.

---

* For the definition of a complete metric space, see H. L. Royden. *Real Analysis*. New York: Macmillan, 1968.

From assumption (3) it follows that if $\mathbf{F}(t_1, \mathbf{x}_0) = \mathbf{x}_1$, then $\mathbf{F}(-t_1, \mathbf{x}_1) = \mathbf{x}_0$. Hence, for given t_1, the correspondence $\mathbf{x}_0 \to \mathbf{x}_1$ is one-to-one and continuous both ways. For every value of t, therefore, $\mathbf{F}(t, \mathbf{x})$ represents a *topological transformation* of X onto X.

We see that the dynamical system can be interpreted as a continuous family of topological transformations which are parametrized by t. Thus

$$\mathbf{x} \to \tau_t[\mathbf{x}] = \mathbf{F}(t, \mathbf{x}).$$

These transformations form a group, since by assumption (3)

$$\tau_{t_2}[\tau_{t_1}[\mathbf{x}]] = \tau_{t_1+t_2}[\mathbf{x}].$$

Hence the composition of transformations corresponds to addition of the corresponding parameters t, and we obtain

$$\tau_{t_2} \cdot \tau_{t_1} = \tau_{t_1+t_2}.$$

For $t - 0$, the identity transformation will result.

For fixed $\mathbf{x}_0 \in X$ and variable $t \in R$, the function $\mathbf{F}(t, \mathbf{x}_0)$ represents a *motion* in the space X, the geometric locus of which is a *trajectory* or *orbit* of the dynamical system.

A *rest point* or *critical point* of a dynamical system is a point $\mathbf{x}_0$ such that

$$\mathbf{F}(t, \mathbf{x}_0) = \mathbf{x}_0$$

for every $t \in R$.

A trajectory is *periodic* if for some $T > 0$,

$$\mathbf{F}(t + T, \mathbf{x}_0) = \mathbf{F}(t, \mathbf{x}_0)$$

for every $t \in R$. In order for this to occur, it is sufficient that $\mathbf{F}(T, \mathbf{x}_0) = \mathbf{x}_0$, according to assumption (3).

Not every system of solutions of a differential equation is a dynamical system. First of all, the differential equation must be autonomous. Furthermore, it should be defined for all t and all x. Finally, no solution is allowed to have a finite escape time; that is, we can rule out the possibility that $\|\mathbf{x}(t)\| \to \infty$ as t tends to some finite t_0.

In some cases one is interested only in the geometric aspect of the trajectory curves, which is quite independent of their parametrization $\mathbf{x}(t)$ (that is, independent of the velocity with which they are described). In that case one defines two dynamical systems to be *equivalent*, if both have the same trajectories.

4.3 Theorem

Every autonomous differential equation is equivalent to another with a bounded second member.

Let

$$\mathbf{x}' = \mathbf{f}(\mathbf{x}) \tag{6.4.1}$$

be a given differential equation. Write

$$\mathbf{x}' = \frac{\mathbf{f}(\mathbf{x})}{1 + \|\mathbf{f}(\mathbf{x})\|}, \tag{6.4.2}$$

Obviously, at every point $\mathbf{x}$, the second members are proportional (parallel vectors). Hence the trajectories coincide.

4.4 Vinograd Theorem

Given the differential equation (6.4.1), *defined in an open region A properly contained in the space X, there is another differential equation, defined on the whole space X, which in A is equivalent to the given one.*

Let B be the complement of A. If $\rho(\mathbf{x}, \mathbf{y})$ is the distance between the points $\mathbf{x}$ and $\mathbf{y} \in X$, the (minimal) distance between a point $\mathbf{x}$ and the closed set B can be defined by

$$\rho(\mathbf{x}, B) = \inf \{\rho(\mathbf{x}, \mathbf{y}) \,|\, \mathbf{y} \in B\}.$$

Then the function

$$g(\mathbf{x}) = \frac{\rho(\mathbf{x}, B)}{1 + \rho(\mathbf{x}, B)}$$

is nonzero at every point of A, identically zero on B, and bounded by 1.

The differential equation

$$\mathbf{x}' = \mathbf{f}(\mathbf{x})g(\mathbf{x}) \tag{6.4.3}$$

therefore satisfies the conditions of the theorem.

By Theorem 4.3, we may assume that $\mathbf{f}(\mathbf{x})$ is bounded. Hence the right member of (6.4.3) is bounded and tends to zero, when $\mathbf{x}$ tends to the boundary of A.

Each point $\mathbf{x} \in B$ is critical because $g(x) = 0$ for it. At every point $\mathbf{x} \in A$, either $\mathbf{x}'$ is zero for (6.4.1) and (6.4.3), or both vectors have the same direction and orientation, since they differ only by the scalar positive factor $g(\mathbf{x})$. Hence the trajectories of both systems coincide in A.

According to these two theorems, it is possible to study the behavior of the solutions of a differential equation [providing it is autonomous and satisfies the conditions for existence and uniqueness of solutions], assuming that the differential equation is defined on the whole space and that it defines a dynamical system.

4.5 *Invariant sets*

Definition

The set $A \subset X$ is called invariant with respect to the dynamical system $\mathbf{F}$, *if*

$$\mathbf{F}(t, A) = A \tag{6.4.4}$$

for all $t \in R$.

Here

$$\mathbf{F}(t, A) = \{\mathbf{F}(t, \mathbf{x}) \mid \mathbf{x} \in A\};$$

that is, $\mathbf{F}(t, A)$ is the image of the set A under the transformation $\tau_t(\cdot) = \mathbf{F}(t, \cdot)$. [Definition (6.4.4) is equivalent to requiring that if $\mathbf{x} \in A$, then $\mathbf{F}(t, \mathbf{x}) \in A$ for every $t \in R$.]

Condition (6.4.4) could be replaced by the apparently more general

$$\mathbf{F}(t, A) \subset A \tag{6.4.5}$$

for every $t \in R$. But the latter is equivalent to the first condition. Indeed, if

$$\mathbf{F}(t, A) = B \subset A,$$

we can show that every point $\mathbf{x}$ of A belongs to B; hence $B = A$. Assume $\mathbf{x} \in A$; then

$$\mathbf{F}(-t, \mathbf{x}) = \mathbf{y} \in A.$$

So

$$\mathbf{F}(t, \mathbf{y}) \in B,$$

and since $\mathbf{F}(t, \mathbf{y}) = \mathbf{x}$, the assertion is proved. [See also Figure 6.8.]

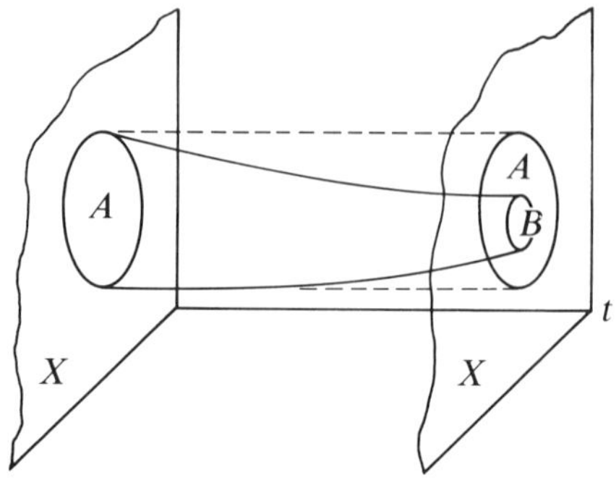

Figure 6.8

Examples of invariant sets are

(a) a critical point,

(b) a trajectory,

(c) any set of trajectories.

Conversely to (c), if an invariant set contains $\mathbf{x}_0$, it must contain the whole trajectory through $\mathbf{x}_0$. Hence any invariant set is a set of trajectories.

4.6 Theorem

The closure of an invariant set is invariant.

Let $\mathbf{x}_0 \in \bar{A}$ = closure of A. Then there exists a sequence

$$\mathbf{x}_1, \mathbf{x}_2, \mathbf{x}_3, \ldots \to \mathbf{x}_0, \qquad \mathbf{x}_i \in A \qquad i = 1, 2, \ldots,$$

and by continuity of $\mathbf{F}$, for every t,

$$\mathbf{F}(t, \mathbf{x}_0) = \lim \mathbf{F}(t, \mathbf{x}_i) \in \bar{A},$$

since $\mathbf{F}(t, \mathbf{x}_i) \in A$.

4.7 Theorem

The set of critical points is closed.

Let $\mathbf{x}_0 = \lim \mathbf{x}_i$ $(i = 1, 2, \ldots)$, where $\mathbf{x}_i$ are critical points. Then, for every t,

$$\mathbf{F}(t, \mathbf{x}_0) = \lim \mathbf{F}(t, \mathbf{x}_i) = \lim \mathbf{x}_i = \mathbf{x}_0 .$$

4.8 Theorem

If $\mathbf{y}_0 = \lim \mathbf{F}(t, \mathbf{x}_0)$ *for fixed* $\mathbf{x}_0$ *as* $t \to +\infty$ *(or* $t \to -\infty$*), then* $\mathbf{y}_0$ *is a critical point.*

Given any sequence $t_1, t_2, \ldots \to +\infty$, then

$$\mathbf{x}_i = \mathbf{F}(t_i, \mathbf{x}_0) \to \mathbf{y}_0 .$$

Hence

$$\mathbf{F}(t, \mathbf{y}_0) = \lim \mathbf{F}(t, \mathbf{x}_i) = \lim \mathbf{F}(t + t_i, \mathbf{x}_0) = \mathbf{y}_0 .$$

The proof for $t \to -\infty$ is similar. [It should be noted that due to the uniqueness of the trajectory through each point, no trajectory can reach a critical point in finite time.]

Calling a point which is not a critical point a *regular point*, the following theorem is a direct consequence of Theorem 4.7.

4.9 Theorem

Given any regular point, there is a neighborhood of it which is without critical points.

Problems

1. Which of the following systems are equivalent [that is, have the same trajectories]?

(a) $\begin{cases} x' = -x \\ y' = -y, \end{cases}$ (b) $\begin{cases} x' = y \\ y' = -x, \end{cases}$

(c) $\begin{cases} x' = \dfrac{1}{x^2 + y^2 + 1} \\ y' = 0, \end{cases}$ (d) $\begin{cases} x' = x^3 + x^2 y \\ y' = x^2 y + y^3, \end{cases}$

(e) $\begin{cases} x' = x^2 y \\ y' = 0, \end{cases}$ (f) $\begin{cases} x' = y(x^2 - y^2) \\ y' = -x(x^2 - y^2), \end{cases}$

(g) $\begin{cases} x' = x^3 y^2 \\ y' = 0, \end{cases}$ (h) $\begin{cases} x' = y(x^4 - y^4) \\ y' = -x(x^4 - y^4), \end{cases}$

(i) $\begin{cases} x' = y/\sqrt{x^2 + y^2} \\ y' = -x/\sqrt{x^2 + y^2}, \end{cases}$ (j) $\begin{cases} x' = y - \sin y \\ y' = 0, \end{cases}$

(k) $\begin{cases} x' = 1 \\ y' = 0, \end{cases}$ (l) $\begin{cases} x' = x^3 \\ y' = y^3. \end{cases}$

2. Prove that the union and intersection of invariant sets are invariant.

3. Prove that if a finite set (a set containing only a finite number of points) is invariant, then each of its points is a critical point.
4. Prove that if an invariant set is not connected, each component must be invariant.
5. Prove that given any set S, there is a smallest invariant set A containing S.

5 *Limit Sets*

5.1 ω- and α-limit points

In this section we will use the following notation:

For a fixed $\mathbf{x}_0 \in X$ and variable $t \in R$, the *trajectory* $\mathbf{F}(t, \mathbf{x}_0)$ will be denoted by $\Gamma_{\mathbf{x}_0}$ (the trajectory through $\mathbf{x}_0$).

The *positive half-trajectory starting at* $\mathbf{x}_0$, that is, the set $\{\mathbf{F}(t, \mathbf{x}_0) \,|\, t \in R^+\}$, will be denoted by $\Gamma^+_{\mathbf{x}_0}$.

The *negative half-trajectory starting at* $\mathbf{x}_0$, that is, the set $\{\mathbf{F}(t, \mathbf{x}_0) \,|\, t \in R^-\}$, will be denoted by $\Gamma^-_{\mathbf{x}_0}$.

The *positive half-trajectory starting at* $\mathbf{F}(t_0, \mathbf{x}_0)$, that is, the set $\{\mathbf{F}(t, \mathbf{x}_0) \,|\, t \geqq t_0\}$, will be denoted by $\Gamma^+_{t_0, x_0}$.

The *negative half-trajectory starting at* $\mathbf{F}(t_0, \mathbf{x}_0)$, that is the set $\{\mathbf{F}(t, \mathbf{x}_0) \,|\, t \leqq t_0\}$, will be denoted by $\Gamma^-_{t_0, x_0}$.

$\mathbf{y}_0$ will be called an *ω-limit point of the trajectory* $\Gamma_{\mathbf{x}_0} = \mathbf{F}(t, \mathbf{x}_0)$, if there exists a sequence

$$t_1 < t_2 < t_3 < \cdots < t_n < \cdots, \qquad \lim_{n\to\infty} t_n = +\infty,$$

such that

$$\lim_{n\to\infty} \mathbf{F}(t_n, \mathbf{x}_0) = \mathbf{y}_0 . \tag{6.5.1}$$

Similarly, $\mathbf{z}_0$ will be an *α-limit point* if there is a sequence

$$t_1 > t_2 > t_3 > \cdots, \qquad \lim_{n\to\infty} t_n = -\infty$$

such that (6.5.1) holds.

5.2 *Definitions of the ω- and α-limit sets*

The ω-limit set of the trajectory $\Gamma_{\mathbf{x}_0}$ *is the set of all ω-limit points of* $\Gamma_{\mathbf{x}_0}$. *It is denoted by* $\Omega(\Gamma_{\mathbf{x}_0})$.

The α-limit set of the trajectory $\Gamma_{\mathbf{x}_0}$ *is the set of all α-limit points of* $\Gamma_{\mathbf{x}_0}$. *It is denoted by* $\mathrm{A}(\Gamma_{\mathbf{x}_0})$.

Since the definitions of ω- and α-limit sets are completely symmetric (replacing t by $-t$), we will consider only the case of ω-limit points and sets. [Everything carries over, changing t into $-t$, to the α-limits.]

An equivalent definition of an ω-limit point is: $\mathbf{y}_0$ is an ω-limit point of $\Gamma_{\mathbf{x}_0}$, if given any neighborhood U of $\mathbf{y}_0$ and any arbitrary real T, there is some $t_1 \geqq T$ such that $\mathbf{F}(t_1, \mathbf{x}_0) \in U$. [The reader should prove the equivalence of both definitions.]

A direct definition of an ω-limit set, given by Lefschetz, is: the ω-limit set of the trajectory $\Gamma_{\mathbf{x}_0}$ is defined by

$$\Omega(\Gamma_{\mathbf{x}_0}) = \bigcap_{t \in R} \overline{\Gamma^{+}_{t, \mathbf{x}_0}}. \tag{6.5.2}$$

This should be interpreted to mean that given $\mathbf{x}_0$, one must consider all points $\mathbf{F}(t, \mathbf{x}_0)$, then take all half-trajectories starting at these points, and then take the closure of each of these half-trajectories and then their intersection (that is, the set of points common to all of these closures). This set is $\Omega(\Gamma_{\mathbf{x}_0})$.

Let us apply this definition to the case of a trajectory tending asymptotically to a point $\mathbf{y}_0$. [See Figure 6.9.] $\Gamma^{+}_{t, \mathbf{x}_0}$ is the arc of trajectory from $\mathbf{x}(t)$ to $\mathbf{y}_0$, excluding $\mathbf{y}_0$. Its closure $\overline{\Gamma^{+}_{t, \mathbf{x}_0}}$ is this same arc plus the point $\mathbf{y}_0$. The set of points common to all these closures reduces to the point $\mathbf{y}_0$. Hence this point is the ω-limit set.

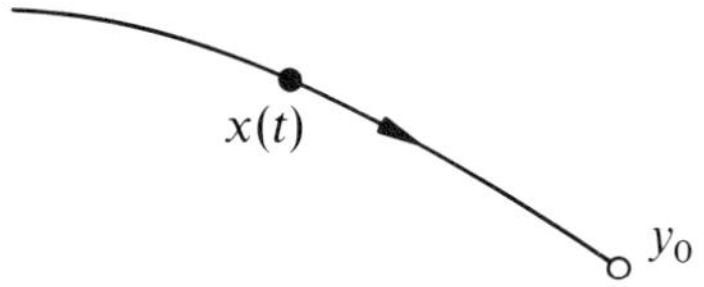

Figure 6.9

In order to prove that every ω-limit point according to the first definition is also in $\Omega(\Gamma_{\mathbf{x}_0})$ according to (6.5.2), we must show that if $\mathbf{y}_0$ is an ω-limit point, then in every neighborhood of $\mathbf{y}_0$ there are points of $\mathbf{F}(t, \mathbf{x}_0)$ with t arbitrarily large; $\mathbf{y}_0$ therefore belongs to the closure of $\Gamma^+_{t, \mathbf{x}_0}$ for every t. This follows from the first definition of ω-limit point given above.

In order to prove that every point in (6.5.2) is a limit point according to the first definition, we will construct the sequence t_n of (6.5.1). To this effect, let us take a sequence $T_1 < T_2 < T_3 < \cdots$ tending to infinity and a sequence of neighborhoods of $\mathbf{y}_0$, $U_1 \supset U_2 \supset U_3 \supset \cdots$ such that $\mathbf{y}_0$ is the only common point of all the U_i and every neighborhood V of $\mathbf{y}_0$ contains some U_n [take U_n as an open ball of radius $1/n$ and center $\mathbf{y}_0$, for example].

Since $\mathbf{y}_0$ belongs to the closure of $\Gamma^+_{T_n, \mathbf{x}_0}$, there is some $t_n > T_n$ such that $\mathbf{F}(t_n, \mathbf{x}_0) \in U_n$. Hence $t_n \to +\infty$ and $\mathbf{F}(t_n, \mathbf{x}_0) \to y_0$ as $n \to \infty$, and the requirements of the first definition are met.

5.3 Theorem

The ω-limit set of a trajectory $\Gamma_{\mathbf{x}_0}$ is a closed invariant set.

We know that it is closed from the second definition of (6.5.2), since the intersection of any number of closed sets is always closed.

To prove that it is invariant, take any $\mathbf{y}_0 \in \Omega(\Gamma_{\mathbf{x}_0})$. According to the first definition of (6.5.2), there is some increasing sequence $t_n \to +\infty$ such that $\mathbf{F}(t_n, \mathbf{x}_0) \to y_0$. For any $T \in R$, then

$$\mathbf{F}(T + t_n, \mathbf{x}_0) = \mathbf{F}(T, \mathbf{F}(t_n, \mathbf{x}_0)) \to \mathbf{F}(T, \mathbf{y}_0).$$

This also proves that

$$\mathbf{F}(T, \mathbf{y}_0) \in \Omega(\Gamma_{\mathbf{x}_0}).$$

Since T was arbitrary, the whole trajectory through $\mathbf{y}_0$ must belong to $\Omega(\Gamma_{\mathbf{x}_0})$. Hence the limit set is a union of trajectories and therefore an invariant set.

5.4 Limit set examples

As examples of limit sets we can mention

(a) a critical point,

(b) a closed (periodic) trajectory.

Note that case (a) is really a particular case of (b). In order to prove that a closed trajectory is the limit set of itself, we may take $t_n = t_0 + nT$ (where T is the period and $n = 1, 2, \ldots$) as the sequence appearing in definition (6.5.1). This proves that every point of the closed trajectory is an ω-limit point. Conversely, the fact that no point outside the trajectory can belong to the ω-limit set follows from the next lemma, which shows that an arc of trajectory is a compact set. Since the entire closed trajectory is the same point set as an arc corresponding to one period, it is obviously a compact set. Hence any point not belonging to it is at a finite nonzero distance from it, and it cannot therefore belong to its limit set.

5.5 Lemma

Any finite arc of trajectory

$$\mathbf{F}([a, b], \mathbf{x}_0) = \{\mathbf{F}(t, \mathbf{x}_0) \mid t \in [a, b]\}$$

is a compact set.

This lemma follows from the fact that the image of a compact set through a continuous mapping is also compact. In this case, the finite arc of trajectory is the image of the interval $[a, b]$, which is compact by the Heine-Borel property,† and the mapping $\mathbf{F}(t, \mathbf{x}_0)$, for fixed $\mathbf{x}_0$, is continuous by the definition of a dynamical system.

5.6 Lagrange stability

A trajectory $\Gamma_{\mathbf{x}_0}$ is positively stable, according to Lagrange, if the positive half-trajectory $\Gamma^+_{\mathbf{x}_0}$ is contained in some compact set.

The trajectory $\Gamma_{\mathbf{x}_0}$ is negatively stable, according to Lagrange, if the negative half-trajectory $\Gamma^-_{\mathbf{x}_0}$ is contained in some compact set.

The trajectory $\Gamma_{\mathbf{x}_0}$ is stable, according to Lagrange, if it is contained in some compact set.

[These types of stability are denoted by L^+, L^-, and L, respectively.]

Note that in the definitions of L^+ and L^-, the half-trajectories from $\mathbf{x}_0$ could have been replaced by half-trajectories starting at $\mathbf{F}(t_0, \mathbf{x}_0)$ for any arbitrary t_0; that is, according to Lemma 5.5, by changing the initial time from 0 to t_0 we only add or subtract a compact set, so the definitions still apply.

The reader should prove that every L^+ stable trajectory has a nonempty ω-limit set Ω. The converse is not true: if the limit set $\Omega(\Gamma)$ is nonempty, Γ is not necessarily L^+ stable, as the next example will show.

---

† See H. L. Royden. *Real Analysis*. New York: Macmillan, 1968.

5.7 Theorem

If $\Gamma_{\mathbf{x}0}$ is L^+-stable, then $\Omega(\Gamma_{\mathbf{x}0})$ is connected.

$\Omega = \Omega(\Gamma_{\mathbf{x}0})$ is closed, and from the assumption of $\Gamma_{\mathbf{x}0}$ being L^+-stable, it follows that Ω is contained in a compact set. Hence it is compact itself.

Let us assume that Ω is not connected. Then $\Omega = A \cup B$, where A and B are disjoint nonempty compact sets. There must be points $\mathbf{y}' \in A$ and $\mathbf{y}'' \in B$ and a sequence

$$t_1' < t_1'' < t_2' < t_2'' < \cdots \to \infty$$

such that

$$\mathbf{F}(t_n', \mathbf{x}_0) \to \mathbf{y}' \in A, \qquad \mathbf{F}(t_n'', \mathbf{x}_0) \to \mathbf{y}'' \in B \quad \text{as} \quad n \to \infty.$$

Since the sets A, B are compact, disjoint, and nonempty, they are at some finite nonzero distance d from each other. Thus

$$\rho(A, B) = \inf \{\rho(\mathbf{a}, \mathbf{b}) \,|\, \mathbf{a} \in A, \mathbf{b} \in B\} = d > 0.$$

Furthermore, for n sufficiently large,

$$\rho(\mathbf{F}(t_n', \mathbf{x}_0), A) < \frac{d}{3}, \qquad \rho(\mathbf{F}(t_n'', \mathbf{x}_0), B) < \frac{d}{3};$$

hence

$$\rho(\mathbf{F}(t_n'', \mathbf{x}_0), A) \geqq \rho(A, B) - \rho(\mathbf{F}(t_n'', \mathbf{x}_0), B) > \frac{2d}{3}.$$

Since the distance function is continuous, there is some

$$t_n''', t_n' < t_n''' < t_n''$$

such that

$$\rho(\mathbf{F}(t_n''', \mathbf{x}_0), A) = \frac{d}{2}.$$

Because of the compactness of the set $\overline{\Gamma^+_{\mathbf{x}0}}$, the sequence $\{\mathbf{F}(t_n''', \mathbf{x}_0)\}$ contains some convergent subsequence whose limit point $\mathbf{z}$ satisfies

$$\rho(\mathbf{z}, A) = \frac{d}{2}, \qquad \rho(\mathbf{z}, B) \geqq \rho(A, B), - \rho(\mathbf{z}, A) = \frac{d}{2}.$$

Hence z, which belongs to Ω, does not belong to either A or B. This gives the desired contradiction which proves the theorem.

5.8 Counterexample for limit sets

We can also show that $\Omega(\Gamma)$ may not be connected, if Γ is not L^+-stable.

Let our space X be the strip $|x| \leqq 1$ in the (x, y)-plane. Let the dynamical system be defined for $|x| < 1$ by

$$\begin{cases} x' = \dfrac{x(1 - x^2) - y(1 - x^2)^2}{(1 + x^2)(1 + r)} \\[2ex] y' = \dfrac{y}{1 + r} + \dfrac{x}{(1 - x^2)(1 + r)}, \end{cases}$$

where $r^2 = x^2 + y^2$. Passing to the limit for $|x| = 1$, we obtain $x' = 0$, $y' = \pm\infty$, but we know that by reparametrization [Theorem 6.4.3] we may avoid this infinity and get $y' = \pm 1$. Completing, in this way, the definition of the dynamical system on the whole strip $|x| \leq 1$, we obtain the trajectories shown in Figure 6.10. These trajectories are expanding spirals tending to both lines $x = \pm 1$ and to infinity. The limit set Ω of such a spiral is the union of these two lines; so it is not connected.

This example also shows a case of trajectories with a nonempty limit set, which are not L^+-stable.

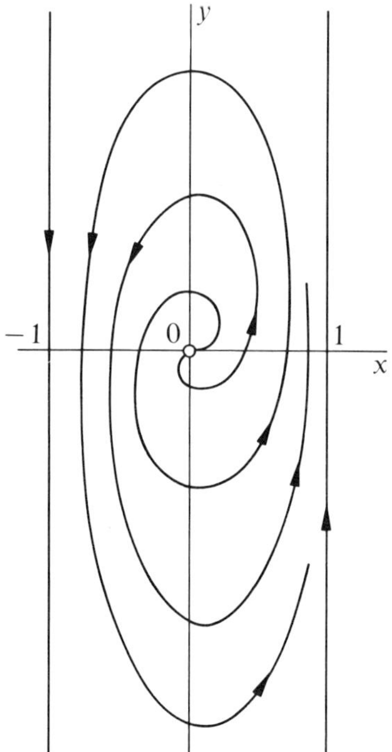

Figure 6.10

Problems

1. A point $\mathbf{x}$ is called *positively stable according to Poisson* (*stable* P^+), if for every neighborhood U of $\mathbf{x}$ and every $T > 0$, there can be found a value $t > T$ such that

$$\mathbf{F}(t, \mathbf{x}) \in U.$$

Prove that if a point $\mathbf{x}$ is stable P^+, then every point of the trajectory through $\mathbf{x}$ is also stable P^+.

2. Given a point $\mathbf{x}_0$ such that in every neighborhood U of $\mathbf{x}_0$ there is some point $\mathbf{x}$ such that the half-trajectory $\Gamma_{\mathbf{x}}^+$ is contained in U, prove that $\mathbf{x}_0$ must be a critical point.

References

An updated, detailed treatment of the general subject of this chapter can be found in

S. Lefschetz. *Differential Equations, Geometric Theory.* New York: Interscience (Wiley), 1957.

V. V. Nemytskii and V. V. Stepanov. *Qualitative Theory of Differential Equations.* Princeton: Princeton Univ. Press, 1960.

Many topics mentioned in this chapter are presented in greater detail in

G. Sansone and R. Conti. *Equazioni Differenziali Non-lineari.* Roma: Edizione Cremonese; translated into English as *Nonlinear Differential Equations.* New York: Macmillan, 1964.

R. Reissig, G. Sansone, and R. Conti. *Nichtlineare Differentialgleichungen höherer Ordnung.* Roma: Edizione Cremonese, 1969.

For a consideration of topological problems dealing with differential equations, which was initiated by Henri Poincaré, we suggest two primary references:

H. Poincaré. Sur les Courbes definies par les equations differencielles. *J. de Math.* (3) **7** (1881): 375–422; **8** (1882): 251–296; (4) **1** (1885): 167–244; **2** (1886): 151–217.

H. Poincaré. *Les Methodes Nouvelles de la Mecanique Celeste*, 3 volumes. Paris: Gauthier-Villars, 1892; 1893; 1899.

A later landmark in the development of the theory of dynamical systems is

G. D. Birkhoff. *Dynamical Systems*, Volume IX. Providence, R.I.: American Mathematical Society (Colloquium Publ.), 1927 (1st ed.); 1966 (2nd ed. rev.).

7

Autonomous Two-dimensional Systems — Critical Points

1 Linear Systems

1.1 Critical points

We will consider linear homogeneous systems of the type

$$\mathbf{x}' = A\mathbf{x}, \tag{7.1.1}$$

where A is a real constant matrix, since the equation is assumed to be autonomous. The vector $\mathbf{x}$ will be thought of as a column vector, although for typographical reasons we may sometimes write it as a row vector. In this chapter we will consider only two-dimensional systems, and for the components of the vector, we will use the traditional coordinates of analytic geometry, $\mathbf{x} = (x, y)$. The phase or state space is the plane, and by drawing the trajectories in the plane, we will get an intuitive insight into many relationships.

The apparently more general nonhomogeneous equation

$$\mathbf{x}' = A\mathbf{x} + \mathbf{b}$$

with constant $\mathbf{b}$ can be reduced to (7.1.1) by translating the coordinates; hence it does not introduce any additional difficulty.

If, in (7.1.1), we make a similarity transformation of coordinates $\mathbf{x} \to \lambda\mathbf{x}$, equation (7.1.1) remains unchanged. Therefore, if we know the behavior of the trajectories in any neighborhood of the origin, we actually know their behavior in the whole plane.

Equation (7.1.1) must necessarily have the origin as a critical point. If any other critical point $\mathbf{x}_1$ exists, then the whole line $\lambda\mathbf{x}_1 (\lambda \in R)$ is critical. If another linearly independent critical point $\mathbf{x}_2$ exists, then all points of the plane are critical and A is the zero matrix. This statement follows directly from linearity and the fact that any vector in the plane is some linear combination of two linearly independent vectors.

In Section 4.2.1 we saw that with an *affine coordinate transformation* $\mathbf{x} = P\mathbf{u}$, where the matrix P is constant and nonsingular, we can transform equation (7.1.1) into another equation of the same type, with its matrix in *canonical form*. [See formulas 4.2.8 and 4.2.9.] Since the coordinate transformation $\mathbf{x} = P\mathbf{u}$ has an immediate geometric meaning, we will discuss the possible shape of trajectories, assuming that the system has already been transformed into canonical form.

In Section 3.1.6 the possible canonical forms of a (2×2) matrix were shown to be

$$(A)\ \begin{pmatrix} \lambda_1 & 0 \\ 0 & \lambda_2 \end{pmatrix}, \quad (B)\ \begin{pmatrix} \lambda & 0 \\ 0 & \lambda \end{pmatrix}, \quad (C)\ \begin{pmatrix} \lambda & 1 \\ 0 & \lambda \end{pmatrix}, \quad (D)\ \begin{pmatrix} \sigma & \omega \\ -\omega & \sigma \end{pmatrix}. \tag{7.1.2}$$

Case (d) corresponds to the complex characteristic roots $\lambda_{1,2} = \sigma \pm i\omega$.

We will discuss these cases under the initial assumption that *all characteristic roots are different from zero; in which case the origin is the only critical point.*

1.2 *Case (A) with* $\lambda_1 > \lambda_2 > 0$

The differential equations are

$$\begin{cases} x' = \lambda_1 x \\ y' = \lambda_2 y \end{cases}$$

with solutions

$$\begin{cases} x = x_0 e^{\lambda_1 t} \\ y = y_0 e^{\lambda_2 t} \end{cases}$$

Eliminating t, we obtain

$$\left(\frac{x}{x_0}\right)^{\lambda_2} = \left(\frac{y}{y_0}\right)^{\lambda_1} = e^{\lambda_1 \lambda_2 t},$$

or

$$y = cx^m$$

with $c = \text{constant}$, $m = \lambda_2/\lambda_1$.

These trajectories are represented in Figure 7.1, where the arrows indicate the direction of motion for increasing values of t.

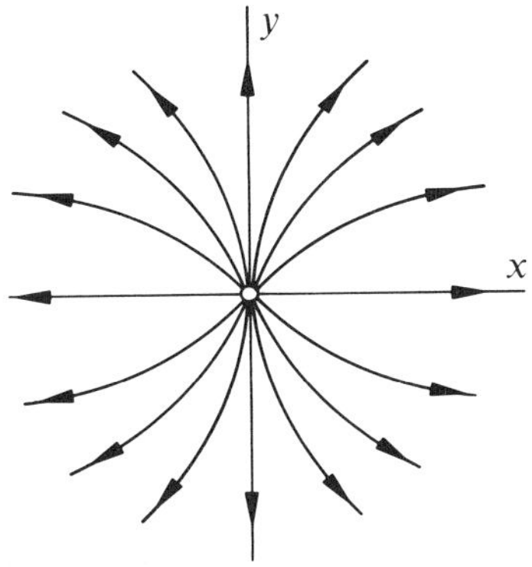

Figure 7.1

1.3 *Case A with* $\lambda_2 > \lambda_1 > 0$

Through the same sort of analysis, we can obtain Figure 7.2.

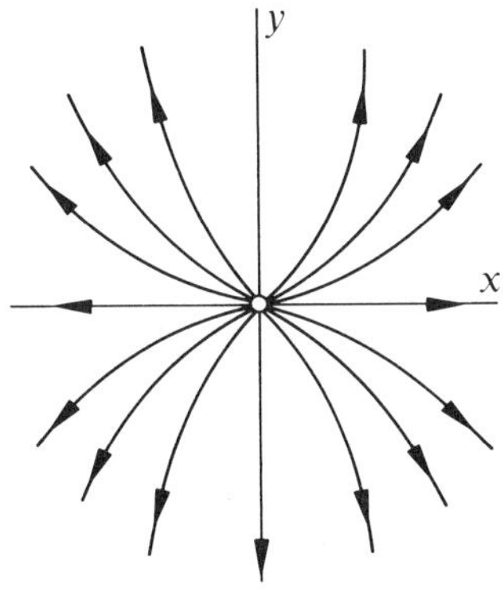

Figure 7.2

1.4 *Case A with* λ_1, λ_2 *negative*

By changing t into $-t$, the differential system is transformed into one of the preceding cases. Therefore the figures remain the same, with the exception of the arrows, which must be reversed.

1.5 Case A with $\lambda_1 > 0 > \lambda_2$

The analysis is still the same, but this time λ_2/λ_1 is negative. Hence the trajectories are like branches of hyperbolas, as shown in Figure 7.3. If λ_1 were negative and λ_2 positive, we could interchange the axes. Without interchanging the axes, the figure would look roughly the same, but the arrows would be reversed.

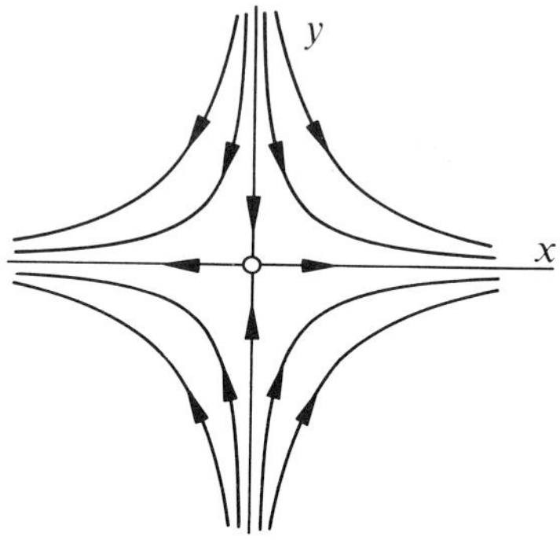

Figure 7.3

1.6 Case B

This coincides with the preceding case for $\lambda_1/\lambda_2 = 1$. The trajectories are half-lines ending at the origin. In Figure 7.4 these are indicated by having the arrows correspond to the case $\lambda_1 = \lambda_2 < 0$.

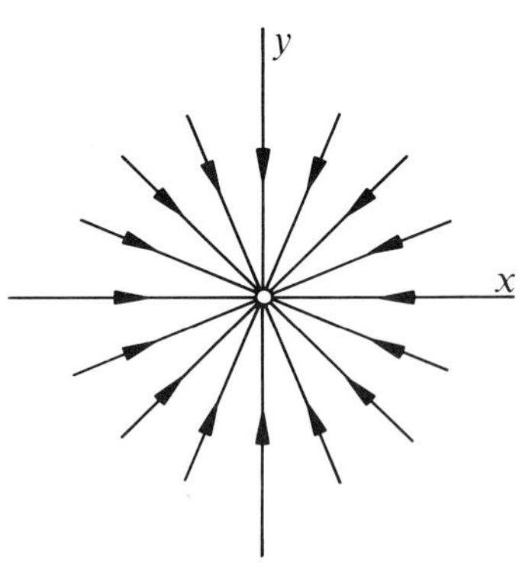

Figure 7.4

1.7 Case C

The differential equations are

$$\begin{cases} x' = \lambda x + y \\ y' = \lambda y. \end{cases}$$

The integration of the second differential equation is immediate, and by substituting in the first, we get the solution

$$\begin{cases} x = e^{\lambda t}(x_0 + y_0 t) \\ y = y_0 e^{\lambda t}. \end{cases}$$

To eliminate $t = (1/\lambda) \ln (y/y_0)$, we substitute this expression in the first equation, obtaining

$$x = \frac{y}{y_0}\left(x_0 + \frac{y_0}{\lambda} \ln \frac{y}{y_0}\right) = k_1 y + k_2 y \ln y.$$

The corresponding trajectories are represented in Figure 7.5. The reader may check the accuracy of the figure by noting that, as $y \to 0$, $x \to 0$ and $y/x = 1/(x/y) \to 0$. The arrows in Figure 7.5 correspond to $\lambda > 0$.

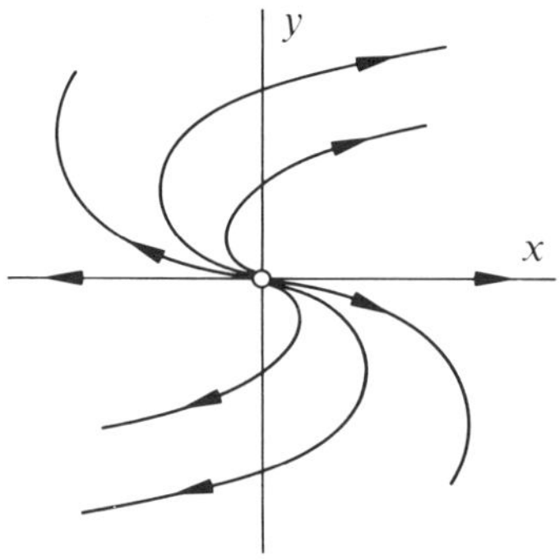

Figure 7.5

1.8 Case D with $\sigma = 0$

The differential equations are

$$\begin{cases} x' = \omega y \\ y' = -\omega x, \end{cases}$$

which can be reduced to the single differential equation

$$x'' = -\omega^2 x.$$

The solutions are

$$x = C \cos(\omega t + \varphi).$$

Hence

$$y = -C \sin(\omega t + \varphi).$$

Eliminating t, we obtain

$$x^2 + y^2 = C^2,$$

which are trajectories represented in Figure 7.6, with the arrows corresponding to $\omega > 0$.

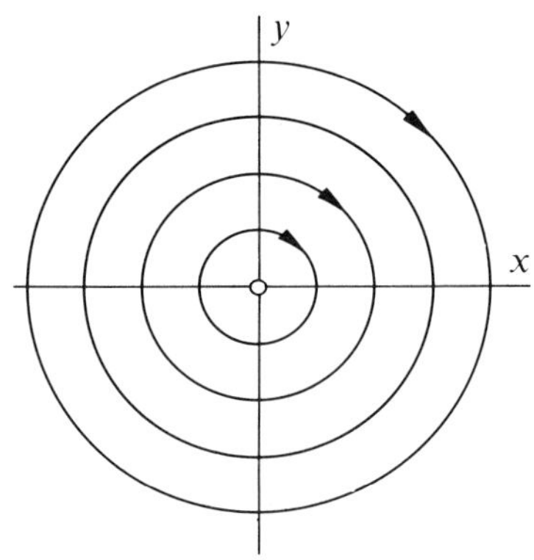

Figure 7.6

1.9 Case D with $\sigma > 0$

If in the differential equations

$$\begin{cases} x' = \sigma x + \omega y \\ y' = -\omega x + \sigma y, \end{cases}$$

we introduce polar coordinates

$$r^2 = x^2 + y^2,\ \theta = \text{arc tan}\,(y/x),$$

we get

$$rr' = xx' + yy' = x^2\sigma + x\omega y + y^2\sigma - y\omega x = \sigma r^2,$$

and

$$\theta' = \frac{1}{1 + (y^2/x^2)} \cdot \frac{y'x - yx'}{x^2} = \frac{-x^2\omega + y\sigma x - y^2\omega - y\sigma x}{x^2 + y^2} = -\omega.$$

Hence the solutions are

$$r = r_0 e^{\sigma t}, \qquad \theta = \theta_0 - \omega t,$$

which are the parametric equations of the logarithmic spirals shown in Figure 7.7.

In case $\sigma < 0$, the spirals tend to the origin for $t \to +\infty$, but they still rotate clockwise because ω did not change its sign.

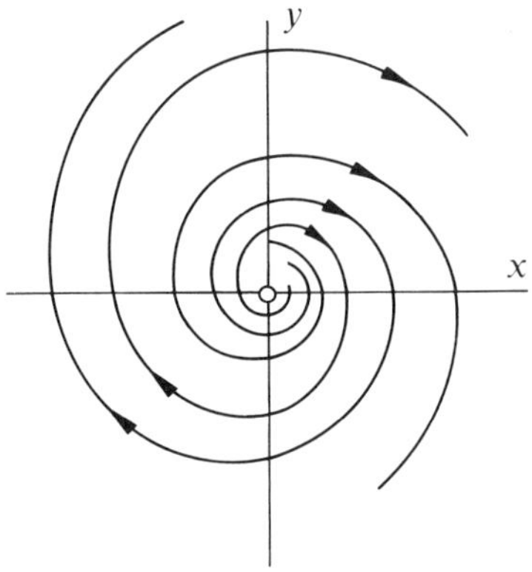

Figure 7.7

1.10 Classification of critical points

In Cases 1.2, 1.3, 1.4, 1.6, and 1.7 the critical point is called a *node* or *nodal point*. A node is characterized by the fact that all trajectories in some neighborhood of it, tend to the critical point [either as $t \to +\infty$ or $t \to -\infty$], and they approach the critical point in a definite direction.

In Case 1.6 [Figure 7.4] each line through the origin contains a trajectory, and the critical point is called a *proper node* or, sometimes, an *isotropic node*. The other nodes are called *improper*.

Figures 7.1 and 7.2 show *improper nodes with two tangents*. There are two lines through the origin which are tangent to the solutions.

Figure 7.5 shows an improper node with one tangent. Here all trajectories are tangent to the same line through the origin.

The critical point of Case 1.5 [Figure 7.3] is called a *saddle point*. There are two trajectories tending to the origin as $t \to +\infty$ and two trajectories tending to the origin as $t \to -\infty$. All the other trajectories remain only a finite time in some neighborhood of the origin.

When the critical point is surrounded by closed trajectories, as in Figure 7.6, it is called a *center*.

In Figure 7.7 the trajectories tend to the origin but not in a definite direction; instead they spiral infinitely many times around the origin. Such a critical point is called a *focus* or a *focal point*.

If the linear system is not in canonical form, these figures must be subjected to an affine transformation; that is, they are referred to some oblique axes. For example, a node with two tangents will look like Figure 7.8. It is easy to imagine the other cases similarly transformed.

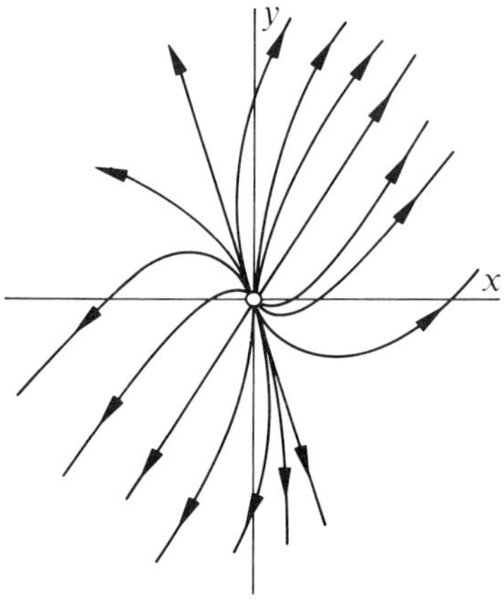

Figure 7.8

For nonlinear differential systems, the situation can be much more complicated, but in many cases the *local picture* of the trajectories (considering only some neighborhood of the critical point) looks topologically the same. We will treat these later.

1.11 Degenerate linear systems $\lambda_1 = \lambda_2 = 0$

If both characteristic roots coincide and are zero, the canonical matrix may be

$$\begin{pmatrix} 0 & 0 \\ 0 & 0 \end{pmatrix} \quad \text{or} \quad \begin{pmatrix} 0 & 1 \\ 0 & 0 \end{pmatrix}.$$

In the first case, all points are critical and all trajectories reduce to points.
In the second case, the solutions will be

$$\begin{cases} x = y_0 t + x_0 \\ y = y_0 = \text{const.} \end{cases}$$

Figure 7.9 shows the corresponding illustration. The x-axis is a line of critical points.

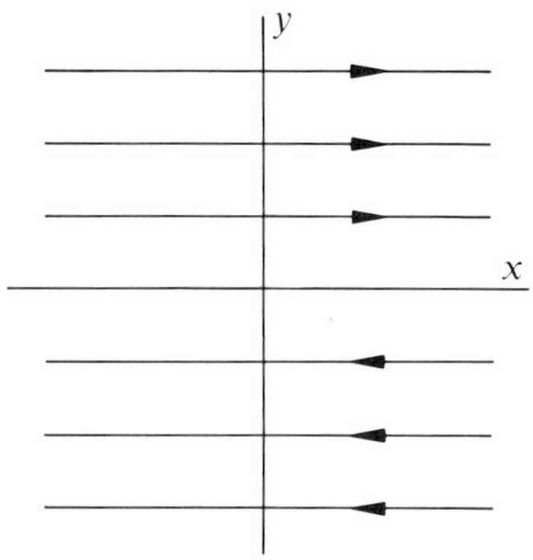

Figure 7.9

1.12 Case A with $\lambda_1 = \lambda, \lambda_2 = 0$

The differential equations are

$$\begin{cases} x' = \lambda x \\ y' = 0. \end{cases}$$

So, the solutions are

$$\begin{cases} x = x_0 e^{\lambda t} \\ y = y_0 = \text{const}, \end{cases}$$

shown in Figure 7.10. The difference between this and the previous case should be noted: in both instances the trajectories are horizontal, but here the y-axis is a critical line and each of the horizontal lines decomposes into the union of two non-degenerate and one degenerate trajectory.

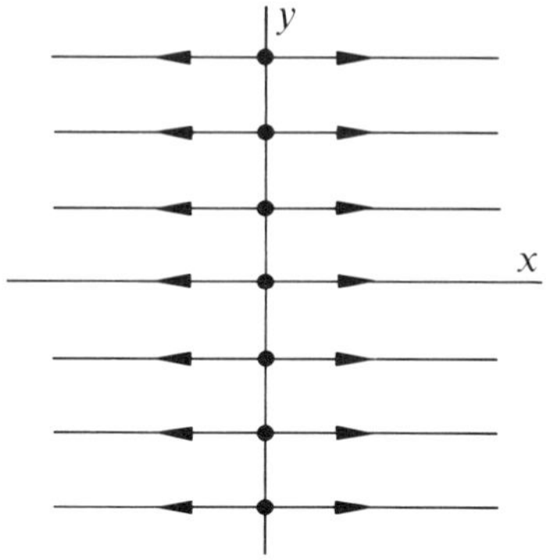

Figure 7.10

Problems

1. Draw the phase portrait [a graph of the trajectories in the phase plane] of the following differential systems:

(a) $\begin{cases} x' = x + y \\ y' = x - y, \end{cases}$ (b) $\begin{cases} x' = x + y \\ y' = x + y, \end{cases}$

(c) $\begin{cases} x' = 3x - y \\ y' = -x + 3y, \end{cases}$ (d) $\begin{cases} x' = x - y \\ y' = -3x - 3y, \end{cases}$

(e) $\begin{cases} x' = 2x - y \\ y' = x. \end{cases}$

2. Study the following piecewise linear systems by drawing the trajectories in the phase plane. Discuss in particular those points where existence and uniqueness of solutions are in doubt. Write the differential equation as a system with $x' = y$.

(a) $$x'' + f(x') = 0, \quad \text{where} \quad f(x') = \begin{cases} +1 & \text{if } x' > 0 \\ -1 & \text{if } x' < 0. \end{cases}$$

[This corresponds to a motion which is slowed down by coulomb friction.]

(b) $$x'' + f(x') + x = 0$$

with $f(x')$, as before. [This corresponds to a harmonic oscillator slowed down by coulomb friction.]

(c) $$x'' + x' - f(x') + x = 0$$

with $f(x')$, as before. [This corresponds to a harmonic oscillator with viscous damping, activated by a unit force acting in the direction of the motion.]

2 Homogeneous Systems

2.1 General considerations

The differential system

$$\begin{cases} x' = X(x, y) \\ y' = Y(x, y) \end{cases} \tag{7.2.1}$$

is *homogeneous* of degree m, if both functions $X(x, y)$ and $Y(x, y)$ satisfy the homogeneity condition

$$X(kx, ky) = k^m X(x, y), \qquad Y(kx, ky) = k^m Y(x, y). \tag{7.2.2}$$

For $m = 1$, a linear system will result.

For homogeneous systems, the lines through the origin are isoclines, since

$$\frac{dy}{dx} = \frac{Y(x, y)}{X(x,y)} \tag{7.2.3}$$

is a homogeneous function of degree zero; thus

$$\frac{Y(kx, ky)}{X(kx, ky)} = \frac{Y(x, y)}{X(x, y)}. \tag{7.2.4}$$

Hence the equations for the trajectories (7.2.3) can be integrated in elementary fashion. [See Section 1.3.3.]

By virtue of (7.2.4), if $\varphi(x, y) = 0$ is a trajectory, then so is $\varphi(kx, ky) = 0$. This means that by knowing the trajectories in some neighborhood of the origin and by using a similarity transformation (expansion), we can ascertain the trajectories in every bounded set of the plane and therefore in the whole plane.

2.2 An example of the homogeneous differential system

Consider

$$\begin{cases} x' = xy \\ y' = x^2 + y^2. \end{cases} \tag{7.2.5}$$

The differential equation of the trajectories is

$$\frac{dy}{dx} = \frac{y}{x} + \frac{x}{y},$$

which was solved in the example in Section 1.3.3, obtaining

$$y^2 = x^2 \ln \left(\frac{x}{x_0}\right)^2.$$

To see how the trajectories are described, we can substitute $y(x)$ in the first equation of the system. Thus

$$\frac{dx}{dt} = x^2\sqrt{\ln (x/x_0)^2}.$$

From the fact that dx/dt is of an order greater than x^2, we can deduce that $x(t)$ will have a finite escape time to infinity. If we wish to avoid this, we may change the differential system to, for example,

$$x' = \frac{xy}{x^4 + y^4}, \qquad y' = \frac{x^2 + y^2}{x^4 + y^4},$$

which has the same trajectories, but its speed of the motion now tends to zero as $t \to \infty$.

2.3 Transformation to polar coordinates

In order to analytically study the homogeneous system, it is convenient to introduce the polar coordinates

$$r^2 = x^2 + y^2, \qquad \theta = \arc\tan \frac{y}{x}. \tag{7.2.6}$$

It follows that

$$rr' = xx' + yy' = xX(x, y) + yY(x, y) = Z(x, y)$$

and

$$\theta' = \frac{1}{1 + (y/x)^2}\,\frac{y'x - x'y}{x^2} = \frac{xY(x, y) - yX(x, y)}{x^2 + y^2} = \frac{N(x, y)}{x^2 + y^2}.$$

Here the so-defined functions $Z(x, y)$ and $N(x, y)$ are homogeneous of degree $m + 1$. We now substitute

$$x = r\cos\theta, \qquad y = r\sin\theta$$

and obtain, by means of the homogeneity relation (7.2.2),

$$\begin{cases} r' = r^m Z(\cos\theta, \sin\theta), \\ \theta' = r^{m-1} N(\cos\theta, \sin\theta). \end{cases} \tag{7.2.7}$$

Observe that the functions $Z(\cos\theta, \sin\theta)$, $N(\cos\theta, \sin\theta)$ depend only on the angle θ. They will be the key for our next discussion.

2.4 *Assume* $N(\cos\theta, \sin\theta) \equiv 0$ *identically*

In this case $\theta' \equiv 0$; hence $\theta =$ constant for all solutions. The origin is a *proper node*. The trajectories are shown in Figure 7.11.

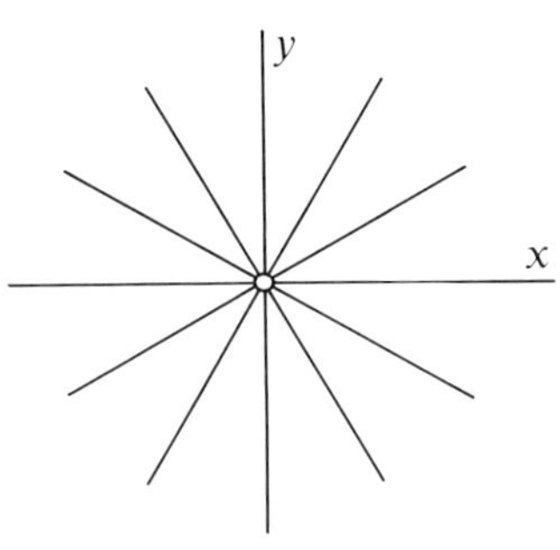

Figure 7.11

2.5 *Assume* $Z(\cos\theta, \sin\theta) \equiv 0$ *identically*

Now $r' \equiv 0$; hence $r =$ constant for all solutions. The origin is a *center*. [See Figure 7.12.]

Note that in the previous case

$$N(x, y) \equiv xY(x, y) - yX(x, y) \equiv 0;$$

hence

$$\frac{dy}{dx} = \frac{Y(x, y)}{X(x, y)} = \frac{y}{x}.$$

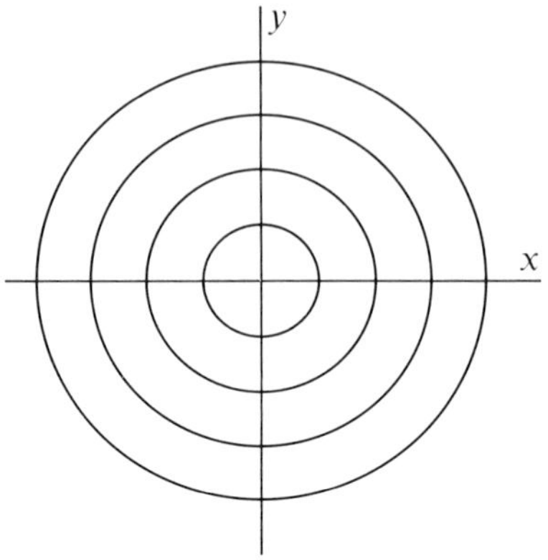

Figure 7.12

In the latter instance

$$Z(x, y) \equiv xX(x, y) + yY(x, y) \equiv 0;$$

hence

$$\frac{dy}{dx} = \frac{Y(x, y)}{X(x, y)} = -\frac{x}{y}.$$

2.6 *Isolated critical points*

For the following we will assume that the origin is an *isolated critical point* and that there is some neighborhood of the origin containing no other critical point. From condition (7.2.4), it follows that there cannot be any other critical point in the whole plane.

Furthermore, we will make the assumption that $X(x, y)$ and $Y(x, y)$ are *polynomials*, which must then be homogeneous of degree m. In this case $N(x, y)$ and $Z(x, y)$ are homogeneous polynomials of degree $m + 1$.

2.7 *Assume the equation* $N(\cos\theta, \sin\theta) = 0$ *has no real roots* (7.2.8)

Then the differential equation for the trajectories is

$$\frac{dr}{d\theta} = r\,\frac{Z(\cos\theta, \sin\theta)}{N(\cos\theta, \sin\theta)}; \tag{7.2.9}$$

hence

$$r = r_0 \exp\left[\int_0^\theta \frac{Z(\cos\theta, \sin\theta)}{N(\cos\theta, \sin\theta)}\, d\theta\right]. \tag{7.2.10}$$

$N(x, y)$ is a polynomial [and thus a continuous function] which does not vanish on the unit circle $x = \cos\theta$, $y = \sin\theta$; hence it has a positive minimum on this circle. The integration in (7.2.10) can therefore be carried out over a complete circuit $[0, 2\pi]$. If K is the value of that integral over $[0, 2\pi]$, then $r(2\pi) = r_0 e^K$.

We remarked earlier that if a trajectory is subject to a similarity transformation (an expansion), then another trajectory is obtained. The second circuit $[2\pi, 4\pi]$ will therefore be similar to the first circuit, and so on.

It follows that if $K \neq 0$, the trajectories will be spirals winding around the origin, and the origin will be a *focus*. [See Figure 7.13.]

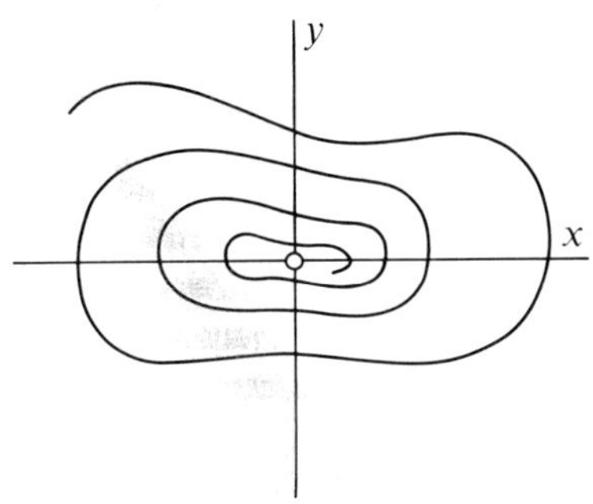

Figure 7.13

In the case $K = 0$ or, explicitly, when

$$K = \int_0^{2\pi} \frac{Z(\cos\theta, \sin\theta)}{N(\cos\theta, \sin\theta)}\, d\theta = 0,$$

then $r(2\pi) = r_0$ and the trajectories will be closed curves around the origin, which then becomes a *center*. [See Figure 7.14.]

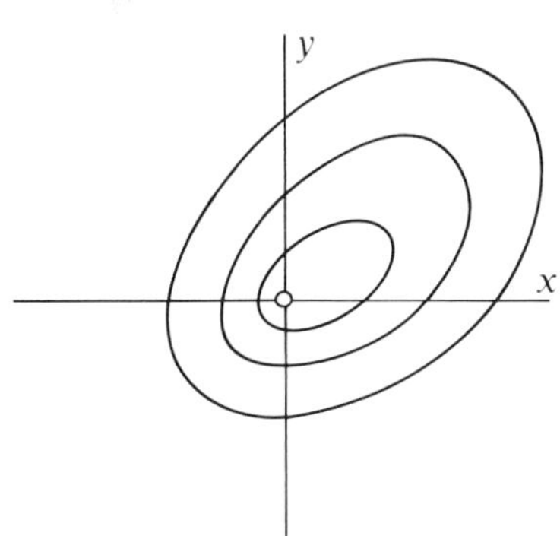

Figure 7.14

2.8 Assume the equation $N(\cos\theta, \sin\theta) = 0$ ***has real roots but is not identically zero*** (7.2.8)

Let θ_0 be a root of equation (7.2.8). Then the half-line $\theta = \theta_0 =$ constant is a trajectory because it satisfies $d\theta/dr = N/Z = 0$. It should be noted that both N and Z cannot vanish simultaneously [except at the origin]; otherwise we would have a whole line of critical points, which would be contrary to the assumption that the origin is an isolated critical point. The half-line $\theta = \theta_0$ is then called an *invariant ray*. We shall study the behavior of trajectories near such an invariant ray.

If, for increasing θ, $N(\cos\theta, \sin\theta)$ changes at θ_0, from positive to negative, then the trajectories in the neighborhood of θ_0 will be such that $\theta \to \theta_0$. If, furthermore, $Z(\cos\theta, \sin\theta) > 0$, then $r(t)$ is increasing and the trajectories will look like those in Figure 7.15(a). If, on the other hand, $Z(\cos\theta, \sin\theta) < 0$, $r(t)$ will decrease, and the result will resemble Figure 7.15(b). [If $N(\cos\theta, \sin\theta)$, for increasing θ, changes from negative to positive at θ_0, the cases $Z > 0$ and $Z < 0$ are reversed.]

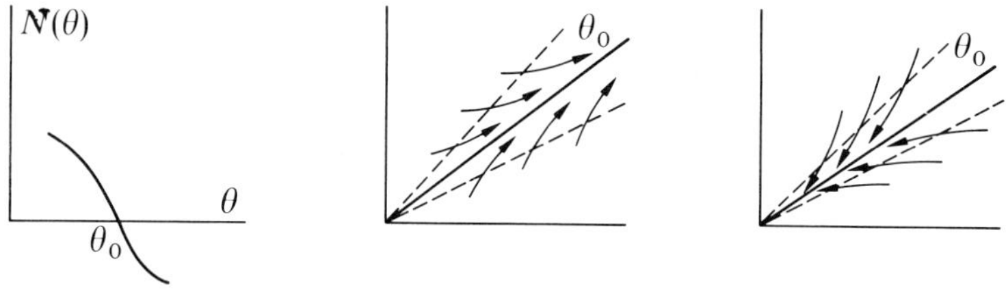

(a) (b)

Figure 7.15

If, at both sides of θ_0, $N(\cos\theta, \sin\theta)$ has the same sign [θ_0 is a root of even multiplicity], then θ will tend to θ_0 for increasing t at one side and for decreasing t at the other. The trajectories will look like those in Figure 7.16, where $Z > 0$ is assumed.

It is possible to pursue the study of concavity and convexity of the trajectories, from their second derivatives, but we will not go into such details here.

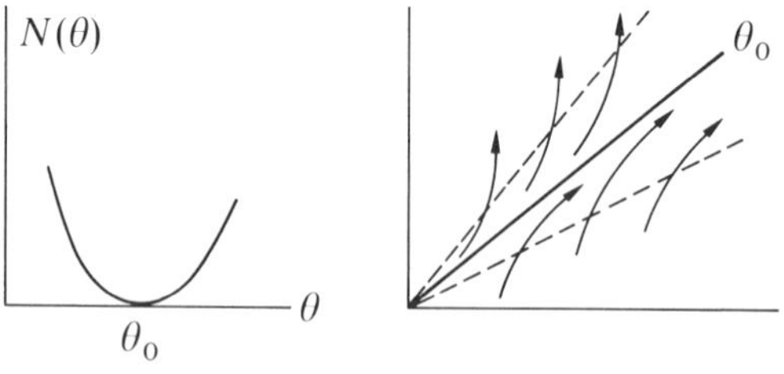

Figure 7.16

Problems

1. Use formula (7.2.10) to check that in

$$\begin{cases} x' = -2y^3 \\ y' = x^3 \end{cases}$$

the origin is a center. [*Hint*: Observe that in (7.2.10),

$$\frac{Z(\cos\theta, \sin\theta)}{N(\cos\theta, \sin\theta)}$$

is an odd function of θ.]

2. Find the invariant rays and study the behavior of the nearby trajectories for the systems

(a) $\begin{cases} x' = y^4 \\ y' = x^4, \end{cases}$ (b) $\begin{cases} x' = -x^3 \\ y' = y^3 - 2xy^2, \end{cases}$

(c) $\begin{cases} x' = xy \\ y' = x^2 + y^2. \end{cases}$

---

3 *Analytic Systems*

3.1 *Assumptions*

We will now consider the system (7.2.1)

$$\begin{cases} x' = X(x, y) \\ y' = Y(x, y), \end{cases}$$

under the hypothesis that X and Y are *analytic functions*, that is, expressible in Taylor series in some region around the origin.

We will also assume that the origin is an *isolated critical point*. Hence we may restrict our considerations to a neighborhood where there are no critical points other than the origin.

We will need the following lemma [the proof of which can be found in Lefschetz, *Differential Equations, Geometric Theory*, Chapter X, pp. 209–210].

3.2 Lemma

Let $f(x, y)$ be a real function, analytic in a neighborhood of the origin, and let $f(0, 0) = 0$. Then, if the neighborhood U [which we may assume circular] is sufficiently small, the curve $f(x, y) = 0$ has in U (at the most) a finite number of branches. These are of the type

$$\begin{cases} x = 0, \text{ or} & (7.3.1a) \\ y = a_k x^{k/n} + a_{k+1} x^{(k+1)/n} + a_{k+2} x^{(k+2)/n} + \cdots & (7.3.1b) \end{cases}$$

where k, n are positive integers and the coefficients a_j are real. Moreover, for U sufficiently small, these branches meet only at the origin.

We will also use the following lemma.

3.3 Definition

Given a differentiable arc $[A, B]$ of a curve, we say that it is an arc without contact, *with respect to a vector field $\mathbf{f}(x, y)$, if at no point of $[A, B]$ the vector $\mathbf{f}(x, y)$ is zero or parallel to the tangent vector to the arc. Analytically speaking, the vector product of the tangent vector to the arc $[A, B]$ and the field vector $\mathbf{f}(x, y)$ at the same point should never be zero.*

3.4 Lemma

Let $[A, B]$ and $[A', B']$ be two arcs without contact for the vector field $\mathbf{f} = (X, Y)$. Assume that $[A, B]$ and $[A', B']$ do not meet, except possibly at the point A, which then coincides with A'. Let the vector field (X, Y) satisfy our assumptions about a differential system (existence, uniqueness) and therefore define a family of trajectories. Let A and A' lie on the same trajectory. Suppose the arc of trajectory AA' does not meet the arcs $[A, B]$, $[A', B']$, except at A, A' and assume the same about BB'. [See Figure 7.17.] Assume, finally, that in the path rectangle $ABB'A'$ there are no critical points. Then every trajectory intersecting $[A, B]$ at some point M intersects $[A', B']$ at a point M' and the correspondence $M \to M'$ is topological, that is, a homeomorphism beween the arcs $[A, B]$ and $[A', B']$.

This lemma is a generalization of the lemma given in (6.3.4) where the same conclusion [stated there in dimension n] was obtained for sufficiently small arcs $[A, B]$, $[A', B']$.

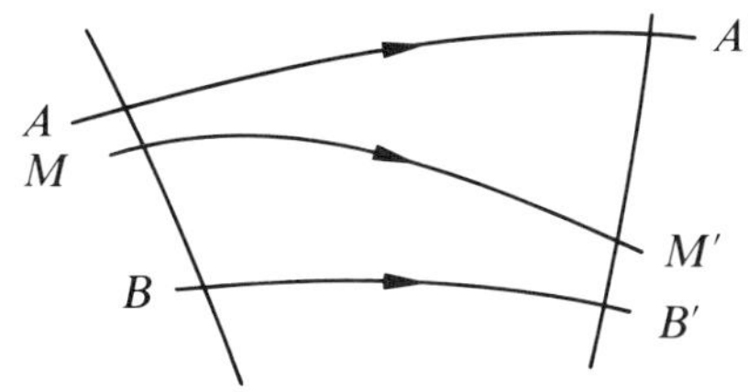

Figure 7.17

Proof

In order to prove this lemma, we must first observe that every trajectory entering the path rectangle $ABB'A'$ at M must leave it somewhere; otherwise a critical point within $ABB'A'$ would exist. This trajectory cannot leave by way of a side formed by the trajectories AA' or BB' because trajectories cannot intersect. If it did leave at the same side $[A, B]$ where M lies, it would be possible to show that $[A, B]$ could not be without contact. So the trajectory through M must leave the path rectangle at some point M' on the opposite side.

Once we have the one-to-one correspondence $M \to M'$, we consider a neighborhood of M and the correspondent neighborhood of M', as in (6.3.4). In those neighborhoods the correspondence is topological; hence the two arcs are homeomorphic.

3.5 *Z-sectors*

Under the assumptions stated in (7.3.1) and following the procedures in (7.2.3), consider the equation

$$rr' = xX(x, y) + yY(x, y) = Z(x, y) = 0. \tag{7.3.2}$$

Since $Z(x, y)$ is analytic, there is a circle of radius R and center at the origin [according to Lemma 7.3.2] such that in its interior

(a) the branches of the curve $Z(x, y) = 0$ are of the type (7.3.1) and do not meet each other except at the origin;

(b) the origin is the only critical point.

Provided R is sufficiently small, we can then state the next lemma.

3.5a Lemma

Each branch of the curve

$$Z(x, y) = 0$$

intersects the circle of radius R at a single point, and the arc of this branch between the origin and this point of intersection is an arc without contact.

Proof

To prove this lemma, consider a particular branch and rotate the coordinate axes in such a way that the tangent to the branch does not coincide with an axis. This rules out such expressions as

$$x = 0 \quad \text{or} \quad y = a_k x^k + \cdots, \quad k \neq 1,$$

for the equation of the branch. Hence the power series expansion of the branch is of the form

$$y = mx + ax^p + \cdots, \quad m \neq 0, \quad p > 1.$$

The tangent to the branch at a general point has the slope

$$\mu = m + pax^{p-1} + \cdots.$$

On the other hand, the vector field (X, Y) has [at the same point] the slope

$$\mu' = -\frac{x}{y} = \frac{-1}{m + pax^{p-1} + \cdots}.$$

[This is true because at this point $Z = xX + yY = 0$.]

As a result*

$$\mu\mu' = -1 + 0(x^{p-1}),$$

and the trajectory is approximately orthogonal to the branch of $Z = 0$ for x sufficiently small. Hence they cannot be tangent to each other, and the branch $Z = 0$ is an arc without contact.

To prove that the branch $Z = 0$ can meet the circle of radius R at only one point, assume the contrary. Thus there would be intersections such as A, B, C in Figure 7.18. But then the arc BC must be tangent at some point to a circle of radius less than R; at this point, the branch would be tangent to the trajectory [which has $r' = 0$]. This provides a contradiction with what we have already seen.

The idea of the following method is the observation that the finite number of branches of the curve $Z(x, y) = 0$ divide the circle of radius R into sectors, as illustrated in Figure 7.19. We call these *Z-sectors*.

In each of these Z-sectors,

$$r' = dr/dt$$

---

* In this and following examples, we will use the notation of infinitesimals. [See, for example, E. C. Titchmarsh, *The Theory of Functions*. London: Oxford University Press, 1939, Chapter I, p. 1.]

By

$$f(x) = 0(g(x))$$

we mean that $|f(x)| < Kg(x)$ for some constant K, if x is sufficiently near to a given limit. In particular, $0(1)$ means a bounded function.

By

$$f(x) = o(g(x))$$

we mean that $f(x)/g(x) \to 0$ as x tends to a given limit. Thus for example, $(1 + x^2) = o(x^3)$ as $x \to \infty$. In particular, $o(1)$ means a function which tends to zero.

From these definitions it follows that

$$0(g(x)) + 0(g(x)) = 0(g(x)), \qquad o(g(x)) + o(g(x)) = o(g(x)),$$
$$0(g(x)) + o(g(x)) = 0(g(x)),$$

and similar relations. If, in particular,

$$r = +\sqrt{x^2 + y^2},$$

then for $r \to 0$ [and therefore $x \to 0$ and $y \to 0$] we have

$$f(x) = o(x) \quad \text{implies} \quad f(x) = o(r).$$

We also have

$$xo(r) = o(r^2) \quad \text{and} \quad yo(r) = o(r^2),$$

in the sense that if $f(r) = o(r)$, then $xf(r) = o(r^2)$ and correspondingly for y.

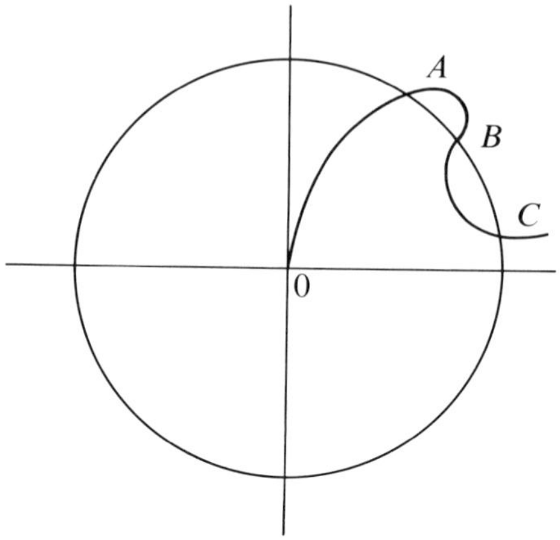

Figure 7.18

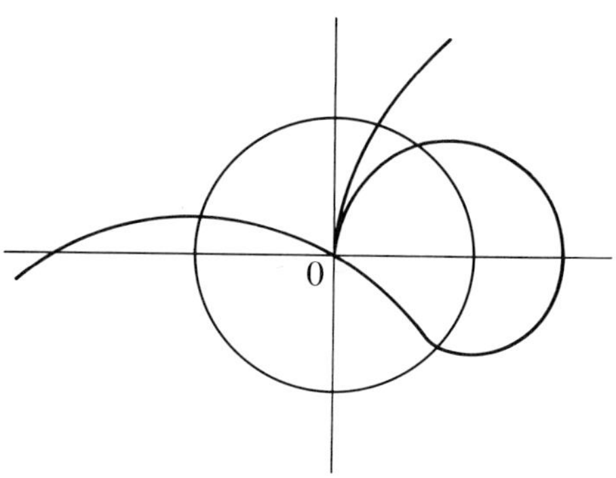

Figure 7.19

has a constant sign. We will find the different possible configurations of the trajectories in such a sector; then, by putting together the different sectors, we will be able to obtain a picture for the whole circle.

Considering a single sector, we may assume that in it

$$r' = dr/dt < 0.$$

The other case is reduced to this one by changing the direction of time ($t \to -t$); this leaves the trajectories unchanged but reverses the arrows on them. Note that for $r' < 0$, all trajectories at points of the circle of radius R enter the sector.

3.6 Theorem

Every trajectory which enters the sector through the circle of radius R leaves through one of the sides or tends to the origin.

Proof

On such a trajectory, $r(t)$ is a monotonous decreasing function because $r' < 0$ in the sector. Hence $r(t)$ tends to some limit r_0. Then the arc of the circle of radius r_0 [$A'B'$ in Figure 7.20] contains some limit point of the trajectory. If this limit point P_0 is interior to the sector, the vector field is at such a point transversal (not tangent) to the circle; hence the trajectory must cross it and r_0 cannot be the limit of $r(t)$. If this limit point is A' or B', the vector field at those points is tangent to the circle ($r' = Z = 0$); hence the trajectory must leave the sector there.

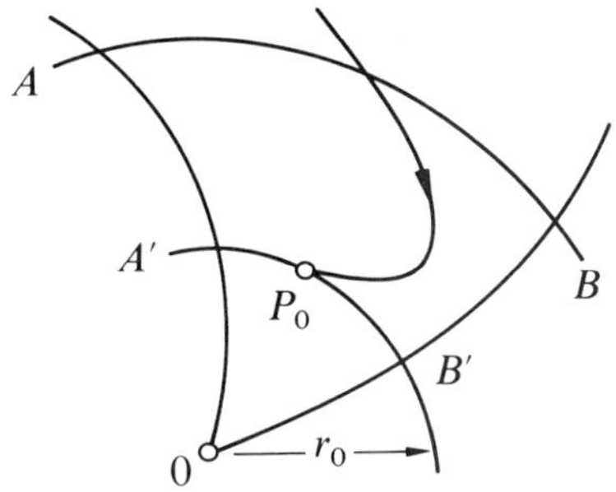

Figure 7.20

Therefore, if the trajectory does not leave the sector by one of the sides, $r(t)$ must tend to zero, and the trajectory, staying within the sector, must tend to the origin. On the other hand, because the origin is a critical point, the trajectory cannot leave the sector at this point.

3.7 Theorem

If the trajectory γ tends to the origin as $t \to +\infty$, then $\theta(t)$ tends to a well-defined limit, either finite or infinite ($+\infty$ or $-\infty$). *In other words, γ is a spiral or approaches the origin with a well-defined tangent.*

Proof

This means that the function $\theta(t)$ cannot have, as $t \to +\infty$, two different limit values, θ_1 and θ_2. In order to prove this theorem, let us assume that this is the case.

According to (7.2.7) and taking into account that the functions are analytic, we may write

$$r^2 \frac{d\theta}{dt} = xY(x, y) - yX(x, y)$$

$$= r^{m+1}[a_{m+1}(\theta) + ra_{m+2}(\theta) + \cdots],$$

where $a_k(\theta)$ is homogeneous of degree k in $\cos\theta$, $\sin\theta$. We may also assume that $xY - yX$ is not identically zero; if this were true, the origin would be a proper node, which certainly satisfies the conditions of this theorem.

As previously stated, we will assume that $\theta(t)$ approaches two limit values, θ_1, θ_2, for $t \to +\infty$. Let

$$\theta_1 < \theta_0 < \theta_2$$

be such that

$$a_{m+1}(\theta_0) \neq 0.$$

[a_{m+1} is the first nonzero coefficient in the above series expansion.] This is possible because a_{m+1} is analytic and not identically zero. Then, for r sufficiently small or (equivalently) for t sufficiently large, whenever $\theta(t)$ crosses θ_0, $d\theta/dt$ must always have the same sign. [It must be equal to the sign of $a_{m+1}(\theta_0)$, which predominates over the remaining terms of the series.] It is therefore impossible that $\theta(t)$ oscillates between θ_1 and θ_2, and the theorem is proved.

It should be noted that the analyticity of the functions plays a decisive role in these results. Without analyticity, configurations such as Figure 7.21 [$\theta(t)$ approaching several limiting values] or Figure 7.22 [infinitely many isolated cycles in every neighborhood of the origin] are quite possible.

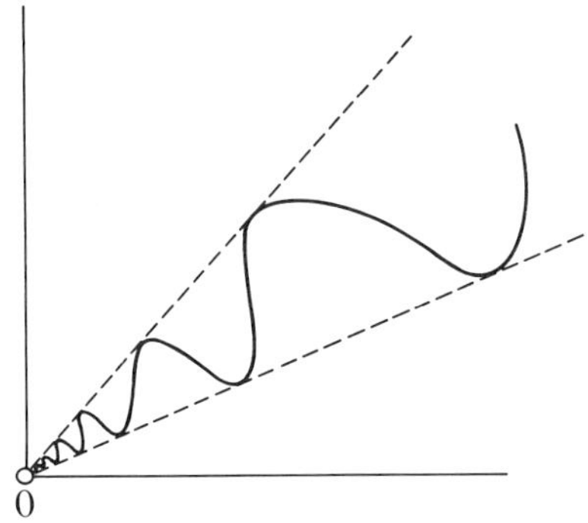

Figure 7.21

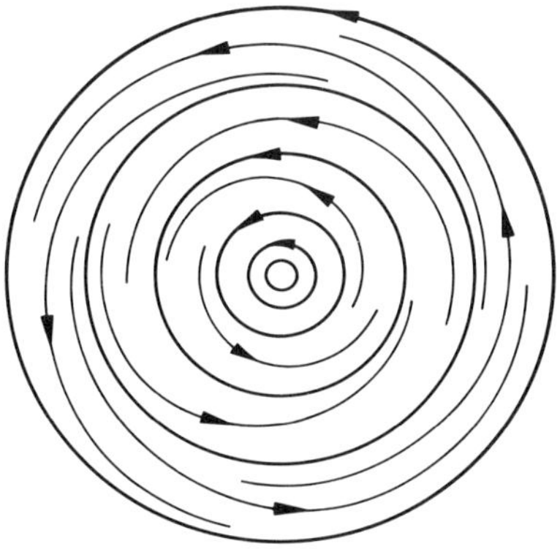

Figure 7.22

3.8 Description of Z-sectors

We have now all the tools we need to describe the kinds of possible behavior of the trajectories in a Z-sector. As previously stated, we will assume that $r' = Z < 0$; the other case is obtained simply by reversing the time direction.

Case 1: The trajectories leave Z on both sides of the sector. The corresponding illustration is given in Figure 7.23(a),(b). At least one trajectory must tend to the origin [see part (b)], but the general case will be that a whole set will tend to the origin [see part (a)]. The configuration of the trajectories is determined by Lemma 7.3.4; the arcs AC and DB [see part (a)] are mapped homeomorphically onto AO and BO. If $C = D$, then the result will be part (b) of Figure 7.23.

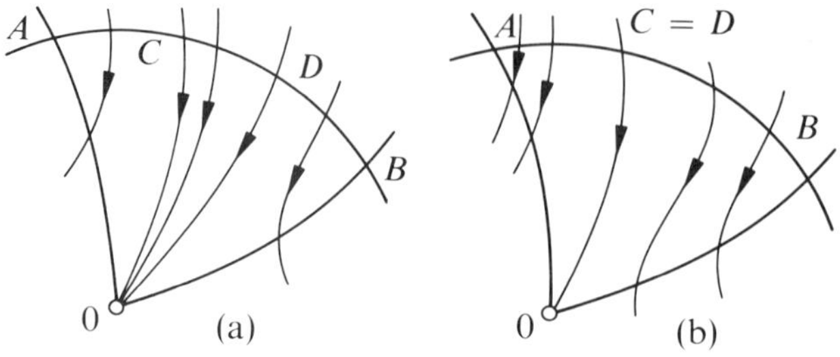

Figure 7.23

Case 2: The trajectories enter Z on both sides of the sector. Then all trajectories must tend to the origin. [See Figure 7.24.]

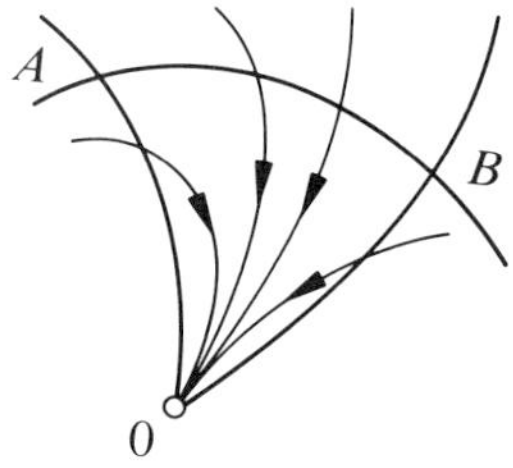

Figure 7.24

Case 3: The trajectories enter on one side of the sector and leave on the other side. Since all trajectories which do not leave must tend to the origin, the configurations shown in Figure 7.25(a),(b),(c) are possible; parts (a) and (b) are essentially the same.

Case 4: The curve $Z = 0$ has no real branches at the origin, so the Z-sector is the entire circle of radius R. The trajectories enter on the circumference of the circle; hence all must tend to the origin. We have seen that they can do this either as spirals ($\theta(t) \to \infty$) or by tending to a definite tangent ($\theta(t) \to \theta_0$). It is easy to see that all trajectories must behave exclusively in one or the other of these two ways, since one trajectory approaching the origin with a definite tangent is sufficient in order to prevent all the other trajectories from spiraling around the origin. Therefore we may have either the node or the focus. [See Figure 7.26.]

Case 5: $Z(x, y) = 0$ identically. In this case $r' = 0$ identically and the origin is a center. [See Figure 7.27.]

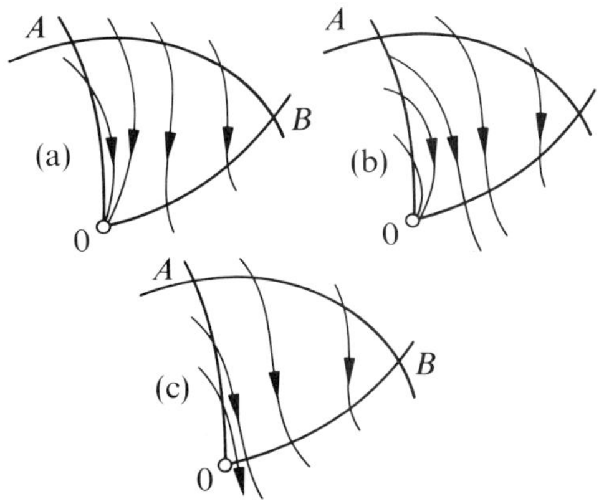

Figure 7.25

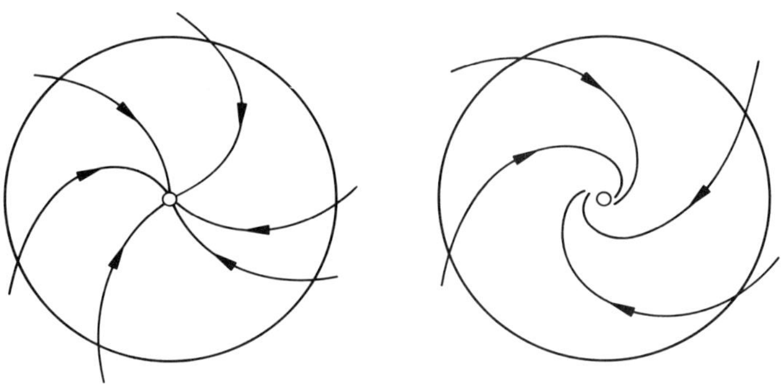

Figure 7.26

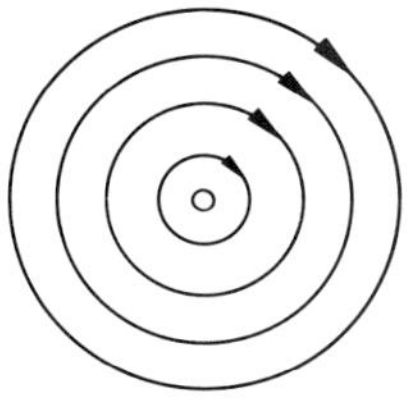

Figure 7.27

3.9 Combination of Z-sectors

To get the whole picture of the trajectories around the origin, we must combine the different Z-sectors, which appear in finite number. The most interesting configurations are the following.

Fan: This is a sector of trajectories, all tending to the origin with definite tangents. [See Figure 7.28.]

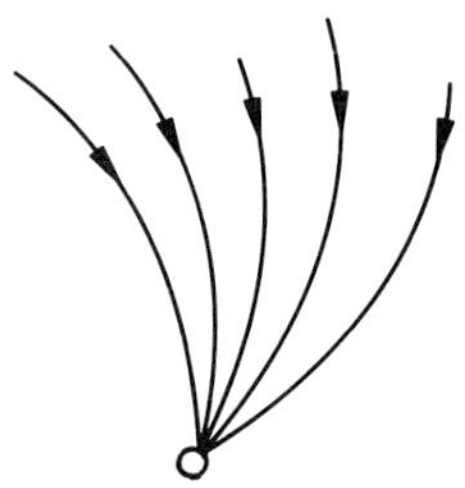

Figure 7.28

Hyperbolic sector: This is a sector limited by two trajectories, one of which leaves the origin and the other of which approaches the origin. All intermediate trajectories pass by, as shown in Figure 7.29.

Figure 7.29

Elliptic sector: This is a sector limited by two trajectories, one leaving the origin and the other approaching it. It is filled with trajectories which tend to the origin for both $t \to +\infty$ and $t \to -\infty$. Figure 7.30 shows the configuration.

In the case of linear differential systems, we have already seen all these configurations of trajectories, with the exception of the elliptic sector. This one can only appear in higher order systems.

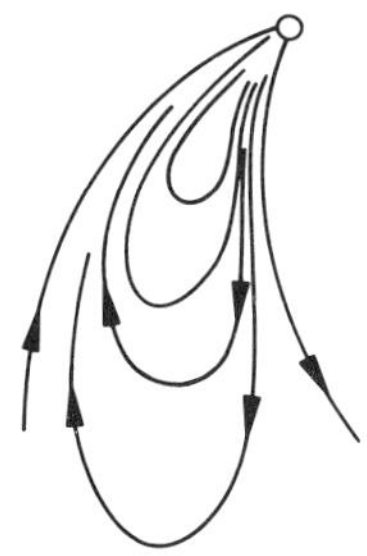

Figure 7.30

We again remind the reader that if the hypothesis of analyticity is dropped, many other, more complicated configurations become possible, as Figures 7.21 and 7.22 illustrate.

Problems

1. Discuss the Z-sectors of problems 1 and 2 on page 217.
2. Study the Z-sectors of the following differential systems. Determine the sign of $N(x, y)$ on the boundary of the Z-sectors and draw, qualitatively, the behavior of the trajectories.

(a) $\begin{cases} x' = x^2 \\ y' = -y. \end{cases}$ (b) $\begin{cases} x' = x^3 \\ y' = -2x^2y + y^3. \end{cases}$

4 Perturbed Linear Systems

4.1 Linear approximation

For a nonlinear system, the analysis of the behavior of the trajectories in the neighborhood of a critical point is quite complicated. By translating coordinates the critical point in question can be moved to the origin; hence we may assume in the following examples that the origin is the critical point under consideration.

If the system is analytic, it may be written

$$\begin{cases} x' = X(x, y) = ax + by + P(x, y) \\ y' = Y(x, y) = cx + dy + Q(x, y), \end{cases} \tag{7.4.1}$$

where $P(x, y)$ and $Q(x, y)$ are power series starting with terms of degree not less than two.

In a more general way, we can consider differential systems of the form

$$\begin{cases} x' = X(x, y) = ax + by + o(r) \\ y' = Y(x, y) = cx + dy + o(r), \end{cases} \tag{7.4.2}$$

where $r^2 = x^2 + y^2$, and $o(r)$ means an infinitesimal of higher order than r. $[o(r)/r \to 0$ for $r \to 0.]$

It is natural, when we consider a small neighborhood of the origin, to neglect the $o(r)$ terms in (7.4.1) and (7.4.2) and consider the *linear approximation*

$$\begin{cases} x' = ax + by \\ y' = cx + dy. \end{cases} \tag{7.4.3}$$

In fact, we will see that for some questions concerning the behavior of the trajectories near the origin, equations (7.4.1), (7.4.2), and (7.4.3) behave in the same way, while for other questions their behavior may be quite different.

One of the basic characteristics of a critical point to be studied is its *stability*. The reader should first review the definitions of stability, asymptotic stability, and instability given in Section 5.2.3, as well as their relation to the characteristic roots of the matrix in the case of a linear system.

4.2 Theorem

If the origin is asymptotically stable for the linear approximation (7.4.3) [*meaning the characteristic roots have negative real parts*], *then it is also asymptotically stable for the nonlinear system* (7.4.2).

The proof of this theorem is really an application of the so-called *Liapunov method*, in which a certain function, the *Liapunov function*, plays the key role.

A remark which we will repeat constantly is that the stability properties of a system are *preserved* under a linear (nondegenerate) transformation of (x, y) coordinates. Hence: *we may assume that the differential system is in its canonical form.*

Proof

Let us first assume that the characteristic roots are real and distinct, and let us denote them by $-\lambda_1$ and $-\lambda_2$. We assume that the linear approximation is in canonical form (7.1.2(a)); hence the system (7.4.2) is

$$\begin{cases} x' = -\lambda_1 x + o(r) \\ y' = -\lambda_2 y + o(r). \end{cases} \tag{7.4.4}$$

Consider the Liapunov function

$$V(x, y) = r^2 = x^2 + y^2. \tag{7.4.5}$$

Along any trajectory, its total derivative dV/dt is

$$V' = \frac{\partial V}{\partial x} x' + \frac{\partial V}{\partial y} y' = -2(\lambda_1 x^2 + \lambda_2 y^2) + o(r^2).$$

In a sufficiently small circular neighborhood of the origin,

$$|o(r^2)| < \lambda_1 x^2 + \lambda_2 y^2$$

and therefore

$$V' < -(\lambda_1 x^2 + \lambda_2 y^2) \leqq -\lambda r^2 = -\lambda V,$$

where $\lambda = \min(\lambda_1, \lambda_2)$. Thus, along every trajectory,

$$V(x(t), y(t)) \leqq V(x_0, y_0) e^{-\lambda t}$$

for $t > 0$. Since $\lambda > 0$, it follows that $r \to 0$ as $t \to +\infty$; hence the origin is asymptotically stable.

If we now consider the case of a double root,

$$\lambda_1 = \lambda_2 = \lambda,$$

the canonical system may be of the same form as (7.4.4), in which case the above proof is valid, or the canonical system may be

$$\begin{cases} x' = -\lambda x + y + o(r) \\ y' = -\lambda y + o(r). \end{cases} \tag{7.4.6}$$

In the latter case we use the Liapunov function

$$V(x, y) = kx^2 + y^2, \tag{7.4.7}$$

where k is a positive constant to be determined later. Note that $V(x, y)$ has the essential property that

$$V(x, y) \to 0 \quad \text{if and only if} \quad r \to 0.$$

The total derivative of $V(x, y)$ along any trajectory is

$$V' = \frac{\partial V}{\partial x} x' + \frac{\partial V}{\partial y} y' = -2(k\lambda x^2 - kxy + \lambda y^2) + o(r^2).$$

The quadratic form in parentheses is positive definite, if we determine k such that

$$k^2 - 4k\lambda^2 < 0,$$

and this is always possible by taking k sufficiently small. Then, for fixed r, the minimum of the quadratic form is proportional to r^2, and for sufficiently small r,

$$|o(r^2)| < k\lambda x^2 - kxy + \lambda y^2.$$

It follows, as in the previous case, that

$$V' \leqq -(k\lambda x^2 - kxy + \lambda y^2),$$

and dividing by V,

$$\frac{V'}{V} \leqq -\frac{k\lambda x^2 - kxy + \lambda y^2}{kx^2 + y^2}.$$

For every $r > 0$, the numerator has a positive minimum proportional to r^2. Similarly, the denominator has a maximum proportional to r^2. Hence there exists a constant $\mu > 0$ such that

$$\frac{V'}{V} \leqq -\mu,$$

and for $t > 0$,

$$V(x(t), y(t)) \leqq V(x_0, y_0)e^{-\mu t}.$$

As we saw earlier, this implies the asymptotic stability of the origin.

Finally, in the case of complex conjugate roots with negative real parts, the Liapunov function to be used may be

$$V(x, y) = r^2 = x^2 + y^2,$$

and the desired result will follow from the same procedure outlined above.

4.3 Theorem

If at least one of the characteristic roots of the linear approximation (7.4.3) *is positive, or if both roots are complex with positive real parts, then the origin is unstable for the nonlinear system* (7.4.2).

Proof

If both characteristic roots are positive or have positive real parts, we use the Liapunov function

$$V(x, y) = x^2 + y^2.$$

As before, we obtain

$$V' = 2(\lambda_1 x^2 + \lambda_2 y^2) + o(r^2) \geqq \lambda_1 x^2 + \lambda_2 y^2 \geqq \lambda V,$$

as λ_1, λ_2 are positive. Thus, for $t > 0$,

$$V(x(t), y(t)) \geqq V(x_0, y_0)e^{\lambda t}.$$

All these relations are true in some neighborhood of the origin; therefore every trajectory must leave this neighborhood after a finite time. Hence the origin is unstable.

If only one characteristic root is positive,

$$\lambda_1 > 0 \geqq \lambda_2,$$

we must use the Liapunov function

$$V(x, y) = x^2 - y^2. \tag{7.4.8}$$

From the expression for the total derivative it follows that

$$\frac{V'}{V} = \frac{2(\lambda_1 x^2 - \lambda_2 y^2) + o(r^2)}{x^2 - y^2} \geqq \frac{\lambda r^2}{r^2} = \lambda > 0,$$

at least in some neighborhood U of the origin.

Now, in any neighborhood of the origin there are points (x_0, y_0) with $V(x_0, y_0) > 0$. Along the trajectories through those points,

$$V(x(t), y(t)) \geqq V(x_0, y_0)e^{\lambda t},$$

for $t > 0$. Hence those trajectories must leave the neighborhood U. Therefore the origin is unstable.

4.4 Lemma

Consider the system of analytic differential equations

$$\begin{cases} x' = X(x, y) = X_m(x, y) + P(x, y) \\ y' = Y(x, y) = Y_m(x, y) + Q(x, y), \end{cases} \tag{7.4.9}$$

where X_m and Y_m are homogeneous polynomials of degree m, and P, Q are power series starting with terms of degree higher than m. This homogeneous approximation of degree m includes the linear approximation as a particular case.

If a trajectory of (7.4.9) tends to the origin in a definite direction θ_0, then θ_0 satisfies the equation

$$\cos\theta_0\, Y_m(\cos\theta_0, \sin\theta_0) - \sin\theta_0\, X_m(\cos\theta_0, \sin\theta_0) = N_0(\cos\theta_0, \sin\theta_0) = 0, \tag{7.4.10}$$

where

$$N_0(x, y) = xY_m(x, y) - yX_m(x, y).$$

Proof

Let $x(t)$, $y(t)$ be the trajectory under consideration or, in polar coordinates, $r(t)$, $\theta(t)$. With the coordinate transformation (7.2.6) and (7.2.7), we obtain

$$\begin{cases} r' = r^{-1}[xX(x, y) + yY(x, y)] \\ \theta' = r^{-2}[xY(x, y) - yX(x, y)]. \end{cases}$$

Substituting (7.4.9) in these expressions, we obtain

$$\begin{cases} r' = r^m[\cos\theta X_m(\cos\theta, \sin\theta) + \sin\theta Y_m(\cos\theta, \sin\theta)] \\ \qquad + \cos\theta P_1(r, \theta) + \sin\theta Q_1(r, \theta), \\ \theta' = r^{m-1}[\cos\theta Y_m(\cos\theta, \sin\theta) - \sin\theta X_m(\cos\theta, \sin\theta)] \\ \qquad + r^{-1}[\cos\theta Q_1(r, \theta) - \sin\theta P_1(r, \theta)]. \end{cases}$$

Here

$$P_1(r, \theta) = P(r\cos\theta, r\sin\theta), \qquad Q_1(r, \theta) = Q(r\cos\theta, r\sin\theta).$$

Note that P_1, Q_1 are power series in r starting with the term r^{m+1}; so we may write

$$\begin{cases} P_1(r, \theta) = r^{m+1}P_2(r, \theta) \\ Q_1(r, \theta) = r^{m+1}Q_2(r, \theta). \end{cases}$$

With this, the formulas become

$$\begin{cases} r' = r^m Z_0(\cos\theta, \sin\theta) + r^{m+1}[\cos\theta P_2 + \sin\theta Q_2] \\ \theta' = r^{m-1}N_0(\cos\theta, \sin\theta) + r^m[\cos\theta Q_2 - \sin\theta P_2]. \end{cases} \tag{7.4.11}$$

If we now imagine r, θ as *cartesian coordinates*, the transformation (7.2.6) maps the punctured (x, y) plane $x^2 + y^2 > 0$ onto the open half-plane $r > 0$, and this correspondence is then analytic. But the point $x = y = 0$ corresponds to the whole line $r = 0$. As long as $r > 0$, each trajectory is transformed into a curve which is a trajectory of the transformed differential equations (7.4.11). Thus every critical point in the (x, y) plane which is different from the origin is mapped into a critical point in the (r, θ) plane.

To study the trajectories in the (r, θ) plane, we may cancel the factor r^{m-1} in (7.4.11). If we do this, then on the line $r = 0$ the points θ with $N(\cos\theta, \sin\theta) = 0$ appear as critical points. This condition is precisely equation (7.4.10).

If, in the (x, y) plane, a trajectory tends to the origin in a definite direction θ_0, then the correspondent trajectory in the (r, θ) plane will tend to the line $r = 0$ such that $\theta \to \theta_0$. Hence, in the (r, θ) plane, the point $r = 0$, $\theta = \theta_0$ will be a critical point and will satisfy condition (7.4.10).

The converse is not true: if θ_0 satisfies condition (7.4.10), there is not necessarily a trajectory tending to $r = 0$ in the direction $\theta = \theta_0$.

4.5 *Theorem*

If, for the linear approximation (7.4.3), *the origin is a center or focus* [*that is, if the characteristic roots are complex*], *then the origin is also a center or focus for the nonlinear system* (7.4.1).

Furthermore, if the origin is a focus for the linear approximation (7.4.3), *then it is also a focus for the nonlinear system* (7.4.1).

Note that it is not necessarily true that if the origin is a center for the linear approximation, then it must also be a center for the nonlinear system. The origin for the nonlinear system may be either a center or a focus.

Proof

If the origin is a focus for the linear approximation, we may assume that it is asymptotically stable, possibly by changing t into $-t$. Then the nonlinear system is also asymptotically stable [Theorem 7.4.2], and in some neighborhood, all trajectories will tend to the origin.

On the other hand, we can show that equation (7.4.10) has no real roots. For the linear system (7.4.3),

$$N_0(x, y) = cx^2 + (d - a)xy - by^2,$$

and a simple computation shows that the discriminant of this quadratic form,

$$(d - a)^2 + 4bc,$$

is precisely the discriminant of the characteristic equation. Hence if the characteristic equation has no real roots, neither will

$$N_0(x, y) = 0$$

have any real roots other than $x = y = 0$.

In conclusion: for the nonlinear system, no trajectory can tend to the origin in a definite direction, and the origin will always be a focus of the same type (stable or unstable) as that in the linear approximation.

If the origin is a center for the linear approximation, we assume a coordinate transformation has been made so that the system is in canonical form. Then the nonlinear system is

$$\begin{cases} x' = \omega y + o(r) \\ y' = -\omega x + o(r). \end{cases}$$

Hence

$$\begin{aligned} \theta' &= \frac{1}{r^2}[xY(x, y) - yX(x, y)] = \frac{-\omega(x^2 + y^2) + o(r^2)}{r^2} \\ &= -\omega + o(1). \end{aligned} \tag{7.4.12}$$

In some neighborhood of the origin, therefore, θ' is approximately equal to $-\omega$; for example

$$\theta' = -\omega + \varepsilon, \ |\varepsilon| < \left|\frac{\omega}{2}\right|.$$

It follows that if, for either $t \to +\infty$ or for $t \to -\infty$, the trajectory does not leave this neighborhood, then in a similar manner $\theta(t)$ will tend to

$+\infty$ or $-\infty$ [according to the sign of ω]. Hence the origin is a center or focus, *provided we can prove that the trajectory does not leave the neighborhood.*

Let the neighborhood under consideration have radius R sufficiently small so that in this neighborhood,

$$|\theta'| > k > 0.$$

[This is possible according to our calculations above.] Let T be a number greater than $2\pi/k$. We can then apply the continuity of solutions in initial conditions to see that if we take a point A on the x-axis sufficiently near the origin, the trajectory through A will remain in the neighborhood of radius R for a time greater than T. Hence θ will change by at least 2π, and the trajectory will meet the x-axis again, as shown in Figure 7.31(a),(b).

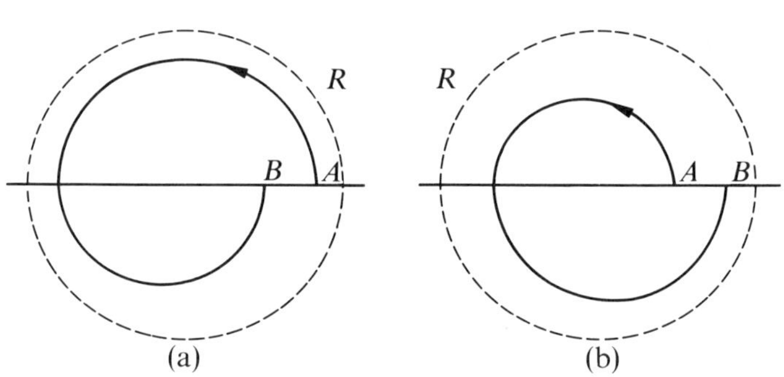

Figure 7.31

Let us consider part (a) of Figure 7.31. Every trajectory starting at $t = 0$ inside the closed curve, formed by the trajectory AB and the segment $\overline{BA}$ of the x-axis, cannot leave this region for $t > 0$. Indeed, it cannot intersect the trajectory AB; while on the segment $\overline{BA}$ of the x-axis, all trajectories enter the region because θ' has constant sign on $\overline{BA}$. Thus we have proved that the trajectory remains forever in the neighborhood of radius R. For part (b) the proof is similar, except $t \to +\infty$ changes into $t \to -\infty$.

4.6 Theorem

If the linear approximation (7.4.3) *has real, distinct, nonzero characteristic roots* λ_1, λ_2, *then the trajectories of the nonlinear system have* [*near the origin*] *the same qualitative configuration as the trajectories of the* *linear approximation. More precisely, the directions of the eigenvectors corresponding to the characteristic roots [and only these directions] are the directions in which the trajectories approach the origin [for either $t \to +\infty$ or $t \to -\infty$], and the time orientation of approach to the origin is the same in both linear and nonlinear cases.*

Proof

If both characteristic roots are of the same sign, all trajectories will tend to the origin for $t \to +\infty$ or for $t \to -\infty$, according to the sign of the characteristic roots. In order to show that $\theta(t)$ tends in the nonlinear case to the same values as in the linear approximation, let us consider the differential equations in canonical form

$$\begin{cases} x' = \lambda_1 x + o(r) \\ y' = \lambda_2 y + o(r). \end{cases}$$

The characteristic directions in which trajectories can approach the origin are given in the linear approximation by

$$N_0(x, y) = (\lambda_2 - \lambda_1)xy = 0;$$

hence they are the directions of the coordinate axes.

Let us consider the x-axis as an example of such a characteristic direction. In an analogous manner to (7.4.12), we now obtain

$$\begin{aligned} \theta' &= \frac{1}{r^2}\,[xY(x, y) - yX(x, y)] \\ &= \frac{1}{r^2}\,[(\lambda_2 - \lambda_1)xy + o(r^2)] \\ &= (\lambda_2 - \lambda_1)\cos\theta\sin\theta + o(1). \end{aligned} \tag{7.4.13}$$

If we consider a sector of a small but fixed angle α on both sides of the x-axis [see Figure 7.32], we may take the radius of the considered neighborhood so small that θ' has a constant sign on each side of the sector. Indeed, formula (7.4.13) shows that for $\theta = \pm\alpha$, θ' is the sum of a fixed nonzero term plus a term tending to zero [the $o(1)$]. This also shows that on both sides of the considered sector, the sign of θ' will coincide for both the linear and nonlinear systems. In the linear case the trajectories either enter on both sides of the sector or leave on both sides. Hence in the nonlinear case the same thing happens, and we then obtain the configuration of trajectories shown in Figure 7.23 and 7.24 [possibly with the arrows reversed.] Since the angle in Figure 7.32 is arbitrarily small, the directions of approach for both the linear and nonlinear systems will coincide.

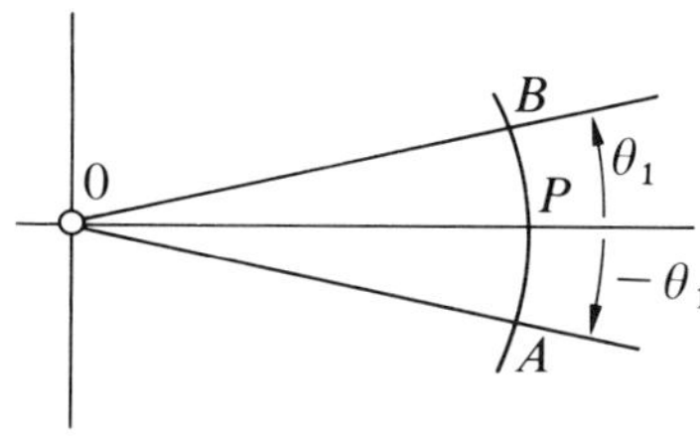

Figure 7.32

If λ_1 and λ_2 are of the opposite sign, the proof concerning a direction of approach to the origin is still valid. By using stability considerations, we see that the kind of approach for $t \to +\infty$ or $t \to -\infty$ is also maintained. Hence we have a saddle point, and the illustration of the nonlinear case will be similar to the one given in Figure 7.33.

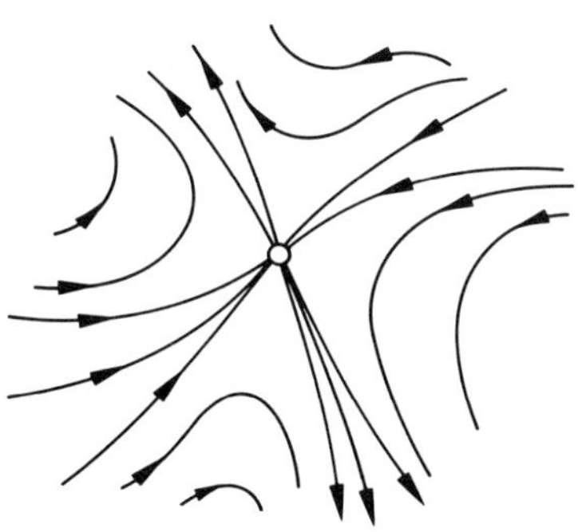

Figure 7.33

Problems

1. Classify the critical points of the following systems. Use linear approximation and state, if possible, the stability properties and the kind of critical point (node, center, focus, or saddle point).

(a) $\begin{cases} x' = x^2 - y^2 \\ y' = 2 - x^2 - y^2. \end{cases}$ (b) $\begin{cases} x' = x^2 - y^2 \\ y' = x^2 - 1. \end{cases}$

(c) $\begin{cases} x' = y^2 - 1 \\ y' = x^2 - 1. \end{cases}$ (d) $\begin{cases} x' = y^2 - 1 \\ y' = 1 - x^2. \end{cases}$

2. Study the system

$$\begin{cases} x' = -x + \dfrac{2y}{\ln(x^2 + y^2)} \\ y' = -y - \dfrac{2x}{\ln(x^2 + y^2)}. \end{cases}$$

Show that, although in linear approximation the origin is a proper node, the nonlinear system has a stable focus at the origin. [*Hint*: Change to polar coordinates.] This will prove that a proper node in linear approximation,

$$N_0(\cos\theta, \sin\theta) \equiv 0,$$

may change into a focus for the nonlinear system. Prove that this is impossible if $N_0(\cos\theta, \sin\theta)$ is not identically zero.

References

Most of the general topics in this chapter are covered extensively in the books by Lefschetz, Nemytskii-Stepanov, and Sansone-Conti, mentioned in the References for Chapter 6. The subject of perturbed linear systems, including perturbations of nonautonomous linear systems, is classical material generally covered in

L. Cesari. *Asymptotic Behavior and Stability Problems in Ordinary Differential Equations*, Volume 16 in the series, *Ergebnisse der Mathematik und ihrer Grenzgebiete*. Berlin: Springer-Verlag, 1963.

R. Bellman. *Stability Theory of Differential Equations*. New York: McGraw-Hill, 1960.

P. Hartman. *Ordinary Differential Equations*, Chapter X. New York: Wiley, 1964.

Of historical importance is the paper by

I. Bendixson. Sur les courbes définies par des équations différentielles. *Acta Math.* **24** (1901): 1–88.

8

Autonomous Two-dimensional Systems — Global Study

1 Differential Equations on Manifolds

1.1 State space different from R^n

Up to this point we have always considered the state space X to be the n-dimensional euclidean space R^n. We mentioned earlier that for some problems an infinite dimensional state space must be considered. Even remaining within finite dimension, however, we should consider n-dimensional manifolds, different from R^n, as state space.

An *n-dimensional differentiable manifold* is defined as a space which is *locally diffeomorphic* [that is, differentiably homeomorphic] to euclidean n-space. By this we mean that every point of the manifold X has a neighborhood which is the topological image, by a transformation T, of a neighborhood of the origin in R^n and that both T and its inverse T^{-1} are differentiable.

Because every point has a neighborhood in which X looks like R^n, it is possible to define a vector field (differential equation) and its integral curves (trajectories) on every such neighborhood and therefore on the whole manifold X. Following these trajectories on adjoining neighborhoods, we obtain a global picture whose properties will be of interest to us.

Here we will consider only such well-known two-dimensional manifolds as the cylinder, torus, and sphere. [The reader may even disregard all generalities and consider only these particular surfaces.]

1.2 Equations on the cylinder

A cylinder can be obtained from the euclidean (x, y) plane, by agreeing that one variable [x, for example] should be considered *modulo* p [written "mod p"]; that is, different values $x, x \pm p, x \pm 2p, \ldots$ should be considered as the same point. [They should be "identified."] This usually happens in the case where x is an angle and therefore $p = 2\pi$.

We define a system of differential equations on this cylinder giving

$$\begin{cases} x' = X(x, y) \\ y' = Y(x, y), \end{cases} \tag{8.1.1}$$

which should satisfy the periodicity condition

$$\begin{cases} X(x + np, y) = X(x, y) \\ Y(x + np, y) = Y(x, y) \end{cases} \tag{8.1.2}$$

for $n = \pm 1, \pm 2, \ldots$.

Once the solutions of this differential system are obtained in the strip $0 \leqq x \leqq p$, we need only match the end points of the trajectories in order to obtain a global picture of them.

1.3 An example of a pendulum equation

The differential equation of the pendulum without friction is

$$x'' = -\frac{g}{l} \sin x,$$

where g is the acceleration of gravity, l the length of the pendulum, and x the angle it makes with the vertical. [See Figure 8.1.] We can write this as a system of the form

$$\begin{cases} x' = y \\ y' = -k^2 \sin x. \end{cases} \tag{8.1.3}$$

The linear approximation is the well-known harmonic oscillator

$$\begin{cases} x' = y \\ y' = -k^2 x. \end{cases}$$

Since x is an angle, it is natural to take x mod 2π and represent the system on the cylinder.

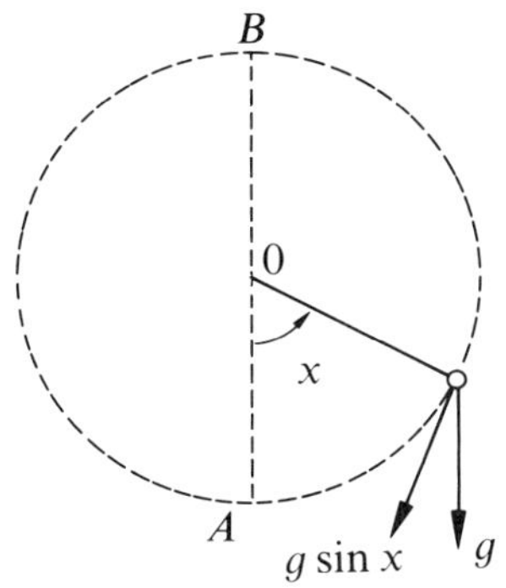

Figure 8.1

The critical points are determined by

$$y = 0, \qquad -k^2 \sin x = 0,$$

or

$$y = 0, \qquad x = 0, \pm\pi, \pm 2\pi, \pm 3\pi, \ldots,$$

which mod 2π are the two points

$$A\colon x = 0, \quad y = 0, \qquad \text{and} \qquad B\colon x = \pi, \quad y = 0.$$

Because there is no friction, it is said to be a *conservative system*; that is, there exists a function $E(x, y)$, having the physical meaning of total energy of the system, which remains constant along any trajectory.

In order to obtain the equations of the trajectories, which are the curves of constant energy, we multiply both members of the second equation of (8.1.3) by $y = x'$, thereby obtaining

$$yy' = -k^2 x' \sin x,$$

from which

$$d\left(\frac{y^2}{2}\right) = -k^2 \sin x\, dx$$

or

$$\frac{y^2 - y_0^2}{2} = -\int_{x_0}^{x} k^2 \sin x\, dx = k^2(\cos x - \cos x_0). \tag{8.1.4}$$

This is the equation of the trajectories. [Physically, the left side represents the gain in kinetic energy and the right side shows the lost potential energy. Both are taken from the initial instant.]

At the point

$$A: x = y = 0,$$

the linear approximation is the one shown above, and its solutions are harmonic oscillations for which the origin is a center. According to Chapter 7 [Theorem 7.4.5], the origin will be a center or a focus for the nonlinear system, but because the system is conservative, the focus is ruled out. Hence the origin must be a center.

At the point

$$B: x = \pi,\ y = 0,$$

we obtain the linear approximation by changing the variable $x - \pi = u$, thereby obtaining

$$\begin{cases} u' = y \\ y' = -k^2 \sin x = +k^2 \sin u \end{cases}$$

for the nonlinear system, and by then replacing $\sin u$ by u,

$$\begin{cases} u' = y \\ y' = k^2 u. \end{cases}$$

The characteristic roots of this linear approximation are

$$\lambda_1 = k,\ \lambda_2 = -k.$$

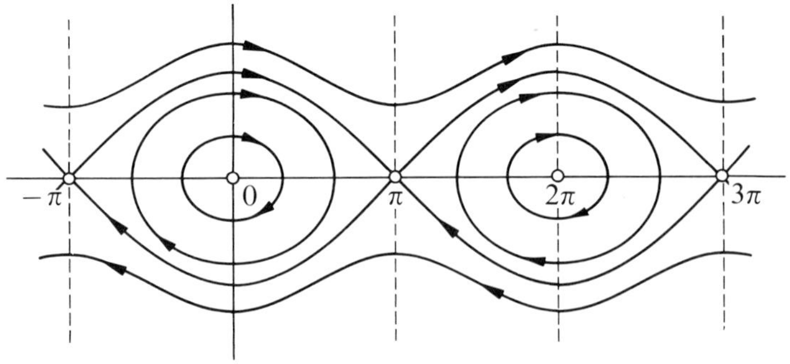

Figure 8.2

It follows that this point is a saddle point for the linear approximation, and therefore a saddle point for the nonlinear system.

In Figure 8.2 the trajectories of (8.1.3) are represented by taking the coordinates x, y in the plane. To obtain the representation on the cylindrical surface, one should wrap this figure around the cylinder in such a way that the point $(x, y) = (0, 0)$ coincides with $(2\pi, 0)$. Any further discussion of properties of the solutions can be easily made by using the equations of the trajectories (8.1.4).

1.4 Equations on the torus

The surface of the torus can be obtained from the (x, y) plane, if we consider both variables x and y to be periodic. We will consider the period of both variables to be equal to one; that is, we will identify the values as x_1 and x_2, providing $x_1 - x_2$ is an integer. Hence the coordinates on the torus are

$$x, y (\mathrm{mod}\ 1). \tag{8.1.5}$$

The differential system will still be (8.1.1), but in order to define uniquely a vector field on the torus, the functions X and Y must satisfy the periodicity condition

$$\begin{cases} X(x+n, y+m) = X(x, y) \\ Y(x+n, y+m) = Y(x, y) \end{cases} \quad \text{for every integer } n, m. \tag{8.1.6}$$

It is convenient to consider the system (8.1.1) also in the plane. In this case the plane will be divided into unit squares,

$$n \leqq x \leqq n+1, \qquad m \leqq y \leqq m+1, \qquad n, m \text{ integers},$$

on which, by periodicity, the trajectories are congruent. So, the basic problem consists of integrating (8.1.1) in the unit square.

This unit square

$$0 \leqq x \leqq 1, \qquad 0 \leqq y \leqq 1$$

is the representation of the torus. The points $(0, y)$ and $(1, y)$ should be identified, and so should $(x, 0)$ with $(x, 1)$. The four vertices $(0, 0)$, $(1, 0)$, $(0, 1)$, and $(1, 1)$ are the same point on the torus.

It should be noted that a trajectory looks quite different when it is considered on the plane (x, y), or when this plane is reduced by periodicity to the unit square representing the torus. Thus, in Figure 8.3(a), (b), the same trajectory is represented —once on the plane and once on the torus. The reader should imagine this same trajectory on the three-dimensional torus: it is a curve which winds around the torus in both directions.

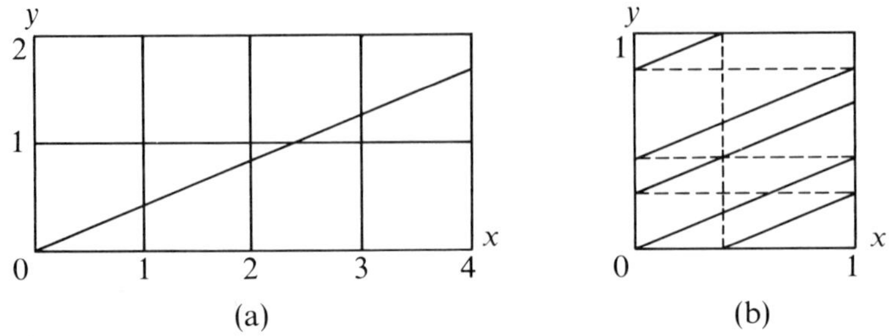

Figure 8.3

One good reason for giving particular attention to the study of the trajectories on the torus, is that this surface [or its diffeomorphically equivalent ones] is the only one on which it is possible to have a vector field without any critical points, as we will prove later. Another reason is that the behavior of the trajectories may be very different from the behavior we have seen in the plane.

1.5 Example

Consider

$$\frac{dy}{dx} = k = \text{const}, \tag{8.1.7}$$

where the variable t has already been eliminated.

On the (x, y) plane, the solutions of this differential equation, are all straight lines of slope k. On the torus they are helices, and therefore two essentially different cases are possible.

Case 1: $k = \text{rational} = q/p$ (p, q integers). The solutions can be written

$$\frac{y - y_0}{q} = \frac{x - x_0}{p}.$$

For $x - x_0 = p$, we have $y - y_0 = q$, and since p, q are integers, the point (x, y) coincides on the torus with (x_0, y_0). Hence the trajectory is closed, and the solution is periodic. All other trajectories are obtained from the one given by translation and are also periodic with the same period.

Case 2: $k = \text{irrational}$. No solution can be periodic because we would have the same situation as before. Indeed, from $x - x_0 = p$, $y - y_0 = q$, we obtain

$$k = \frac{y - y_0}{x - x_0} = \frac{q}{p} = \text{rational}.$$

We will prove that each trajectory has, in this case, the remarkable property that it passes arbitrarily close to every point of the torus. More precisely, given any point on the torus and any neighborhood of it, the trajectory intersects this neighborhood. To prove this we need the following lemma.

1.5a Lemma

If k is irrational, then the set C of numbers of the form

$$a_{p,q} = p + kq, \quad p, q \text{ integers},$$

is a dense subset of the real line R; that is, in every interval, no matter how small, there is some number of the set C.

We notice first that the correspondence between the pairs (p, q), with p, q integers, and the points of the set C is one-to-one. Indeed, if we assume that two pairs $(p, q), (p', q')$ correspond to the same number,

$$p + kq = p' + kq',$$

we immediately get

$$k = \frac{p - p'}{q' - q} = \text{rational}.$$

Another important property is that by addition and subtraction of numbers belonging to C, we always obtain numbers of C:

$$a_{p,q} \pm a_{p',q'} = a_{p \pm p', q \pm q'};$$

hence C is an additive group.

Proof of the lemma

For arbitrarily large p we may write

$$k = \frac{q}{p} + r,$$

with p, q integers and $0 < r < 1/p$. Thus

$$kp - q = pr < 1,$$

and this proves that in the interval (0, 1) there are infinitely many numbers of C. This implies, however, that there are pairs of numbers of C as close to each other as we want; that is, given any $\varepsilon > 0$, there are p, q, p', q' such that

$$|a_{p,q} - a_{p',q'}| < \varepsilon.$$

Then

$$|a_{p-p',q-q'}| < \varepsilon,$$

and all its multiples belong to C, which proves the lemma.

1.5b Theorem

Given a trajectory of (8.1.7) *on the torus, with k irrational, and an arbitrary point* (x_1, y_1), *there are points belonging to the trajectory as near to* (x_1, y_1) *as we please.*

Proof

By translating the coordinate system we may evidently assume that the trajectory passes through the origin $(x_0, y_0) = (0, 0)$. Its equation is then

$$y = kx$$

in the plane. We consider it on the torus by identifying (x, y) with all points $(x + p, y + q)$, where p, q are integers.

We will prove that for every $\varepsilon > 0$, there are integers p, q such that

$$x_1 = x + p$$
$$|y_1 - y + q| < \varepsilon$$
$$y = kx.$$

In other words, (x, y) belongs to the trajectory; x coincides with x_1 [on the torus]; and y almost coincides [with error ε] with y_1.

The stated conditions are indeed equivalent to

$$|y_1 - k(x_1 - p) + q| = |y_1 - kx_1 + kp + q| < \varepsilon,$$

and we can always accomplish this with a proper choice of p, q. [For example, in the preceding lemma the number $y_1 - kx_1$ can be approximated as closely as we wish with numbers of the form $kp + q$.]

1.6 Rotation number

Consider again the general differential equation (8.1.1) with the periodicity condition (8.1.6). We shall assume that *the functions* $X(x, y)$ *and* $Y(x, y)$ *are lipschitzian* and, furthermore, that $X(x, y)$ *is never zero*. It follows that $X(x, y)$ has constant sign which we may assume is positive. Hence $X(x, y)$ and $Y(x, y)$ have some finite maxima and minima, and the minimum of $X(x, y)$ is positive. [Remember that, by periodicity, it is sufficient to consider the values of both functions on the unit square.]

Under this hypothesis, it is enough to consider the differential equation

$$\frac{dy}{dx} = \frac{Y(x, y)}{X(x, y)} = f(x, y), \tag{8.1.8}$$

where $f(x, y)$ is lipschitz-continuous and periodic of period one in x and y.

In the following we will compare the solutions of (8.1.8) in the (x, y) plane and on the torus. In order to distinguish the coordinates on these two manifolds, we will consistently use the notation

$$x = [x] + (x) = \text{integer part of } x + \text{fractional part of } x.$$

Integer part of x = greatest integer $\leq x$.
Hence, if x and y are the coordinates on the plane, then (x) and (y) are the corresponding coordinates on the torus.

We designate the solution of (8.1.8) [in the plane], starting at the point $x_0 = 0$, $y(0) = y_0$, by

$$y(x) = u(x, y_0) = u(x, x_0, y_0). \tag{8.1.9}$$

Since $f(x, y)$ is bounded, the solutions of (8.1.8) can be extended to the whole interval $-\infty < x < +\infty$.

By periodicity, if $L(y_0)$ represents the trajectory passing through the point $x = 0$, $y = y_0$, then after translating $L(y_0)$ along the x- and y-axes by integers p and q, we will obtain another trajectory. On the other hand, all trajectories of (8.1.8) meet the line $x = 0$. Hence all trajectories are represented by $L(y_0)$ with appropriate choices of y_0.

Consider the mapping

$$y \to u(p, y) \quad \text{for fixed integer } p.$$

This is a mapping of the line $x = 0$ onto the line $x = p$. By uniqueness [the fact that solutions do not intersect], this mapping is one-to-one, continuous, and order-preserving. [$y_1 < y_2$ implies $u(p, y_1) < u(p, y_2)$.]

The lines $x = 0$ and $x = p$ (p integer) represent the same circle on the torus; the mapping

$$(0, y) \to (0, (u(p, y))), \qquad 0 \leqq y < 1,$$

is a mapping of this circle onto itself, which is one-to-one, continuous, and order-preserving. Hence

$$u(p, 0) < u(p, y) < u(p, 1) = u(p, 0) + 1 \quad \text{for} \quad 0 < y < 1.$$

Let us define the operator u by

$$u(y) = (u(1, y)), \qquad 0 \leqq y < 1; \tag{8.1.10}$$

then its powers are

$$u^n(y) = (u(n, y)). \tag{8.1.11}$$

These powers are defined for integer n by products and inverses of operators, and the ordinary rules of operation are applicable. Thus

$$u^n \cdot u^m(y) = u^{n+m}(y) = (u(n+m, y)),$$
$$u^{-n}(y) = (u(-n, y)),$$
$$u^0(y) = y.$$

$L(y_0)$ is a closed trajectory on the torus if and only if for some integers n, m,

$$u(n, y_0) = y_0 + m. \tag{8.1.12}$$

Hence

$$u^n(y_0) = y_0, \tag{8.1.13}$$

and the transformation u^n has a *fixed point*. We will see that in that case the number m/n has a very important meaning and is called the *rotation number*. Even if there is no closed trajectory, we can still define a rotation number according to the following theorem.

1.7 *Theorem*

Under the assumptions of continuity and periodicity stated, the solutions of (8.1.8) *have the property that for each of them, the quotient*

$$\frac{\Delta y}{\Delta x} = \frac{y(x) - y(0)}{x}$$

has the same well-defined limit, as $x \to \pm\infty$. *This number will be denoted by* k *and will be called the rotation number* of* (8.1.8).

Proof

Let us call $L(n, m)$ (n, m integers) the trajectory solution of (8.1.8) through the point $x = n$, $y = m$ [in the plane]. For example, $L(0, 0)$ is the trajectory through the origin.

We will say that the point with integer coordinates $x = p$, $y = q$ is *above* $L(n, m)$ if, with the notation (8.1.9),

$$u(p, n, m) < q.$$

Similarly, we say that (p, q) is *below* $L(n, m)$ if $u(p, n, m) > q$.

In the same sense we say that a trajectory is *above* or *below* another trajectory. Since trajectories cannot intersect, this relationship is well defined.

Another fact to be noted is that any two trajectories $L(n, m)$ and $L(p, q)$ (n, m, p, q integers) are congruent by a translation of components $p - n$ and $q - m$.

We may classify all points having integer coordinates (p, q) with $p > 0$ into two classes:

Class (a): those points (p, q) which belong to or are above the trajectory $L(0, 0)$.

Class (b): the remaining points (p, q) which are then below $L(0, 0)$.

If we now assign to each point (p, q) the fraction q/p, then we shall prove that the Classes (a) and (b) both form a Dedekind cut defining a real number.

If (p, q) belongs to Class (a), then the same is true for the points $(2p, 2q)$, $(3p, 3q)$, ... This is obvious if (p, q) belongs to $L(0, 0)$; if not, then (p, q) is above $L(0, 0)$ and hence all of the trajectory $L(p, q)$ is above

---

* It was introduced by Poincaré.

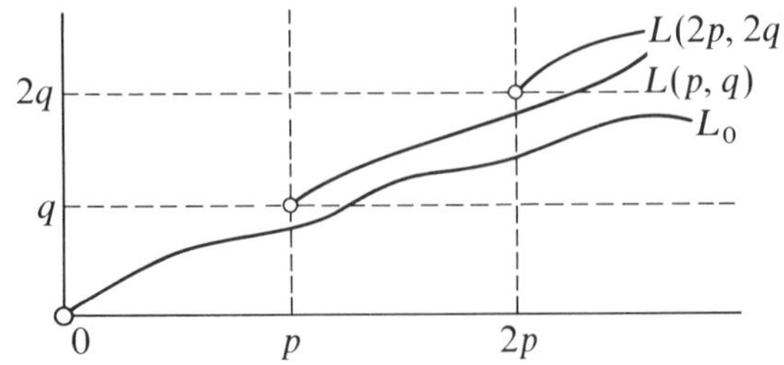

Figure 8.4

$L(0, 0)$. [See figure 8.4.] Since $L(p, q)$ is congruent to $L(0, 0)$, it follows that $L(2p, 2q)$ will be above $L(p, q)$ and hence above $L(0, 0)$.

Now let (p, q) belong to Class (a). We will show that if $q'/p' > q/p$, then (p', q') also belongs to Class (a). Indeed, assuming the contrary [(p', q') belongs to Class (b)], then (pp', qp') must belong to the same class as (p, q), and $(p'p, q'p)$ to the same class as (p', q'). Hence, if (p', q') belongs to Class (b), then $q'p < qp'$ and therefore $q'/p' < q/p$—a contradiction. This proves that Class (a) and (b) form a Dedekind cut.

This Dedekind cut defines the real number k. What we still have to prove is that this same number can be obtained as the limit of $\Delta y/\Delta x$ as $\Delta x \to \infty$ on any trajectory.

It is no restriction to assume that, as $\Delta x \to \infty$, $\Delta x = x - x_0 = x$; in other words, assume $x_0 = 0$. Indeed, if this limit exists,

$$\lim_{x\to\infty} \frac{y(x) - y(0)}{x} = \lim_{x\to\infty} \frac{y(x) - y(x_0) + y(x_0) - y(0)}{(x - x_0) + x_0}$$

$$= \lim_{x\to\infty} \frac{y(x) - y(x_0)}{x - x_0}, \tag{8.1.14}$$

and it is independent of x_0.

Let $L(y_0)$ be the trajectory through the point $(0, y_0)$ and $u(x, y_0)$ its y-coordinate for the corresponding x-value. We will consider only trajectories with $0 \leqq y_0 < 1$, since all other trajectories are congruent to these by integer translations.

From $0 \leqq y_0 = u(0, y_0) < 1$ it follows that

$$u(p, 0) \leqq u(p, y_0) < u(p, 0) + 1,$$

Hence

$$u(p, 0) < u(p, y_0) - y_0 + 1,$$

and

$$u(p, y_0) - y_0 < u(p, 0) + 1.$$

Finally,

$$u(p, 0) - 1 < u(p, y_0) - y_0 < u(p, 0) + 1. \tag{8.1.15}$$

Dividing by $p = \Delta x$ and letting $p \to +\infty$, both extreme members tend to the same limit

$$\lim_{p \to +\infty} \frac{u(p, 0)}{p}.$$

By the way the Dedekind cut determining k was defined, it is easily seen that this limit is precisely k. Formula (8.1.15) proves, therefore, that along *any trajectory*

$$\frac{u(x, y_0) - y_0}{x} \to k \quad \text{as} \quad x \to +\infty.$$

That the restriction $x = p =$ integer is not essential is proven in a way similar to (8.1.14).

Finally, if $x \to -\infty$, it is possible to make exactly the same argument, thereby obtaining a limit k', which obviously cannot be different from k. One can also consider the translated trajectory

$$\lim_{p \to +\infty} \frac{u(-p, y_0) - y_0}{-p} = \lim_{p \to +\infty} \frac{\{y_0 - m\} - \{u(-p, y_0) - m\}}{p},$$

where the integer $-m$ is the integer part of $u(-p, y_0)$ so that the second member is the quotient of increments corresponding to the trajectory starting at $y(0) = u(-p, y_0) - m$. [We leave the details of the proof to the reader.]

Corollary

If there exists a periodic solution [*closed trajectory*], *then the rotation number*, $k = m/n$, *is rational. In this case, if we take the origin on the closed trajectory, then* (n, m) *is the first point with integer coordinates on it.*

1.8 Theorem

Under the conditions of Theorem 8.1.7, if the rotation number k is rational, there exists at least one periodic solution.

Proof

Let

$$k = \frac{m}{n}, \qquad m \text{ and } n \text{ integers,}$$

and consider the trajectory L_0 passing through the origin. If L_0 also passes through the point (n, m), we have already found a periodic solution. Otherwise the point (n, m) is above or below L_0. Assume (n, m) is *above* L_0, the other case being similar. Then the whole trajectory $L(n, m)$ is above L_0.

Between the trajectories L_0 and $L(n, m)$ there cannot be any other trajectory $L(p, q)$ passing through a point (p, q) with integer coordinates. To show this, we first rule out that (p, q) is of the form (rn, rm) (r integer). Indeed, the relative position of $L(rn, rm)$ with respect to $L(n, m)$ is the same as that of $L(n, m)$ with respect to L_0, so $L(rn, rm)$ is above $L(n, m)$ and not between L_0 and $L(n, m)$. The case that (p, q) is of the form (rn, rm) with $r < 0$ implies that the relative position of L_0 with respect to $L(rn, rm)$ is the same as $L(n, m)$ with respect to L_0. So $L(rn, rm)$ is below L_0 and again not between L_0 and $L(n, m)$. [Compare Figure 8.4.]

So, if $L(p, q)$ is between L_0 and $L(n, m)$, it must be that $q/p \neq m/n$. Since $L(p, q)$ is above L_0, it follows from the definition of rotation number that the point (p, q) is above the line $y/x = m/n = k$, and hence $q/p > m/n$.

Consider two cases.

Case 1: If $n < p$, then $q/p > m/n$ implies $(q - m)/(p - n) > k$. If we imagine the origin translated to the point (n, m) and the rotation number k defined by the new trajectory through the new origin, which is the old $L(n, m)$, then this last inequality implies that the point of new coordinates $(p - n, q - m)$ is above the new trajectory through the new origin; in the old coordinates this means that the point (p, q) is above $L(n, m)$. [See Figure 8.5.]

Case 2: If $n > p$, then $(m - q)/(n - p) < k$ and the trajectory through (n, m) must be below $L(p, q)$. In Figure 8.5 the coordinates of the point in these conditions are called p', q'.

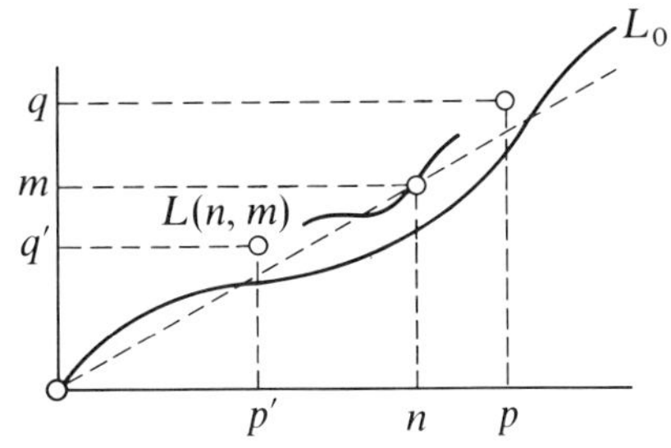

Figure 8.5

Hence in both cases $L(p, q)$ is above $L(n, m)$ and cannot be between $L(n, m)$ and L_0, proving that between these two last trajectories there cannot by any point of integer coordinates (p, q).

Now we will consider the intersection y_1 of $L(n, m)$ with the y-axis:

$$\begin{aligned} y_1 &= u(0, n, m) = (u(-n, 0, 0)) \\ &= \text{fractional part of } u(-n, 0, 0). \end{aligned}$$

With the notation of (8.1.11), we may write

$$y_1 = u^{-n}(0).$$

Define the sequence

$$y_0 = 0, \qquad y_1 = u^{-n}(0), \qquad y_2 = u^{-2n}(0), \ldots, \qquad y_r = u^{-rn}(0), \ldots.$$

Our previous result states that in the (x, y) plane, between L_0 and $L(n, m) = L(0, y_1)$, there are no trajectories passing through a point with integer coordinates. Hence no such trajectory intersects the open interval $(0, y_1)$ on the y-axis. On the torus, therefore, L_0 never meets one of the two arcs of the circle corresponding to the y-axis, which is determined by the points $y = 0$, $y = y_1$. Let us call this arc $\theta = (0, y_1)$.

The transformation u^{-n} transforms the arc $\theta = (0, y_1)$ into

$$u(\theta) = \theta_1 = (y_1, y_2),$$

which is disjoint from θ. Similarly, the end point y_2 cannot coincide with an end point, 0 or y_1, of θ because this would result in a closed trajectory [if this were the case, our proof would be already complete].

For the same reasons,

$$u^{-n}(\theta_1) = \theta_2 = (y_2, y_3)$$

is disjoint from $(0, y_2)$. Furthermore, the transformation u^{-n} preserves the order on the circle; thus the sequence $y_1, y_2, y_3, \ldots$ is monotone and cannot make a complete turn around this circle. This means it is bounded above by $y = 1$ [or bounded below by 0]; it therefore has a limit point $z = \lim y_r$, as $r \to \infty$. It follows then from the continuity of the transformation u^{-n} that

$$u^{-n}(z) = u^{-n}(\lim y_r) = \lim u^{-n}(y_r) = \lim y_{r+1} = z.$$

Hence z is an invariant point of the transformation u^{-n}, and the trajectory $L(0, z)$ is closed. This proves the theorem.

Problems

1. Indicate the behavior of the trajectories of the following differential systems on the cylinder (y, θ), $0 \leqq \theta \leqq 2\pi$.

   (a) $y' = dy/d\theta = y$. (b) $y' = y \sin \theta$.
   (c) $y' = y \cos^2 \theta$. (d) $y' = y^2 \sin \theta$.
   (e) $y' = y^2$.

2. Comparing with the frictionless case treated in the text, study the pendulum with friction

$$x'' + kx' + \sin x = 0.$$

   Hint: take the trajectories in the frictionless case as curves of equal total energy, and take into account that in the case with friction, this energy is always decreasing during the motion.

3. Indicate the behavior of the trajectories of the following differential systems on the torus (x, y), $0 \leqq x \leqq 1$, $0 \leqq y \leqq 1$. Indicate the critical points.

   (a) $y' = dy/dx = 4\pi \cos 2\pi x$. (b) $y' = \dfrac{\cos 2\pi x}{\cos 2\pi y}$.

   (c) $y' = \dfrac{\sin 2\pi x}{\sin 2\pi y}$. (d) $y' = \dfrac{\cos 2\pi x}{\sin 2\pi y}$.

   (e) $y' = 2\pi \cot 2\pi x$. (f) $y' = \dfrac{\cos 2\pi nx}{\cos 2\pi my}$ n, m integers.

   (g) $y' = \dfrac{1 + \cos 2\pi x}{1 + \cos 2\pi y}$.

2 *The Index*

2.1 *Index of a curve*

Many problems can be solved elegantly by using the concept of index and the corresponding theory developed by Henri Poincaré. This notion of index can be generalized to higher dimensions, but in what follows we will consider only the euclidean plane, designating its points in a generic way by $\mathbf{x}$. A *Jordan curve* is a homeomorphic [topological] image of the circle

$$\mathbf{x} = \mathbf{x}(t), \qquad 0 \leqq t \leqq 1, \qquad \mathbf{x}(0) = \mathbf{x}(1). \tag{8.2.1}$$

This curve will be called Γ. Note that since a homeomorphism is a one-to-one correspondence, Γ does not intersect itself.

We will also consider a *vector field* $\mathbf{f}(\mathbf{x})$ defined in some region of the plane which includes all the points of Γ. We assume $\mathbf{f}(\mathbf{x})$ to be *continuous* and to be *different from zero at all points of* Γ. Then, on Γ,

$$\mathbf{f}(\mathbf{x}) = \mathbf{f}(\mathbf{x}(t))$$

is a continuous function of the variable t, and this vector function is never zero. When t varies from 0 to 1, $\mathbf{x}(t)$ describes the curve Γ. At the same time, $\mathbf{f}(\mathbf{x}(t))$ varies so that

$$\mathbf{f}(\mathbf{x}(1)) = \mathbf{f}(\mathbf{x}(0)).$$

Consider any fixed directed line [for example, the axis of abscissas] and the angle from this line to the vector $\mathbf{f}(\mathbf{x}(t))$. Denote this angle by $\varphi(t)$; from the fact that $\mathbf{f}(\mathbf{x}(1)) = \mathbf{f}(\mathbf{x}(0))$, it follows that

$$\varphi(1) = \varphi(0) + 2\pi k,$$

where k is an integer.

Definition

Under these conditions, the number k [that is, the number of revolutions which the vector $\mathbf{f}(\mathbf{x}(t))$ makes as $\mathbf{x}(t)$ describes Γ] is called the index of the curve Γ with respect to the vector field $\mathbf{f}(\mathbf{x})$ and denoted by

$$J = J(\Gamma, \mathbf{f}) = \frac{\Delta\varphi}{2\pi}. \tag{8.2.2}$$

Note that the index J is necessarily an integer; its sign depends on the orientation of Γ [the sense of increasing t] and on the sense chosen as positive for measuring angles.

2.2 Theorem

If the Jordan curves Γ_0 and Γ_1 can be continuously deformed one into the other without passing any critical point [*that is, a point where* $\mathbf{f}(\mathbf{x}) = 0$], *then both curves have the same index*

$$J(\Gamma_0, \mathbf{f}) = J(\Gamma_1, \mathbf{f}).$$

In order to give a precise meaning to this statement, we must define such a *continuous deformation*: there exists a function $\mathbf{g}(t, s)$, defined and continuous for $t \in [0, 1]$, $s \in [0, 1]$, such that for every fixed s, $\mathbf{g}(t, s)$ is a Jordan curve and for $s = 0$, $\mathbf{g}(t, 0)$ represents Γ_0, while for $s = 1$, $\mathbf{g}(t, 1)$ represents Γ_1. One can imagine that for fixed $s \in [0, 1]$, $\mathbf{g}(t, s)$ represents a Jordan curve which changes continuously, as s changes from $s = 0$ to $s = 1$. In the language of modern mathematics, one says that Γ_0 and Γ_1 are *homotopic*, and $g(t, s)$ is a *homotopy* from Γ_0 to Γ_1.

Proof

While $\Gamma = \Gamma_s$ changes continuously with $s \in [0, 1]$, the index will also be a continuous function of s. But this index is an integer; hence it must be constant.

In this proof it is essential that the variable curve Γ_s never meets a critical point, where $\mathbf{f}(\mathbf{x}) = 0$, because the index for Γ_s would not be defined.

2.3 Theorem

If Γ is a Jordan curve and $\mathbf{f}(\mathbf{x})$ and $\mathbf{f}_1(\mathbf{x})$ are two vector fields defined and never zero on Γ, and if on Γ the angle between $\mathbf{f}(\mathbf{x})$ and $\mathbf{f}_1(\mathbf{x})$ is constant, then

$$J(\Gamma, \mathbf{f}) = J(\Gamma, \mathbf{f}_1).$$

The proof is immediate from the definition.

2.4 Theorem

If Γ is a Jordan curve and $\mathbf{f}(\mathbf{x})$, $\mathbf{f}_1(\mathbf{x})$ are two vector fields, defined and never zero on Γ, and if $\mathbf{f}(\mathbf{x})$ and $\mathbf{f}_1(\mathbf{x})$ on Γ, are never opposed [their angle is never $180°$], then

$$J(\Gamma, \mathbf{f}) = J(\Gamma, \mathbf{f}_1).$$

Proof

Denote by φ and φ_1 the angles of $\mathbf{f}(\mathbf{x})$ and $\mathbf{f}_1(\mathbf{x})$ with some fixed direction and also by $\theta = \varphi - \varphi_1$. Then, defining φ and φ_1 properly, we have everywhere on Γ

$$-\pi < \theta < +\pi.$$

From $\theta(t = 0) \equiv \theta(t = 1) \bmod 2\pi$ and the continuity of $\theta(t)$ the desired result follows.

2.5 Analytic expression of the index

Assume that the vector field is given by

$$x' = X(x, y), \qquad y' = Y(x, y),$$

where these functions are differentiable. Assume also that the curve Γ is differentiable and rectifiable so that it is possible to integrate along Γ. Then $\Delta/2\pi$ is given by

$$J(\Gamma) = \frac{1}{2\pi}\int_\Gamma d\left(\text{arc tan}\,\frac{Y}{X}\right) = \frac{1}{2\pi}\int_\Gamma \frac{X\,dY - Y\,dX}{X^2 + Y^2}. \tag{8.2.3}$$

2.6 Poincaré's formula for the index

Assume Γ to be differentiable and denote the tangent vector to Γ by T. It is then easy to see that considering T as a vector field on Γ, the index of Γ will be $+1$, if the sense in which Γ is described for increasing t values is the sense in which angles are considered positive [normally counterclockwise; see Figure 8.6].

For a general vector field, we say that the point A [belonging to the curve Γ] is a point of *exterior contact* of Γ, if the trajectory of the vector field through A is tangent and exterior to Γ [as is the case for the points A and A' in Figure 8.7. Similarly, the points B and B' in the same figure are called points of *interior contact*.

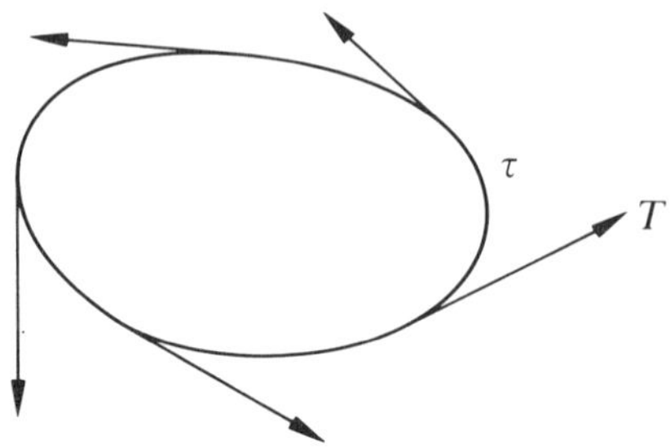

Figure 8.6

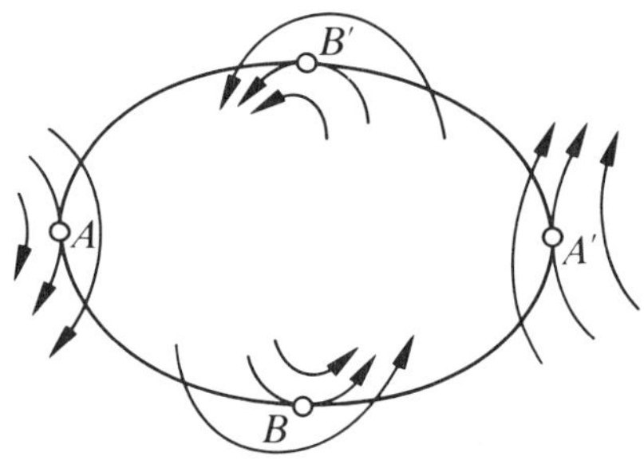

Figure 8.7

In case the Jordan curve Γ has only a finite number of points of contact with the vector field, we denote by E the number of points of exterior contact of Γ and by I the number of points of interior contact of Γ. Then the following formula of Poincaré holds true:

$$J = \frac{I - E + 2}{2}. \tag{8.2.4}$$

To establish this formula, we compare the vector $\mathbf{f}(\mathbf{x})$ with the tangent vector $\mathbf{T}$. If on Γ the directions of both vectors never coincide, then according to Theorem 8.2.4, the index for both vector fields will be the same, $+1$.

Each time the point $\mathbf{x}(t)$ on Γ passes a point of contact, the angle φ from $\mathbf{T}$ to $\mathbf{f}(\mathbf{x})$ passes through the value 0 or π (mod 2π). This happens with increasing values of φ, if the point is of interior contact [such as B, B'], and with decreasing values of φ, if the point is of exterior contact [such as A, A']. Hence, for every two exterior contacts, the index decreases by one, and for every two interior contacts, it increases by one, thus giving formula (8.2.4).

2.7 *Index of a point*

Let $\mathbf{x}_0$ be a regular or isolated critical point so that in some neighborhood U of $\mathbf{x}_0$ there are no critical points of the vector field, with the possible exception of $\mathbf{x}_0$ itself. Let Γ be a Jordan curve described in a positive [counterclockwise] sense, surrounding $\mathbf{x}_0$ such that no other critical point is inside or on Γ. Such a curve does exist, since we need only to take it in U to satisfy this condition.

Then *the index of* Γ *with respect to the vector field under consideration does not depend on the particular curve* Γ *chosen*. This means that for any two curves Γ_0 and Γ_1 satisfying the requirements, the index is the same. Indeed, we can change Γ_1 continuously into Γ_0 without meeting any critical point.

Definition

The index of such a curve Γ, *which does not contain in its interior any other critical point than possibly* $\mathbf{x}_0$ *itself, is called the index of* $\mathbf{x}_0$ *with respect to the vector field* $\mathbf{f}(\mathbf{x})$.

2.8 *Index of a regular point*

If $\mathbf{x}_0$ is a regular point, then its index is $J(\mathbf{x}_0) = 0$.

Indeed, the curve Γ can be taken sufficiently near $\mathbf{x}_0$ so that the angle between the variable vector $\mathbf{f}(\mathbf{x})$ on Γ and the fixed vector $\mathbf{f}(\mathbf{x}_0)$ is less than ε. Hence $\mathbf{f}(\mathbf{x})$ cannot make a whole revolution on Γ and $J(\Gamma) = 0$.

2.9 *Index of a curve containing a finite number of critical points in its interior*

Assume that in the interior of Γ there are three critical points $\mathbf{x}_1$, $\mathbf{x}_2$ and $\mathbf{x}_3$, as shown in Figure 8.8. Enclosing these critical points in small circles Γ_1, Γ_2 and Γ_3 and making the indicated cuts, we obtain a new Jordan curve Γ^* containing no

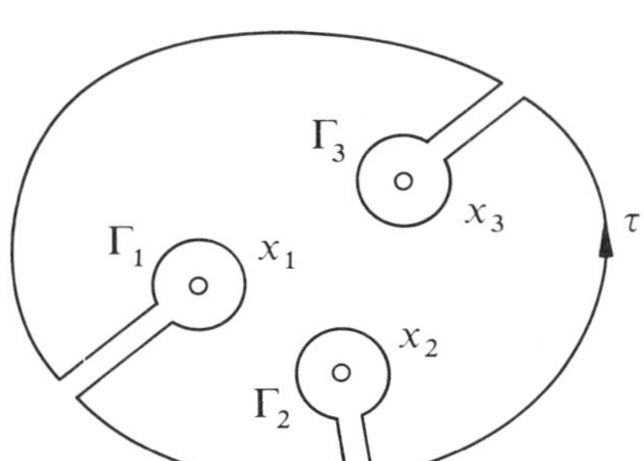

Figure 8.8

critical point in its interior. From Sections 8.2.7–8.2.8 it follows that the index of Γ^* is zero.

The curve Γ^* is made up by the curve Γ, described in the positive sense, plus the curves Γ_1, Γ_2 and Γ_3, described in the negative sense, plus the cuts, each of which is described twice in opposite directions. Thinking of the index as the line integral (8.2.3) therefore, we may write

$$J(\Gamma^*) = J(\Gamma) - J(\Gamma_1) - J(\Gamma_2) - J(\Gamma_3) = 0,$$

or

$$J(\Gamma) = J(\Gamma_1) + J(\Gamma_2) + J(\Gamma_3).$$

The following generalization is immediate:

The index of a curve containing in its interior a finite number of critical points is equal to the sum of the indices of these points.

And we can add the corollary:

If the index of a curve is different from zero, then in its interior there is necessarily some critical point.

2.10 Index of the elementary critical points

By elementary critical points, we mean those arising in linear differential systems. They are the node, center, focus and saddle point, as represented in Figure 8.9. The index can be obtained directly or with the aid of Poincaré's formula, considering a Jordan curve enclosing the origin. The values obtained are as follows:

Index of the node $= +1$

Index of the center $= +1$

Index of the focus $= +1$

Index of the saddle point $= -1$.

Note that the index does not change if the sense of the vector field is reversed. For example, the stable and the unstable node both have index $+1$.

Another nice result, which is easy to prove, is the following theorem.

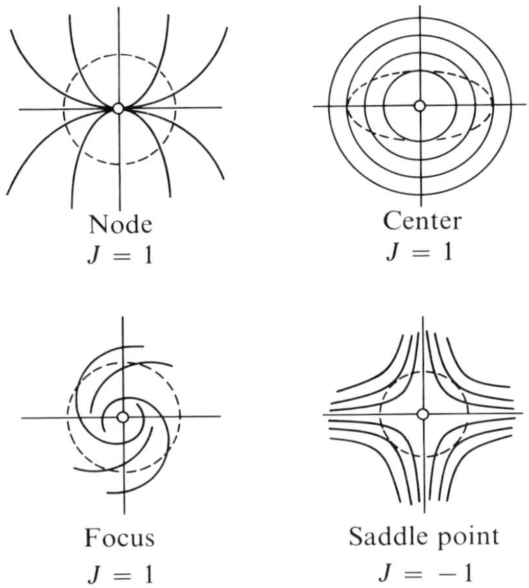

Figure 8.9

2.11 Theorem

In the interior of every closed trajectory there is some critical point.

Indeed, the vector field is always tangent to this curve. Hence its index is $+1$, and the corollary in Section 2.9 can be applied.

Note: Not all the critical points interior to a closed trajectory can be saddle points. Their sum must be $+1$, and each saddle point must have index -1.

2.12 Index of a surface

By *surface* we mean a two-dimensional, differentiable manifold, that is, a topological space which is locally diffeomorphic [differentiably homeomorphic] to a disk in the euclidean plane. In other words, every point has a neighborhood which can be described by coordinates x, y such that the coordinate systems in overlapping neighborhoods are related in a differentiable way.

We also assume that this manifold V is compact and can be *triangulated* [decomposed into a finite number of pieces called *cells*, each diffeomorphic to a plane triangle]. Each of these cells is then equivalent, as in the definition of index, to a plane triangle.

We can define a vector field on this manifold V through each of the neighborhoods, aided by a local coordinate system. Let us assume that this vector field $\mathbf{f}(\mathbf{x})$ has on V only a finite number of critical points, where $\mathbf{f}(\mathbf{x}) = 0$.

We can therefore define the index of the surface V in the following way:

(a) *Decompose V into a finite number of triangular cells, such that the edges (dividing lines) of these do not pass through any critical point. Let $\{V_i\}$ be these cells.*

(b) *Compute the index of the Jordan curve Γ_i, as boundary of the cell V_i, for every i.*

(c) *The sum of the indices of these Γ_i is defined to be the index of V with respect to the vector field* $\mathbf{f}(\mathbf{x})$.

But, for this definition to make sense, it must be independent of the particular triangulation chosen. Hence we must give the following proof.

If any triangulation is replaced by a *finer* one [obtained by adding more dividing lines to the dividing edges], then any cell V_i of the first triangulation will be subdivided into cells $V_{i1}, V_{i2}, \ldots, V_{ip_i}$. Since these behave like plane triangular subdivisions, the index of V_i will be equal to the sum of the indices of the V_{ip}'s. Hence the total sum remains the same. Now, if we want to compare two different triangulations [neither one being a refinement of the other], we can consider a third triangulation which is a common refinement of them both. For this, it is sufficient to take all the edges of both triangulations and add suitable dividing lines to form all triangular cells. [The cells obtained by taking all edges of both triangulations may be polygons of different numbers of sides.] As seen, this common refinement has the same index as each of the initial triangulations; hence these have the same index.

This definition is applicable in a natural way to all *orientable* surfaces (surfaces with *two* sides), taking a coherent orientation as positive sense on all cells V_i. But this is not essential. Even if the surface is *not orientable*, it is sufficient to take *in each cell* the same sense as positive for the angles and for describing Γ_i. This determines uniquely the index of each cell, and by summation the index of the whole surface.

2.13 Poincaré's Theorem

The index of a surface V with respect to a differentiable vector field $\mathbf{f}(\mathbf{x})$, *defined on V, which has on V only a finite number of critical points, is independent of the vector field and equal to the Euler-Poincaré characteristic of that surface.*

We will prove this theorem only for the particular case of V being the sphere.

Proof

In order to calculate the index of the sphere, we decompose it into two cells by a small circle. This circle cuts the sphere into two pieces [a big one and a small one], each equivalent to a circular disk. We can take the circle in such a way that it does not meet any critical point and that there is also no critical point in the small disk. Then, if the circle is small enough so that the vector field is almost constant on the small disk, we obtain a figure like 8.10.

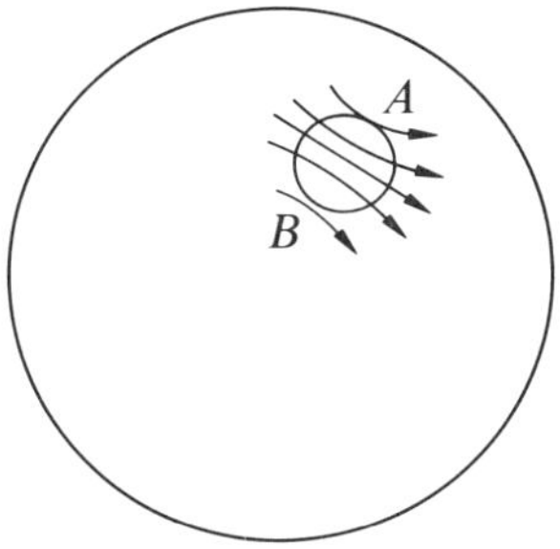

Figure 8.10

We now evaluate the index of the small and the large disk. Under the previously mentioned conditions, the vector field on the small disk is almost constant; thus the circle has only two exterior contacts. By Poincaré's formula, the index is zero. [We could have asserted this initially, since this disk does not contain any critical point.]

The vector field on the large disk may be extremely complicated, but we can again evaluate its index by using the Poincaré formula. Only now, for the large disk, the contacts of the vector field with the circle are interior contacts (see Figure 8.10, points A and B). Hence

$$J = \frac{I - E + 2}{2} = \frac{2 + 2}{2} = 2.$$

Adding the indices of the small and the large disk, the index of the sphere comes out to be $0 + 2 = 2$.

The Euler-Poincaré characteristic of a surface can, in general, be computed by triangulating the surface and counting:

$$\begin{aligned}\text{Euler-Poincaré characteristic} &= \text{number of vertices}\\ &\quad - \text{number of edges}\\ &\quad + \text{number of faces (cells)}\end{aligned}$$

For the sphere, such a triangulation can be a tetrahedron, with 4 vertices, 6 edges and 4 faces, so that again $J(\text{sphere}) = 2$ results.

An immediate result of this is that there cannot exist a vector field on the sphere without critical points. Particular examples of vector fields on the sphere are:

(1) A vector field such that its trajectories are meridians (geographical longitude = constant). Both poles are critical points of the node type.
(2) A vector field such that its trajectories are parallel circles (geographical latitude = constant). The poles are critical points of the center type.
(3) A vector field such that the trajectories are intersections of the sphere with the planes containing a fixed line tangent to the sphere. These trajectories are represented in Figure 8.11. The only critical point is not elementary, and it necessarily has index 2.

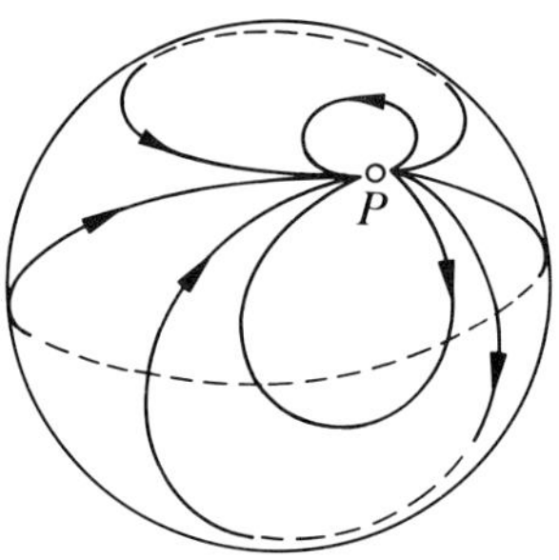

Figure 8.11

As a result of Poincaré's theorem, the only surfaces on which a vector field without critical points can exist are the surfaces of index zero. These are the torus and its diffeomorphic images [equivalent by continuous differentiable deformation]. This makes the torus unique among all surfaces.

Problems

1. Compute the index of the critical points in the preceding problems (page 256, problem 3) and check that the sum of the indices is zero.
2. Consider the sphere as the one-point-compactification of the plane (i.e., consider one single point P_∞ at infinity), and study the trajectories on the sphere corresponding to the linear systems (node, focus, etc.).

3. Consider the sphere divided by the equator and two complete meridian circles into eight octants. Consider the boundary of each octant as a limit cycle. Where four octants meet we have a saddle point. How many such saddles are there and where are the remaining critical points, so that the sum of their indices is two?

3 *Trajectories on the Projective Plane*

3.1 *Compactification of the euclidean plane*

Until now, when we have spoken about the trajectories on the plane, we have meant the *finite* or *euclidean plane*; that is, the collection of *ordinary* points, each of which is at a finite distance from the origin. For many problems it is convenient to add some *ideal* points, which may be thought of as being at an infinite distance from the origin. This really amounts to transforming the euclidean plane into a *compact* manifold by adding limit points, defined by sequences which tend to infinity.

There are several different *compactifications* of the euclidean plane. One possibility is to add just one point at infinity as the limit for all sequences tending to infinity. This transforms the plane into a manifold equivalent topologically to a sphere; hence it is sometimes called *Gauss' sphere*. This compactification is commonly used in complex variables, where it is also known as the *Riemann sphere*.

Another compactification is the *projective plane*, obtained by adding infinitely many *ideal* points [points at infinity] in such a way that every straight line has exactly one point at infinity. [By tending to infinity in both senses on a given line, we approach the *same* point at infinity.] *Parallel lines* are assumed to have the *same* point at infinity [they intersect at infinity], but *nonparallel lines* have *different* points at infinity. In this way, every two lines in the projective plane intersect in exactly one point.

3.2 *Representation of the projective plane*

Let us project the projective plane from a point 0, exterior to the plane, on a sphere Σ of center 0 [see Figure 8.12]. Every [ordinary or ideal] point of the projective plane determines exactly one projecting line through 0, and this line again meets the sphere Σ in two opposite points. The points of the projective plane are therefore in a one-to-one relation with the pairs of diametrically opposite points on the sphere.

Hence:

> *The projective plane is equivalent to a sphere on which we have identified every two diametrically opposite points.*

Another way to look at the representation of the projective plane on the sphere Σ is to consider only the lower hemisphere in Figure 8.12. Then:

> *Every ordinary point of the projective plane corresponds to exactly one point of the hemisphere. But the ideal points* [*at infinity*] *of the projective plane still have two corresponding points on the equator, boundary of the hemisphere.*

Still another representation of the projective plane can be obtained on a circular disk, since this is equivalent to the hemisphere.

> *To obtain a manifold equivalent to the projective plane, take a circular disk and identify every two diametrically opposite points of the boundary.*

When we draw the full picture of the projective plane, this latter representation is preferable for obvious reasons.

Among the properties of the projective plane, a particularly curious one is the fact that *the projective plane is one-sided, that is, nonorientable* [the Möbius strip, for example].

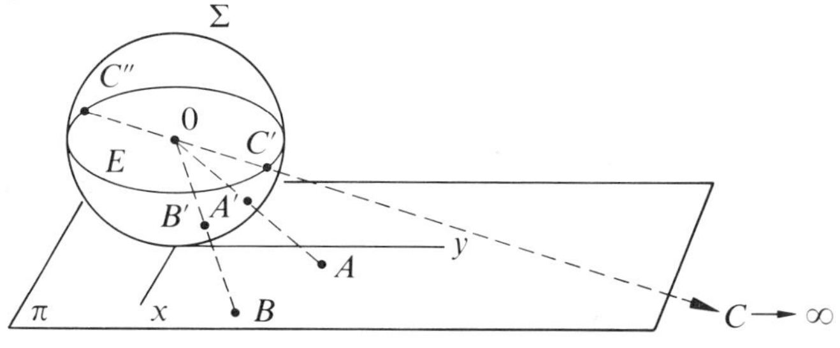

Figure 8.12

3.3 Projective form of the differential equation

Let x, y be the usual coordinates on the euclidean plane. We now introduce the *homogeneous coordinates* u, v, w by the relations

$$x = \frac{u}{w}, \quad y = \frac{v}{w}.$$

The differential equation

$$\frac{dy}{dx} = \frac{Y(x, y)}{X(x, y)} \tag{8.3.1}$$

can be written

$$\begin{vmatrix} dx & dy \\ X(x, y) & Y(x, y) \end{vmatrix} = 0.$$

From now on we will assume that $X(x, y)$ and $Y(x, y)$ *are polynomials without common factors.* [Possible common factors have been cancelled out.] Let n and m be the degrees of X and Y, respectively. If, for example, $n \geqq m$, then

$$\begin{cases} X_0(u, v, w) = w^n X\left(\dfrac{u}{w}, \dfrac{v}{w}\right) \\[2ex] Y_0(u, v, w) = w^n Y\left(\dfrac{u}{w}, \dfrac{v}{w}\right) \end{cases}$$

are homogeneous polynomials of degree n in u, v, w. In case $n \leqq m$, we would multiply by w^m in order to get X_0, Y_0.

We can write the differential relation

$$\begin{vmatrix} du & dv & dw \\ u & v & w \\ X_0(u, v, w) & Y_0(u, v, w) & 0 \end{vmatrix} = 0, \tag{8.3.2}$$

which, developed, is

$$A(u, v, w)\, du + B(u, v, w)\, dv + C(u, v, w)\, dw = 0. \tag{8.3.3}$$

We shall prove that in the euclidean plane [the finite part of the projective plane] this relation is equivalent to the differential equation (8.3.1).

If $x(t)$, $y(t)$ satisfy the differential equation (8.3.1) and if we substitute $u = x(t)$, $v = y(t)$, $w = 1$ [hence $dw = 0$], it can be immediately verified that (8.3.2) holds [since two rows are proportional]. Conversely, if the curve given in homogeneous coordinates by $u(t)$, $v(t)$, $w = 1$ satisfies (8.3.2), it is sufficient to put $x = u(t)$ and $y = v(t)$ to have a solution of (8.3.1).

Equation (8.3.2) satisfies the following homogeneity condition: if $u(t)$, $v(t)$, $w(t)$ satisfy this relation and if $r(t)$ is any differentiable factor, then the functions $u_1 = r(t)u(t)$, $v_1 = r(t)v(t)$, $w_1 = r(t)w(t)$ also satisfy (8.3.2). Indeed, in

$$\begin{vmatrix} d(ru) & d(rv) & d(rw) \\ ru & rv & rw \\ X_0(ru, rv, rw) & Y_0(ru, rv, rw) & 0 \end{vmatrix}$$

one can factor out r in the second row and r^n in the third row; while each term in the first row can be written as a sum [e.g., $d(ru) = r\,du + u\,dr$.] The rest of the proof follows by applying elementary properties of determinants.

If the curve given in homogeneous coordinates, $u(t)$, $v(t)$, $w(t)$, satisfies the relation (8.3.2), one can change to ordinary coordinates by dividing by $w(t)$, which then plays the role of the factor $r^{-1}(t)$. As long as $w(t)$ is not zero [this happens on all finite parts of the projective plane], the correspondent curve in ordinary coordinates will satisfy (8.3.1).

Note that if we want to have the differential equation of the curves on the unit sphere, which represents the projective plane, it is sufficient to add the condition $u^2 + v^2 + w^2 = 1$ or, in differential form, $u\,du + v\,dv + w\,dw = 0$.

We can therefore state that:

> *Where the differential equation* (8.3.1) *is defined* [*that is, on the finite parts of the plane*], *it is equivalent to* (8.3.2). *Hence* (8.3.2), *which makes sense also at the ideal points* [*at infinity*] *of the projective plane, can be taken as extending* (8.3.1) *to those points, where* $w = 0$.

The *ideal line* of the projective plane [corresponding to the equator of the equivalent sphere] has the equation $w = 0$; hence $dw = 0$. It follows that the ideal line always satisfies (8.3.2) and is therefore a trajectory.

Critical points are those points (u, v, w) for which the relation $du\colon dv\colon dw$ is not determined; hence those for which, in (8.3.3),

$$A(u, v, w) = B(u, v, w) = C(u, v, w) = 0.$$

Given an ideal critical point, we can easily study it by first making a change of coordinates, which takes that point into an ordinary one [into the origin, for example]. We then have no difficulty in classifying it as node, center, focus, or saddle point [in linear systems, for example]. Note that since the ideal line [equator] is a trajectory, a point at infinity cannot, in general, be a center or focus. Occasionally, this may happen when, in some cases, the solution $w = 0$ can be canceled out in (8.3.3). Note also that there is no difficulty in defining the index of a critical point at infinity, since it can be changed, by a coordinate transformation, into an ordinary point.

In the central projection of Figure 8.12 of the projective plane onto the sphere Σ, each critical point on the projective plane is projected on two critical points on the sphere. Since the index of the sphere is $+2$, it follows that *the index of the projective plane* is $+1$.

Example 1

Consider the differential equation $dy/dx = -x/y$. Equation (8.3.3) then becomes

$$\begin{vmatrix} du & dv & dw \\ u & v & w \\ v & -u & 0 \end{vmatrix} = uw\, du + vw\, dv - (u^2 + v^2)\, dw = 0.$$

The only critical point is (0, 0, 1), the origin. Figure 8.13 represents the projective plane on the circular disk; diametrically opposite points on the circumference should be identified. The trajectories are the circles, and the straight lines are the coordinate axes.

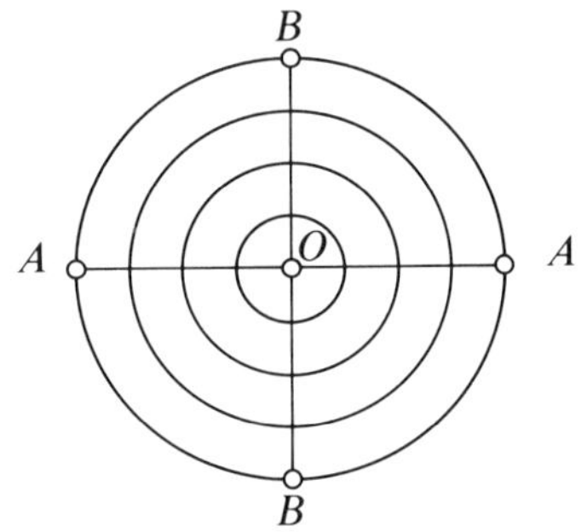

Figure 8.13

Example 2

Consider the differential equation $dy/dx = y/x$. Equation (8.3.3) then becomes

$$\begin{vmatrix} du & dv & dw \\ u & v & w \\ u & v & 0 \end{vmatrix} = -vw\, du + uw\, dv = 0.$$

Critical points are the origin (0, 0, 1) and all ideal points $w = 0$. But, if we cancel the common factor w in the differential equation, the points of the circumference of the circle [the ideal line] are not critical anymore, and the circumference itself is not a trajectory. This representation is shown in Figure 8.14.

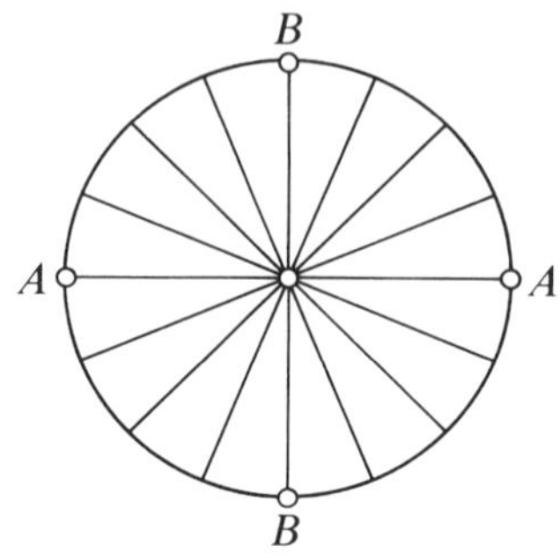

Figure 8.14

Example 3

Consider the differential equation $dy/dx = 0$. We may write it $dy/dx = 0/1$, obtaining

$$\begin{vmatrix} du & dv & dw \\ u & v & w \\ w & 0 & 0 \end{vmatrix} = w^2\, dv - vw\, dw = 0.$$

Cancelling the factor w, the only critical point is the ideal point $(1, 0, 0)$. The circumference $w = dw = 0$ is a trajectory. On the other hand, if we do not cancel the factor w, then all points of the circumference are critical. The corresponding figure is 8.15.

That $(1, 0, 0)$ is a proper node can be seen by writing the equation as $dv/dw = v/w$ and imagining v, w as cartesian coordinates.

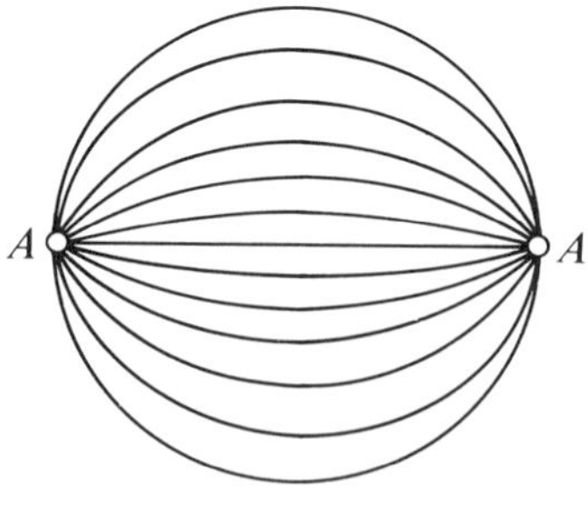

Figure 8.15

Example 4

Consider the differential equation $dy/dx = 2y/x$. We then obtain

$$\begin{vmatrix} du & dv & dw \\ u & v & w \\ u & 2v & 0 \end{vmatrix} = -2vw\,du + uw\,dv + uv\,dw = 0.$$

There are three critical points: (0, 0, 1), (0, 1, 0) and (1, 0, 0). The first one is the origin, which we already know. To examine the second, we put $v = 1$, $dv = 0$ and obtain $du/dw = u/2w$. This point is therefore an improper node with two tangents. To examine the third point, make $u = 1$, $du = 0$ and obtain $dv/dw = -v/w$. This point is therefore a saddle point. The corresponding figure is 8.16.

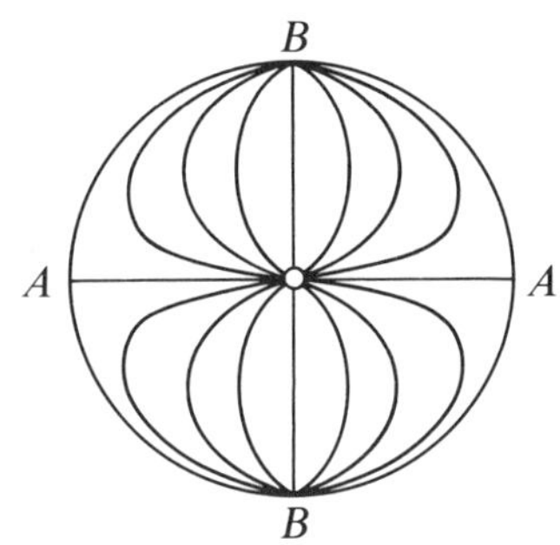

Figure 8.16

Example 5

Consider the differential equation $dy/dx = (2xy)/(x^2 - y^2)$. We change to polar coordinates $dr = r \cot\theta\, d\theta$, from which we get $r = C \sin\theta$. The trajectories are given in Figure 8.17. On the projective plane, we obtain

$$\begin{vmatrix} du & dv & dw \\ u & v & w \\ u^2 - v^2 & 2uv & 0 \end{vmatrix} = -2uvw\,du + (u^2 - v^2)w\,dv + (u^2 + v^2)v\,dw = 0.$$

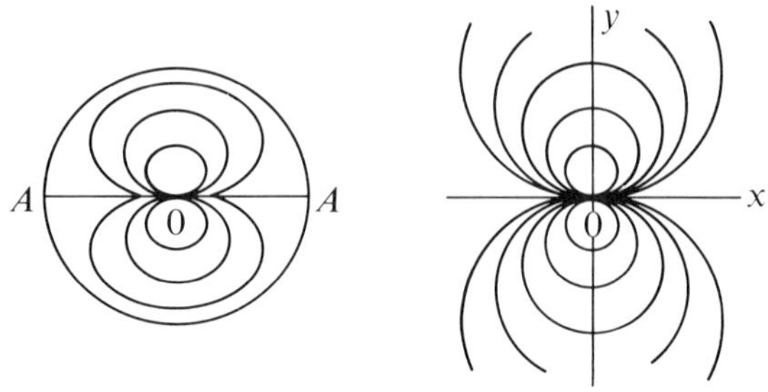

Figure 8.17

The critical points are (0, 0, 1) and (1, 0, 0). The first one, which is an ordinary point, is a critical point of higher order, index +2. The second one is a saddle point [see Figure 8.17].

Problem

1. Study the behavior of the trajectories of the following differential systems on the projective plane:

   (a) $x' = x(x + y)$, $y' = y(x - y)$.
   (b) $x' = y(x + y)$, $y' = x(x - y)$.
   (c) $x' = y^3$, $y' = x^2$.
   (d) $x' = -y^3$, $y' = x^3$.

---

4 *Limit Sets in the Plane*

4.1 Limit cycles

We again consider the differential system

$$\begin{cases} x' = X(x, y) \\ y' = Y(x, y). \end{cases} \tag{8.4.1}$$

The word *cycle* is just another name for a closed trajectory, which we know corresponds to a periodic solution.

A cycle is *isolated* if it has a neighborhood which is free of other closed trajectories. This is the opposite to what happens, for example, with the differential equation $dy/dx = -x/y$, where all trajectories are closed. We shall see that for an isolated cycle, in some neighborhood of it all trajectories tend asymptotically, for either $t \to +\infty$ or $t \to -\infty$, to that cycle. For this reason, it is called a *limit cycle*. Let us first look at some examples.

Consider, in polar coordinates, the system

$$\begin{cases} r' = r(1 - r) \\ \theta' = 1. \end{cases}$$

Obviously, $r = 1$ is a cycle. If $r > 1$, then $r' < 0$; so r is decreasing. If $r < 1$, then $r' > 0$; so r is increasing. Hence, as $t \to +\infty$, all neighboring trajectories tend to the cycle, which is therefore a *stable limit cycle* [see Figure 8.18].

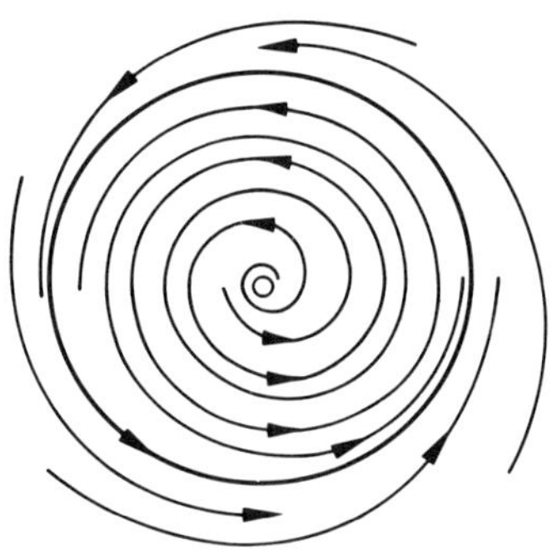

Figure 8.18

Now consider

$$\begin{cases} r' = -r(1 - r) \\ \theta' = 1. \end{cases}$$

Here all neighboring trajectories tend to the cycle as $t \to -\infty$. It is an *unstable limit cycle* [see Figure 8.19].

Finally, consider

$$\begin{cases} r' = r(1 - r)^2 \\ \theta' = 1. \end{cases}$$

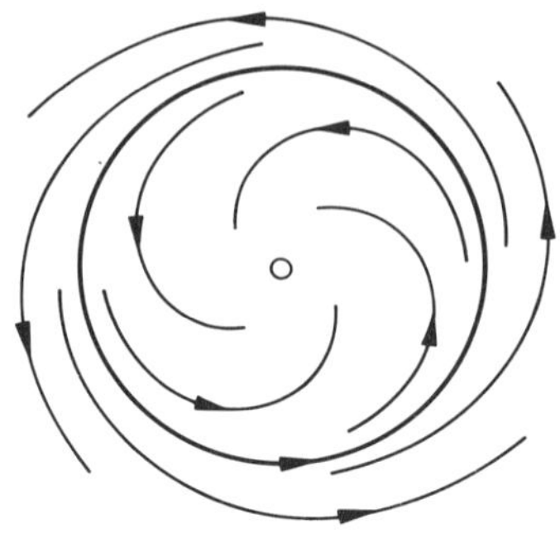

Figure 8.19

Here the behavior of the neighboring trajectories is different on opposite sides of the cycle. Since $r' \geqq 0$, the trajectories approach the cycle from the interior, but they spiral away in the exterior. This cycle is therefore stable from the inside and unstable from the outside. It is sometimes called a *semistable cycle* [see Figure 8.20].

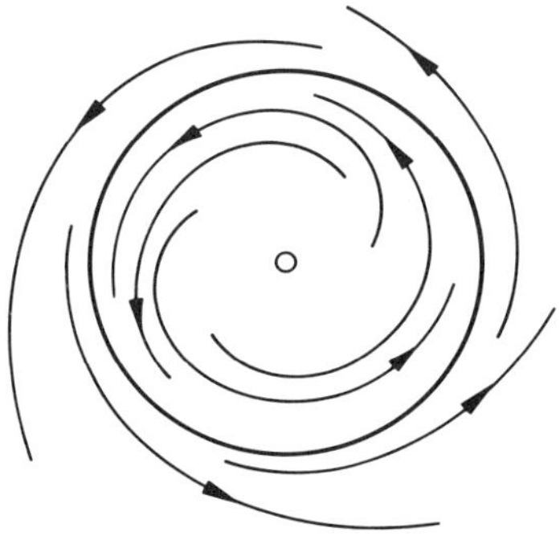

Figure 8.20

4.2 Poincaré–Bendixson Theorem

Let us assume that in (8.4.1) *X and Y are continuous and therefore satisfy conditions which insure the uniqueness of the solutions and their continuous dependence on initial conditions. Furthermore, assume that*

$$\Gamma_0 : x = \varphi(t), \qquad y = \psi(t)$$

is a periodic solution of period T and that, in some neighborhood of it, there are no other periodic solutions [hence Γ_0 is an isolated cycle]. Then there exists $\varepsilon > 0$ such that every trajectory which comes within a distance less than ε from Γ_0 tends asymptotically to Γ_0 as $t \to +\infty$ or as $t \to -\infty$ [spiraling around Γ_0]. Furthermore, on each side of the cycle, all neighboring trajectories behave the same way in that either all approach Γ_0 as $t \to +\infty$ or all approach Γ_0 as $t \to -\infty$.

Proof

Let

$$P_0: x_0 = \varphi(t_0), \qquad y_0 = \psi(t_0)$$

be a point of Γ_0. Since it is a regular point, there is a circular neighborhood of radius r_0, without critical points, such that the angle between the vectors (X, Y) corresponding to any two points of this neighborhood is less than some fixed α, $0 < \alpha < \pi/4$.

Let v be a segment of the normal to the trajectory Γ_0 at P_0, and let the length of v be r_0 on each side of Γ_0. Then v is an arc without contact, and all trajectories which intersect it cross it in the same sense. Furthermore, we can determine a neighborhood V of P_0 such that every trajectory through V intersects v, before leaving the circle of radius r_0 and center P_0 [see Figure 8.21].

As the trajectory through P_0 comes back to P_0 after a time T, it follows [from the continuous dependence of the solution on the initial conditions] that there is a neighborhood U of P_0 such that every trajectory Γ, passing at time $t = 0$ through a point P_1 of U, is again in the neighborhood V of P_0 at the time $t = T$. U can be taken sufficiently

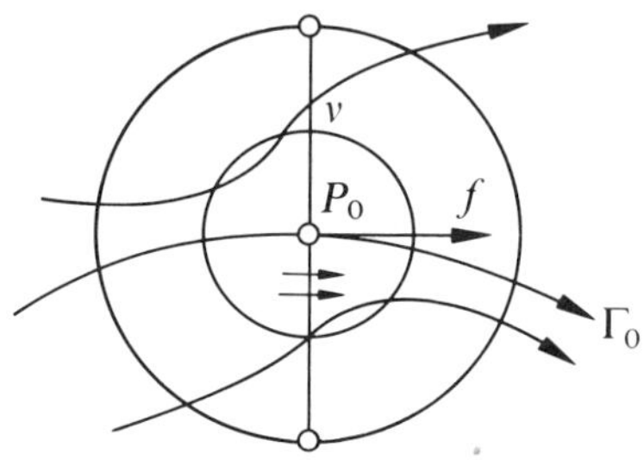

Figure 8.21

small so that in the whole interval $t \in [0, T]$, the distance between the points $x(t)$ describing Γ_0 and Γ is less than a given $\delta > 0$.

Take a point P_1 of v belonging to the neighborhood U. Then the trajectory Γ through P_1 stays close to Γ_0 on the entire revolution for $0 \leqq t \leqq T$, after which Γ again meets v in some point P_2 [see Figure 8.22].

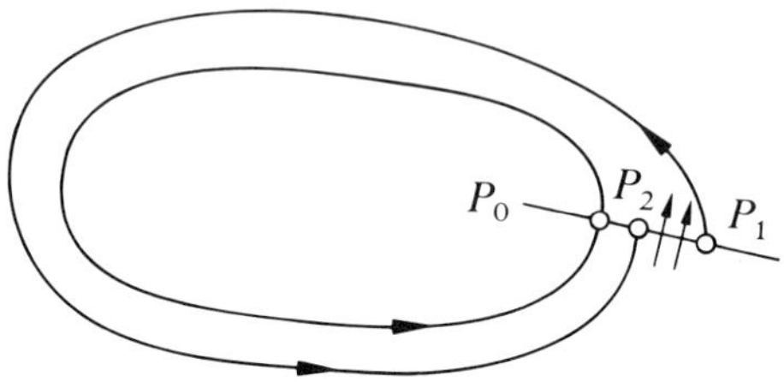

Figure 8.22

We now need the following topological property of the plane. *Every Jordan curve in the plane [that is, every homeomorphic image of a circle] divides the plane in three disjoint point sets*

(a) *the Jordan curve J itself;*
(b) *the interior region I, which is bounded;*
(c) *the exterior region E, which is unbounded.*

The regions I and E are connected. Given any two points in one of these regions, it is possible to join them with a continuous curve completely belonging to the same region. On the other hand, the curve J "separates" the two regions I and E. Given any point of I and any point of E, every continuous curve joining these points must necessarily intersect J.

Γ_0 is a Jordan curve. The trajectory Γ cannot intersect Γ_0, since trajectories do not meet. Hence, if P_1 belongs to the exterior region determined by Γ_0, so must P_2. We shall consider this to be the true case, as illustrated in Figure 8.22.

The case where both P_1 and P_2 are interior is similar. If P_1 has been taken sufficiently near P_0, Γ cannot be a closed trajectory, since Γ_0 is assumed to be an isolated cycle. Hence P_1 and P_2 cannot coincide. So P_2 can be nearer to or farther from P_0 than P_1.

Assume that P_2 is between P_0 and P_1, as indicated in Figure 8.22; then P_2 also belongs to the neighborhood U. For the same reasons as before, the trajectory Γ, after passing P_2, winds around Γ_0 and again intersects v at a point, P_3. Then P_3 must be between P_2 and P_0. Indeed, consider the region bounded by the arc of trajectory Γ: P_1P_2, plus the arc of normal v: P_2P_1. The trajectory Γ then enters this region at P_2. It cannot leave crossing the arc of trajectory P_1P_2, nor can it leave by the way of v, since trajectories only enter the region there. Thus it must still be inside this region when it intersects v again, and P_3 must therefore be between P_2 and P_0.

The same situation arises again and again. After P_3, Γ must wind around Γ_0 and intersect v successively at points $P_4, P_5, \ldots$. This sequence of points is monotone and can never reach P_0; hence it has some limit point Q. We shall show that Q must coincide with P_0.

If Q were different from P_0, the trajectory through Q could not be closed. Using the same reasoning as before, this trajectory would then wind around Γ_0 and intersect v at Q_1, which would be nearer to P_0 than Q. But then Γ, which comes arbitrarily close to Q, would also come arbitrarily close to Q_1 [by the continuous dependence of solutions on initial conditions]. But this means that Q could not be the limit point of the sequence $P_1, P_2, P_3, \ldots$. Hence Q must coincide with P_0.

All this proves that if P_2 is between P_1 and P_0, the trajectory Γ winds around Γ_0 approaching it asymptotically as $t \to +\infty$. In the other case, where P_1 lies between P_2 and P_0, reversing the sense of increasing time we have the same situation as before. Thus Γ tends asymptotically to Γ_0 as $t \to -\infty$. It is easy to see that a similar situation arises on the other side of Γ_0. So, in some neighborhood of P_0, we have proved that all trajectories behave as described.

For each point $P_0 \in \Gamma_0$, such a neighborhood can be determined. The collection of these neighborhoods covers the compact set Γ_0; hence a finite number of them covers Γ_0. Now ε can be chosen to be the minimum of the distance from the compact set Γ_0 to the points outside this finite number of neighborhoods [i.e., to the complement of the set $\bigcup_i U_i$, where U_i are those neighborhoods and this complement is a closed set]. The value of ε so determined then satisfies the requirements of the theorem.

Finally, from this discussion we can also see that on each side of Γ_0, the manner of approach must be the same for all trajectories Γ.

4.3 Limit set configurations

In Chapter 6, Section 5, the definition and some general properties of limit sets were given. In the plane a more detailed study is possible.

It is convenient to compactify the plane [that is, to replace it by Gauss' sphere or the projective plane] in order to have a compact manifold. In Chapter 6 we saw some nice properties of limit sets on compact manifolds. On such manifolds every trajectory has a nonvoid limit set [not true in the euclidean plane] which is closed and connected. We now present some typical configurations.

If P belongs to $\Omega(\Gamma)$ and if P is a critical point, then $\Omega(\Gamma)$ may coincide with P or be a larger set.

If P belongs to $\Omega(\Gamma)$ and P is a regular point, then the whole trajectory through P belongs to $\Omega(\Gamma)$. If this trajectory is closed, then it coincides with $\Omega(\Gamma)$. Indeed, we have seen that the way Γ must approach this closed trajectory precludes the limit set $\Omega(\Gamma)$ containing other points than those of the limit cycle.

If P belongs to $\Omega(\Gamma)$, if P is regular, and if the trajectory through P is not closed, then $\Omega(\Gamma)$ must contain other points not belonging to this trajectory. A typical case is the so called "path polygon" [see Figure 8.23,a]. The behavior of Γ is similar to the approach to a closed cycle, but this cycle is replaced by a sequence of open trajectories and critical points. A special case of this configuration is shown in Figure 8.23,b.

The limit set can be more complicated, containing, for example, infinitely many critical points. Similar considerations can be made for the $A(\Gamma)$ limit sets.

(a) (b)

Figure 8.23

Problem

1. Study the stability of the closed orbits of:

(a) $\theta' = 1$, $r' = \sin r$ (r, θ polar coordinates). (b) $\theta' = 1$, $r' = r \sin \dfrac{1}{r}$.

(c) $\theta' = 1$, $r' = r\left(1 - \sin \dfrac{1}{r}\right)$. (d) $\theta' = \cos \dfrac{1}{r}$, $r' = r \sin \dfrac{1}{r}$.

References

General discussion of material in this chapter will again be found in the books by Lefschetz, Nemytskii-Stepanov, and Sansone-Conti, mentioned in the References for Chapter 6. For a readable, modern introduction to differentiable manifolds, see

I. M. Singer and J. A. Thorpe. *Lecture Notes on Elementary Topology and Geometry.* Glenview, Ill.: Scott, Foresman and Co., 1967.

A classical, rather weighty reference for the topics of combinatorial topology, classification of surfaces, and the theory of the index, including the case of higher dimension, is

P. Alexandroff and H. Hopf. *Topologie*. Berlin: Springer-Verlag, 1935; New York: Chelsea, 1965 (reprint).

The proof of Poincaré's theorem (2.13) can be found in L. Bieberbach. *Theorie der Differentialgleichungen*. Berlin: Springer-Verlag, 1930.

9

Second-order Equations

1 Conservative Systems

1.1 Types of equations

The linear equation

$$A(t)x'' + B(t)x' + C(t)x = e(t) \tag{9.1.1}$$

has been considered in Chapters 3, 4, and 5. If $A(t)$ is identically zero in some interval, the equation is not really of the second order there. If $A(t)$ is zero at particular points, they are then called *singular points*. The behavior of the solutions of a differential equation near a singular point can be quite complicated and is best studied, for analytic equations, in the complex domain through power series expansions [i.e., Taylor, Laurent, and Puiseux or fractional power series]. Here we shall not pursue the problem of singular points and shall therefore assume that $A(t)$ is different from zero.

When equation (9.1.1) corresponds to a time-invariant physical system, the left member will be independent of t; hence A, B, and C will be constants. On the other hand, the right member, $e(t)$, usually represents an exterior action upon the physical system [forcing term] and will be time-dependent, even if the physical system is time-independent.

A generalization of (9.1.1) is the differential equation

$$Ax'' + Bx' + g(x) = e(t), \tag{9.1.2}$$

with A and B constants or even with A and B functions of x and x'. These are *nonlinear equations*, and as long as the left member does not depend explicitly on t, they represent time-invariant physical systems with forcing term $e(t)$.

1.2 *Integral of motion*

If, in the general case of equation

$$\mathbf{x}' = \mathbf{f}(\mathbf{x})$$

where $\mathbf{x} \in R^n$, there is a function $E(\mathbf{x})$ with the properties

(a) $E : D \to R$ is defined and continuously differentiable on an open region $D \subset R^n$,

(b) $E(\mathbf{x})$ is not constant on any open subset of D,

(c) $E(\mathbf{x})$ is constant along every trajectory $\mathbf{x}(t)$ of the differential equation,

then $E(\mathbf{x})$ is called an *integral* of the differential equation also. [It may be referred to as a *first integral* or an *integral of motion*.]

If an integral exists for which the domain of definition is all of R^n, then we say that the differential system $\mathbf{x}' = \mathbf{f}(\mathbf{x})$ is *conservative*. This certainly happens if the differential equation represents a physical system for which the total energy [potential + kinetic energy] is constant. If no such integral exists, the system is nonconservative. *Dissipative systems* are those nonconservative systems for which the total energy is nonincreasing along every trajectory $\mathbf{x}(t)$.

In the case of the second-order equation, an integral is of the form $E(x, x')$. Its existence allows one to analyze the solutions of the differential equation more easily.

1.3 *Free oscillations of conservative systems*

Consider the differential equation

$$x'' + g(x) = 0. \tag{9.1.3}$$

Multiplying both members by $dx = x'\,dt$ and integrating, we obtain

$$\int \frac{dx'}{dt} x'\,dt + \int g(x)\,dx = \int x'\,dx' + \int g(x)\,dx = \text{const.}$$

Letting

$$G(x) = \int_0^x g(x)\,dx, \tag{9.1.4}$$

we get

$$\frac{x'^2}{2} + G(x) = u = \text{const} = \text{total energy}. \qquad (9.1.5)$$

This is already the equation of the trajectories in the phase plane, $(x, y = x')$. The complete solution of (9.1.3) is immediate by separation of variables:

$$t - t_0 = \int_{x_0}^{x} \frac{dx}{\sqrt{2(u - G(x))}}. \qquad (9.1.6)$$

In phase plane the motion $x(t)$, $y(t) = x'(t)$ will take place along a curve (9.1.5) and will be restricted to an x interval, where $G(x) \leqq u$ [see Figure 9.1.]

If, for example, $xg(x) > 0$ for $x \neq 0$, then $G(x) \geqq 0$. Furthermore, if $G(x) \to +\infty$ as $x \to \pm\infty$, then for any total energy u the motion is bounded and the trajectories are ovals around the origin, as represented in Figure 9.1. The period of oscillation is obtained from (9.1.6) if we substitute the limits of integration x_0, x by the limits of oscillation a and b. Thus

$$T = 2 \int_{b}^{a} \frac{dx}{\sqrt{2(u - G(x))}}. \qquad (9.1.7)$$

In the linear case, $g(x) = kx$, and the period of oscillation T is independent of the amplitude $A = a - b = 2a$.

In the nonlinear case, if $g(x)$ increases more than linearly with x [for example, if $g(x) = x + x^3$], we say that it corresponds to a *hard spring*. $g(x)$ is taken to represent the *restoring force* of a spring in a mechanical system represented by the differential equation. In formula (9.1.7), $G(x)$ will then increase faster than x^2 with respect to x [with the latter corresponding to the linear case]. Hence *for increasing amplitude of oscillation, the period will decrease.*

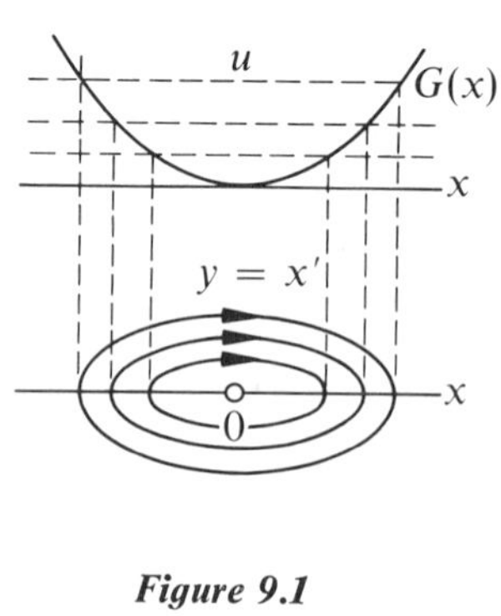

Figure 9.1

If $g(x)$ increases less than linearly, we say that it corresponds to a *soft spring*, and *the period of oscillation will increase with increasing amplitude.*

If we represent the relation between the period of oscillation and the amplitude, we obtain a graph like that shown in Figure 9.2. The pendulum $x'' = -k \sin x$ with

$$g(x) = k \sin x = k\left[x - \frac{x^3}{3!} + \frac{x^5}{5!} - \cdots\right]$$

is an example of a soft spring.

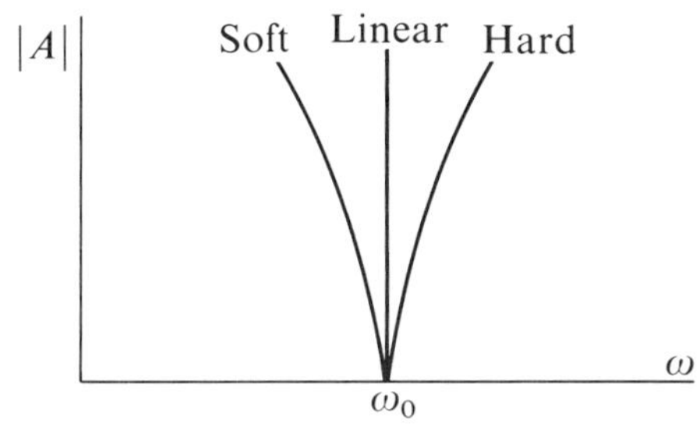

Figure 9.2

1.4 Forced oscillations

The differential equation

$$x'' + g(x) = e(t), \tag{9.1.8}$$

in spite of its apparent simplicity, is very difficult to discuss in general. A long and complicated theory can be developed in order to describe the behavior of the solutions. But we shall restrict our attention to some particular cases and results which are already of importance in their applications.

Consider $g(x) = ax + bx^3$, which is the simplest nonlinear odd function, and consider the forcing term to be $e(t) = F \cos \omega t$, where F is a constant. The corresponding differential equation is called *Duffing's equation*

$$x'' + ax + bx^3 = F \cos \omega t. \tag{9.1.9}$$

We will assume, without mathematical justification, that this equation has periodic solutions of angular frequency ω. This seems reasonable, at least for small values of the nonlinear term [i.e., of the parameter b]. The coefficient a is assumed positive; hence the corresponding linear equation [for $b = 0$] has oscillatory solutions. If $b > 0$, we have a hard spring; if $b < 0$, we have a soft spring.

Assuming its existence, let us try to determine the periodic solution by a method of successive approximations in which we will be satisfied with the second approximation.

As a first approximation, we take the solution of the linear equation $[b = 0]$:

$$x_1'' + ax_1 = F \cos \omega t.$$

According to Section 5.4, the solution will be

$$x_1 = A \cos \omega t,$$

where

$$A = \frac{F}{a - \omega^2}.$$

In Figure 9.3, A is represented as a function of ω. Note that for $F = 0$, the ω-axis and the vertical line $\omega = \sqrt{a}$ result. This vertical line is the same as the one in Figure 9.2.

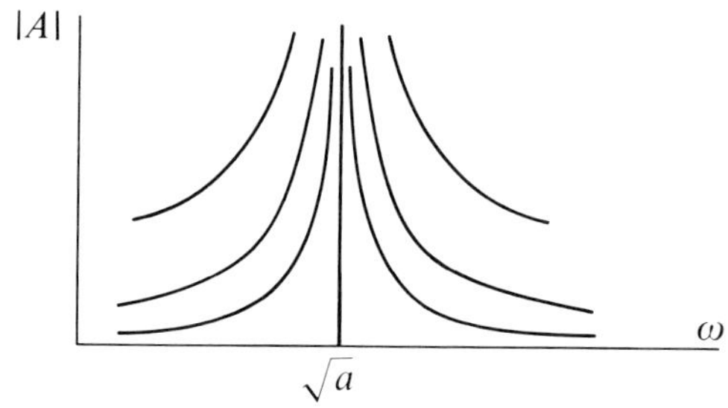

Figure 9.3

In order to obtain the second approximation $x_2(t)$, we write (9.1.9) in the form

$$x'' + \omega^2 x = (\omega^2 - a)x - bx^3 + F \cos \omega t. \tag{9.1.10}$$

We now replace x by the first approximation x_1 in the second member, obtaining the second approximation as a solution. Using the formula

$$\cos^3 \omega t = \tfrac{3}{4} \cos \omega t + \tfrac{1}{4} \cos 3\omega t$$

from trigonometry, we obtain

$$x_2'' + \omega^2 x_2 = [(\omega^2 - a)A - \tfrac{3}{4}bA^3 + F]\cos\omega t - \tfrac{1}{4}bA^3\cos 3\omega t. \qquad (9.1.11)$$

This is a linear differential equation, and the forcing term of angular frequency ω gives resonance. Hence a periodic solution for this equation can exist only if this term cancels out:

$$\omega^2 = a + \tfrac{3}{4}bA^2 - \frac{F}{A}. \qquad (9.1.12)$$

This relation is therefore a necessary condition for the existence of a periodic solution [within the approximation of our calculation]. The curves of ω as function of A [for fixed F] are given in Figure 9.4,a, where $b > 0$ is assumed [hard spring]. Since only the absolute value $|A|$ has the physical meaning of an amplitude of oscillation, the graph should be drawn like that in Figure 9.4,b. Note that this graph is really a deformation of the linear one in Figure 9.3.

For $b < 0$ [soft spring], the curves look like the ones in Figure 9.5.

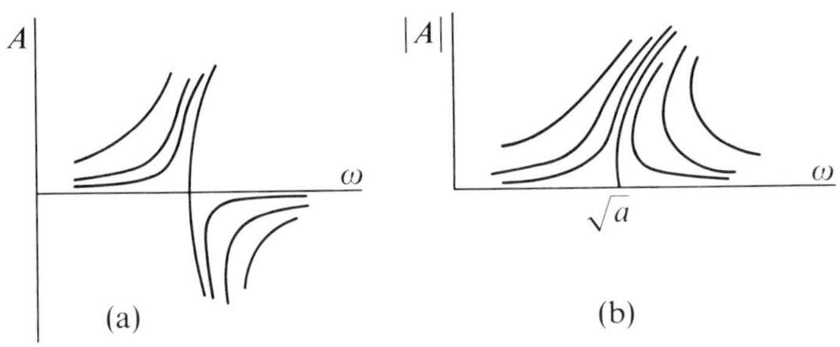

Figure 9.4

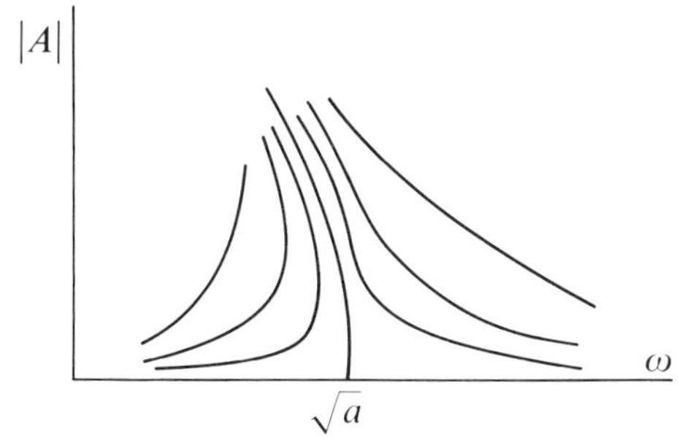

Figure 9.5

1.5 Subharmonics

In Duffing's equation, a new feature appears which is impossible in linear vibrations. This is the existence of *subharmonics—oscillations whose frequencies are fractions of the forcing frequency*. We shall not try to give the whole theory, but, as a simple example, we shall establish a condition for the existence of subharmonics of order 1/3 for equation (9.1.9).

Assume the wanted solution $x(t)$ developed in Fourier series. By symmetry, all the sine terms and the even cosine terms will cancel out in the calculations; hence we will not write them from the very beginning:

$$x(t) = A_1 \cos \frac{\omega t}{3} + A_3 \cos \omega t + A_5 \cos \frac{5\omega t}{3} + \cdots. \tag{9.1.13}$$

Substituting this expression in (9.1.9), it is necessary to express products and powers of cosines by the following formulas:

$$\cos^3 \frac{\theta}{3} = \frac{3}{4} \cos \frac{\theta}{3} + \frac{1}{4} \cos \theta$$

$$\cos^2 \frac{\theta}{3} \cos \theta = \frac{1}{4} \cos \frac{\theta}{3} + \frac{1}{2} \cos \theta + \cdots$$

$$\cos \frac{\theta}{3} \cos^2 \theta = \frac{1}{2} \cos \frac{\theta}{3} + \cdots$$

$$\cos^3 \theta = \frac{3}{4} \cos \theta + \cdots,$$

where the dots mean terms in multiples of θ. In this way, we formally obtain the conditions

$$\left(a - \frac{\omega^2}{9}\right) A_1 + \frac{3}{4} b(A_1^3 + A_1^2 A_3 + 2A_1 A_3^2) = 0 \tag{9.1.14}$$

and

$$(a - \omega^2) A_3 + \frac{1}{4} b(A_1^3 + 6A_1^2 A_3 + 3A_3^3) = F. \tag{9.1.15}$$

Again, we apply a method of successive approximations. As first approximation $x_1(t)$, we take the solution of the linear equation with $b = 0$. But, if $b = 0$, equation (9.1.14) shows that $A_1 \neq 0$ implies $\omega = 3\sqrt{a}$. If this condition is satisfied, A_1 is arbitrary and $A_3 = -(F/8a)$. Here A_1 corresponds to a free oscillation [which can exist because there is no damping]; while A_3 is the usual harmonic response.

For the next approximations, we rewrite equations (9.1.14–9.1.15) in the form

$$\omega^2 = 9a + \tfrac{27}{4} b(A_1^2 + A_1 A_3 + 2A_3^2) \tag{9.1.16}$$

and

$$-8aA_3 = F + (\omega^2 - 9a)A_3 \\ - \tfrac{1}{4}b(A_1^3 + 6A_1^2A_3 + 3A_3^3). \qquad (9.1.17)$$

We will not calculate further approximations, but we will discuss the significance of these formulas.

In the first approximation we got $A_3 = -(F/8a)$, and (9.1.16) represents a relation between frequency ω and the amplitude A_1 of the subharmonic. The curve (9.1.16) represents an ellipse or hyperbola [depending upon b being negative or positive]. For $b > 0$, ω^2 has a minimum. For $b < 0$, ω^2 has a maximum, located at $A_1 = -1/2(A_3)$; the corresponding value of ω^2 is

$$\omega^2 = 9(a + \tfrac{21}{16}bA_3^2).$$

In conclusion [up to second approximation], there can be a subharmonic of frequency $\omega/3$ only if

$$\text{for}\begin{cases} b < 0 \\ b > 0 \end{cases}, \qquad \omega^2 \begin{matrix} < \\ > \end{matrix} 9(a + \tfrac{21}{16}bA^2),$$

where $A = A_3$ is the amplitude of the component of the forcing frequency ω. [A rigorous and detailed treatment of the subject is given in the References at the end of this chapter.]

Problems

1. Classify the following oscillatory systems into "hard" and "soft" spring cases:

   (a) $x'' + 3 \sin x = 0$. (b) $x'' + \sin x - \dfrac{1}{4} \sin 2x = 0$.

   (c) $x'' + \sin x - \dfrac{1}{8} \sin 2x = 0$. (d) $x'' + \tan x = 0$.

   (e) $x'' + g(x) = 0$, where $g(x) = x$ for $|x| \leqq 1$, $g(x) = -1$ for $x < -1$, $g(x) = 1$ for $x > 1$.

   Obtain the algebraic equations of and draw the trajectories in this last case.

2. For the following oscillatory systems, find the first integral, find an integral expression of the period of oscillation (as function of the amplitude or equivalent parameter), in particular find the limit of this period when the amplitude tends to zero, and describe qualitatively the oscillations of large amplitude.

   (a) $x'' + \sin x = 0$. (b) $x'' + \tan x = 0$.
   (c) $x'' + x + x^2 = 0$. (d) $x'' + x + x^3 = 0$.

   (e) $x'' + \dfrac{x}{x+1} = 0$. (f) $x'' + xe^{-x^2} = 0$.

   (g) $x'' + x^3 = 0$.

3. Consider the discontinuous differential equation

$$x'' + k(x)x = 0, \quad \text{where} \quad 0 < k(x) = \begin{cases} k_1 \text{ small for } x > 0, \\ k_2 \text{ large for } x < 0. \end{cases}$$

   Find the period of oscillation. Make $k_2 \to +\infty$. This corresponds to the bouncing of a perfectly elastic body on a rigid wall. Draw the graph of the trajectories in the phase plane.

4. As in the preceding problem, consider the equation of a perfectly elastic bouncing ball subject to gravity:

$$x'' + k(x) = 0, \quad \text{where} \quad k(x) = \begin{cases} k_1 > 0 & \text{for} \quad x > 0, \\ k_2 < 0 & \text{for} \quad x < 0, \end{cases}$$

   and $|k_2|$ is very large ($k_2 \to -\infty$).
   Find the period and draw the graph of the trajectories in the phase plane.

5. Check that formula (9.1.12) is correct from the point of view of the physical dimensions of the magnitudes involved.

6. Write equation (9.1.12) for the case of the differential equation of the pendulum

$$x'' + k^2 \sin x = F \cos \omega t$$

   for the approximation up to the term in x^3.

7. Disregarding the approximation written in the preceding problem, discuss the qualitative behavior of the solutions, drawing (qualitatively) the curves of amplitude $= f(\omega)$ (as in (9.1.12)). Discuss particularly these curves when the amplitude approaches and surpasses π. Show that a reasonable first approximation for very large F (and amplitude) would be $x'' = F \cos \omega t$. Compare with Figure 8.2.

---

2 *Nonconservative Systems*

2.1 Liénard's equation

The equation

$$x'' + f(x, x')x' + g(x) = 0, \tag{9.2.1}$$

describing free nonlinear oscillations, can be applied to a wide range of physical problems. A less general but still important differential equation is

$$x'' + f(x)x' + g(x) = 0, \tag{9.2.2}$$

known as *Liénard's equation*. [*Note*: Liénard considered the case $g(x) = x$]. In this equation, the friction term is linear in x' but depends on the position x.

Let us introduce the following notation:

$$\begin{cases} F(x) = \int_0^x f(x)\,dx \\ G(x) = \int_0^x g(x)\,dx. \end{cases} \tag{9.2.3}$$

It is convenient to introduce the phase plane as

$$\begin{cases} x' = y - F(x) \\ y' = -g(x). \end{cases} \tag{9.2.4}$$

The total energy of the system is given by

$$E = \frac{x'^2}{2} + G(x).$$

Differentiating we obtain

$$E' = x'x'' + g(x)x' = x'(x'' + g(x)) = -x'^2 f(x).$$

Therefore, if $f(x)$ is positive, the energy decreases and the system is dissipative. If $f(x)$ is negative, however, the energy increases and there is a kind of "self-excitation." But for this to happen there must be an exterior source of energy. In electrical circuits, for example, we speak of a "negative resistance" in such cases.

Instead of the energy we may consider the function

$$u(x,y) = \frac{y^2}{2} + G(x). \tag{9.2.5}$$

It follows that

$$du = F(x)\,dy.$$

If the system is dissipative, the solutions will tend to the origin, which then will be asymptotically stable. A more interesting case is when, for small values of x, the function $f(x)$ [and the damping] is negative, and therefore the energy of the system is increased. The oscillations [we assume that the solutions are of this type] will then increase in amplitude. But, if for large x, $f(x)$ is positive, the [positive] damping will not allow the oscillations to grow excessively, and a stable intermediate oscillation will result. This is the typical behavior of some systems which are capable of self-sustained oscillations [electronic oscillators, for example]. Sufficient conditions to cause this type of solution to appear are given in the following theorem.

2.2 *Theorem*

Assume that

(a) $f(x)$ *is an even function,* $g(x)$ *is odd,* $xg(x) > 0$ *for every* $x \neq 0$, *and* $f(0) < 0$;

(b) $f(x)$ *and* $g(x)$ *are continuous for every* $x \in R$, *and* $g(x)$ *satisfies a Lipschitz condition;*

(c) $F(x) \to \pm\infty$ *as* $x \to \pm\infty$;

(d) $F(x)$ *has only one positive zero* $x = a$ *and is monotone increasing for* $x \geqq a$.

Then, equation (9.2.2) *has a unique periodic solution [a limit cycle], and all other solutions tend asymptotically to it.*

Proof

From the assumptions, it follows that existence and uniqueness of solutions are guaranteed.

In the x, y plane, let us draw the curve $y = F(x)$ [Figure 9.6]. Above this curve $x' > 0$, and below this curve $x' < 0$. On the other hand, $y' < 0$ for $x > 0$, and $y' > 0$ for $x < 0$.

A trajectory starting at a point such as A [Figure 9.6], must necessarily move to the right ($x' > 0$) and down ($y' < 0$). Thus it must meet the curve $y = F(x)$ at some point B. At B the trajectory has vertical tangent, and below this curve the trajectory moves to the left and down. Hence it will meet the y-axis at a point C. By similar reasoning the trajectory will then move upwards and meet the y-axis again at some point A'. [Note that the trajectory cannot recross the curve $y = F(x)$ in the fourth quadrant, since if it did so, it would cross with a vertical tangent which would force $y' > 0$ for $x > 0$.]

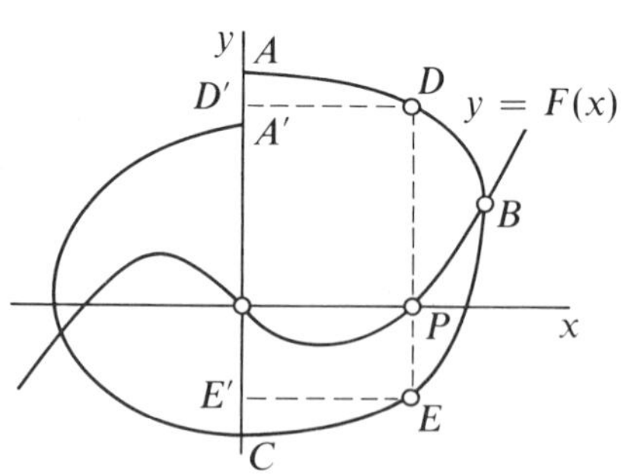

Figure 9.6

If, by chance, A' coincides with A, we have found a periodic solution. By symmetry, this will hold true if and only if $OA = -OC$, which then implies that the function $u(x, y)$ takes the same value at A and C.

If the entire arc ABC of the trajectory does not meet the vertical through $P(x = a)$, then

$$u(C) - u(A) = \int_{ABC} F(x)\,dy > 0;$$

hence $-OC > OA$.

If the abscissa of the point B is greater than the abscissa of the point P, we separate the line integral of $F(x)\,dy$ along the arc ABC into two parts:

$$\int_{ABC} F(x)\,dy = \left\{\int_A^D + \int_E^C\right\} F(x)\,dy + \int_{DBE} F(x)\,dy = \phi_1 + \phi_2 .$$

[Note that $\varphi_1 = \int_A^D + \int_E^C$ is positive and $\varphi_2 = \int_{DBE}$ is negative.] From the uniqueness of solutions, we know that when OA increases, CO also increases. We shall prove that if OA increases, $\varphi_1 + \varphi_2$ decreases monotonically. Indeed, on the arcs AD and EC,

$$du = \frac{F\,dy}{dx}\,dx = \frac{-Fg}{y - F}\,dx.$$

On AD, y increases with OA; hence $\int du$ decreases [F is negative]. On EC, y decreases, while OA increases. But now $dx < 0$, and the integral again decreases. Hence φ_1 decreases.

On DBE, $F(x)$ increases with x, and it is possible to show that, as the abscissa of the point B tends to $+\infty$, the absolute value of $\int_{DBE} F(x)\,dy$ tends to infinity. But this integral is negative [$dy < 0$]; hence it tends to $-\infty$. [The reader can fill in the details of this proof; see Problem 1.]

From all this, it follows that when OA increases from zero to infinity, $u(C) - u(A)$ starts being positive and decreases [monotonically] to $-\infty$. There is therefore exactly one value of OA for which $u(C) = u(A)$, and this corresponds, as we have already seen, to the periodic solution.

Furthermore, if OA is less than the value corresponding to the periodic solution, $u(C) - u(A)$ is positive; hence this [nonperiodic] solution increases in size and tends to the periodic one. Indeed, the successive intersections A', A'', A''', ... of this trajectory with the positive y-axis form a monotone sequence, having a limit point A°. The solution through A° is then periodic by the Poincaré-Bedixson theorem, which proves the assertion.

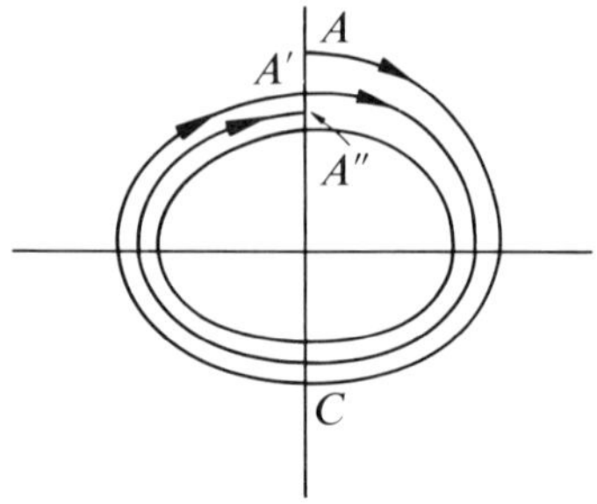

Figure 9.7

For OA greater than the value of the periodic solution, a similar reasoning holds, showing that from the outside the trajectories also tend to the limit cycle (Figure 9.7).

2.3 *Forced oscillations*

The differential equation

$$x'' + f(x)x' + g(x) = e(t) \tag{9.2.6}$$

is sometimes called *Cartwright-Littlewood's equation*, due to these authors' extensive studies of it. One basic problem is to give sufficient conditions for the existence of periodic solutions, assuming that $e(t)$ is periodic.

Particular cases are *Duffing's equation with damping*

$$x'' + cx' + g(x) = F\cos \omega t, \tag{9.2.7}$$

and

$$x'' + cx' + ax + bx^3 = F\cos \omega t. \tag{9.2.8}$$

For $c = 0$ there is no damping, and a description of the system may be made by the curves A as function of ω, as shown in Figure 9.4.

For $b = 0$, the system is linear, and in the case of positive damping, the "resonance curves" in Figure 9.3 are converted into those in Figure 5.7. It is therefore reasonable that, at least for small nonlinearity b, the curves in Figure 5.7 will be deformed, as they were in Figure 9.4,b and 9.5. This is actually the case, and the corresponding curves are shown in Figure 9.8, where (a) represents the linear case, (b) the hard spring, and (c) the soft spring. [For a more detailed treatment, we refer the reader to the References at the end of this chapter.]

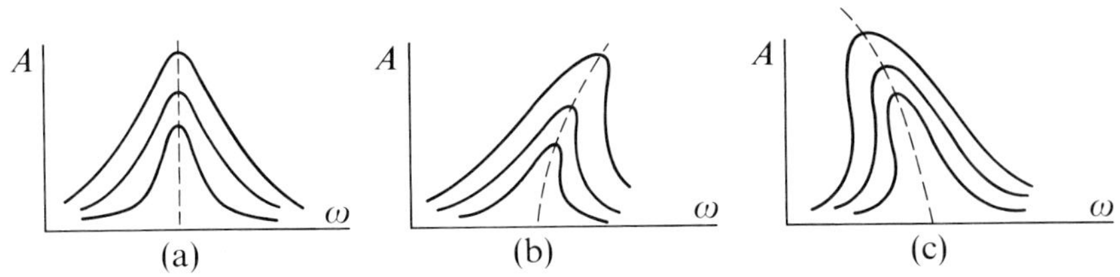

Figure 9.8

Problems

1. Fill in the missing details in the proof of Theorem 9.2.2.
2. Consider, in the assumptions of Theorem 9.2.2, the case when, instead of condition (d),
   (a) $F(x) > 0$ for all $x > 0$. (Hence $f(0) > 0$.)
   (b) $F(x) < 0$ for all $x > 0$. (Hence here $f(0) < 0$.)
   (c) $F(x)$ has several zeros.
   (d) Consider the case when (a), (b), and (d) of Theorem 2.2 are satisfied, but $F(x)$ is bounded as $x \to \pm\infty$.

   In all these cases try to draw a picture of the possible qualitative behavior of the trajectories.
3. Study the qualitative behavior of the solutions of the differential equation
   $$x'' + f(x)x' + x = 0, \quad \text{where } f(x) = -1 \text{ if } |x| \leqq 1, \text{ and } f(x) = +1 \text{ if } |x| > 1.$$
4. Study similarly the differential equation $x'' + x' + g(x) = 0$, where $g(x) = 0$ if $|x| \leqq 1$, $g(x) = 1$ if $x > +1$ and $g(x) = -1$ if $x < -1$.

---

3 *Relaxation Oscillations*

3.1 Van der Pol's equation

A very curious type of oscillation is obtained from second-order differential equations in which the term in x'' is multiplied by a parameter which tends to zero. In the limit, the differential equation is of the first order and has a completely different character [the kind of initial conditions, for example].

Let us now consider *Van der Pol's equation*

$$x'' + \mu(x^2 - 1)x' + x = 0 \qquad \mu > 0. \tag{9.3.1}$$

For $\mu = 0$, it is the equation of the linear harmonic oscillator. For small μ, it is a "perturbed" linear equation of Liénard's type, studied in Section 2.1. Now we are interested in $\mu \to \infty$.

Letting $\mu = 1/\varepsilon$, the equation becomes

$$\varepsilon x'' + (x^2 - 1)x' + \varepsilon x = 0. \tag{9.3.2}$$

Setting $\varepsilon = 0$, we obtain

$$(x^2 - 1)x' = 0$$

with all solutions of the form $x(t) = \text{constant}$.

Now set

$$\mu y = x' + \mu\left(\frac{x^3}{3} - x\right);$$

then equation (9.3.1) becomes the system

$$\begin{cases} x' = \mu\left[y - \left(\dfrac{x^3}{3} - x\right)\right] \\ y' = -\dfrac{x}{\mu} \end{cases} \tag{9.3.3}$$

Since the differential equation satisfies the conditions of Theorem 9.2.2, it has exactly one limit cycle. Let us see how this limit cycle looks for large μ.

Consider, in the (x, y) plane, the curve Γ:

$$y = \frac{x^3}{3} - x.$$

It follows from equation (9.3.3) that, except in the vicinity of Γ, x' is very large in absolute value, while y' is very small [we are assuming μ very large]. Hence the trajectories are nearly horizontal lines, in the sense of increasing values of x above Γ and of decreasing values of x below Γ.

A trajectory starting at a point P [Figure 9.9] will proceed almost horizontally until it nears Γ. There the trajectory will curve downwards and must meet Γ. From this point on, the trajectory will proceed to the left and down, but it cannot cross Γ again at a nearby point. Hence it must proceed in the neighborhood of Γ until reaching the vicinity of point A. Here, proceeding to the left and somewhat down, the trajectory gets away from Γ and becomes practically horizontal. It will meet Γ again at point D and, in a similar manner, will proceed to point C. There it will enter between Γ and its own former arc. Hence we see that the trajectory will approach the limit cycle more and more closely.

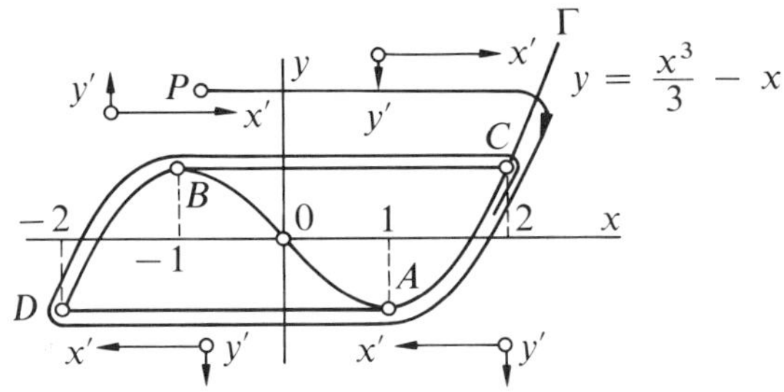

Figure 9.9

The limit cycle will consist approximately of the arcs CA and DB of the curve Γ, plus the horizontal parts BC and AD. The velocity with which the horizontal lines BC and AD are described is very large [of the order of μ]. The velocity on the Γ-arcs CA and DB is very small [of the order of $1/\mu$].

In calculating the period T of the oscillation, we will neglect the time correspondent to the horizontal lines and, by symmetry, write

$$T = 2\int_C^A dt.$$

Since the motion takes place almost on the curve,

$$dt = -\frac{\mu}{x}\,dy = -\mu\,\frac{x^2-1}{x}\,dx.$$

The coordinates of A and C are $(1, -2/3)$ and $(2, 2/3)$. Hence

$$T = 2\int_C^A dt = 2\mu\int_1^2 \frac{x^2-1}{x}\,dx$$

$$= \mu(3 - 2\ln 2) = 1.614\mu. \qquad (9.3.4)$$

In Figure 9.10 the shape of a trajectory is shown for values of $\mu = 0.1$, 1, 10. The trajectory starts near the origin and grows, approaching the limit cycle. It should be noted that for small μ, the oscillations look like harmonic oscillations, but for increasing values of μ, they become more and more distorted.

This type of oscillation occurs in physical systems which, during certain periods of time, undergo a slow change [an accumulation of energy, for example] and then suddenly experience a fast change [a quick release of energy]. The quick change is called *relaxation*, and this type of oscillation is termed a *relaxation oscillation*.

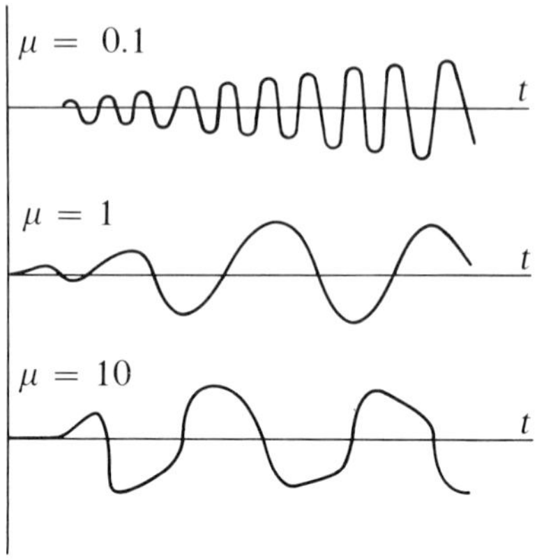

Figure 9.10

3.2 Hysteresis phenomena in oscillations

This is another type of relaxation oscillation, characterized by the stability or instability of different branches of the resonance curves.

Consider Duffing's equation with damping (9.2.8) and assume, for example, $b < 0$. We have already seen that resonance curves showing the amplitude of oscillation as a function of the driving frequency ω for fixed F are like those shown in Figure 9.11.

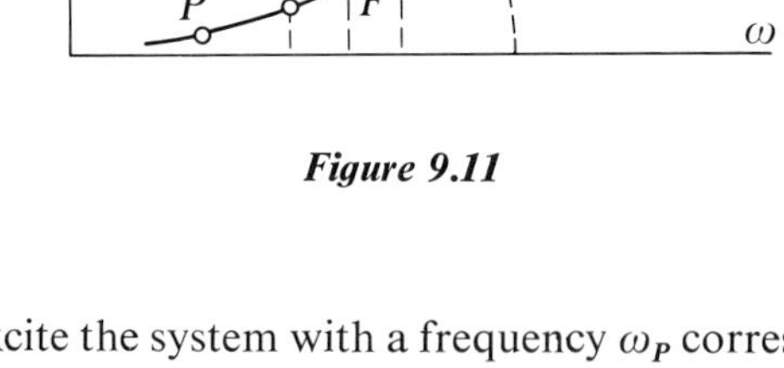

Figure 9.11

Assume that we excite the system with a frequency ω_P correspondent to the point P in the figure. An oscillation of amplitude $A(P)$ will become established in the system. Assume now that we change the frequency ω, but we do this so slowly that at each moment the system is in its steady-state motion [remember that in considering only the periodic solution, we neglect the transients]. This means that on the graph the point of operation will change from P, along the resonance curve, to A, F, and finally B. If we continue to increase the frequency, the amplitude must suddenly change from that corresponding to point B, to that corresponding to point D. This discontinuity of the amplitude as function of the frequency is a kind of relaxation.

Assume we have a frequency larger than that of point D, and we decrease it. On the graph, we will pass the point D, proceed to H, and finally to C. There the amplitude has a discontinuity, jumping to point A again.

In the frequency interval between points C and B, three possible values of the amplitude of oscillation [e.g. F, G, and H] actually correspond to every frequency. Developing the theory further, it can be seen that between the resonance curve and the ω-axis, the transients are such that the amplitude of oscillation grows, while in the region above the resonance curve, the amplitude decreases. From this it follows that the upper and lower branches, AB and CD, of the resonance curve are stable, while the middle branch, BC, is unstable. For example, if at a point like F the amplitude of oscillation increases a little, the point will enter the region of decreasing amplitude and return to F. The same happens if the amplitude decreases slightly from F. But, if the same thing happens at point G, the outcome is different. If, for example, we increase slightly the amplitude from G, the point will enter a region of increasing amplitude and grow, until it reaches H. By decreasing slightly from G, the amplitude would immediately decrease more and more, until it reaches F. Hence, in practice, the branch BC cannot be observed directly.

From an experimental viewpoint, it may be easier to maintain ω constant and change the amplitude of excitation F. For such an experiment, we should consider the curves of A as functions of F. Figure 9.12 shows how this graph can be obtained from that of Figure 9.11. Part (a) shows the resonance curves of A as a function of ω, each curve for a labeled value of F. If we want to maintain $\omega = \omega_1$, we should draw the vertical line $\omega = \omega_1$ in (a) and represent, point by point, the labels F of the resonance curves against the amplitude A. In this way we obtain part (b). Here, also, there is a relaxation oscillation: in the interval where there are three branches, only two are stable, and the operating point will be on one or the other of them, depending upon its past positions.

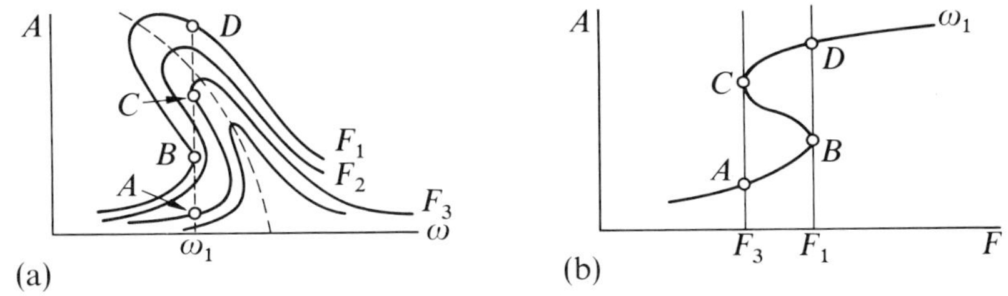

Figure 9.12

The name *hysteresis* is given to the behavior of a system in which, for a given input, the output depends on how this input is reached [by increasing or by decreasing values of the input]. Sometimes, one also speaks of *hereditary* or *memory* systems.

Problems

1. Consider the relaxation oscillations determined by $x'' + \mu f(x)x' + x = 0$, where $f(x) = -1$ if $|x| \leqq 1$, $f(x) = +1$ if $|x| > 1$. Find the period of oscillation for a very large μ.
2. Do the same for the relaxation oscillations corresponding to the same differential equation, with $f(x) = -1$ if $|x| \leqq 1$, $f(x) = k > 0$ very large, if $|x| > 1$. Draw the picture of the trajectories and imagine physically the motion described by them.

References

Besides the general material contained in the books by Lefschetz, Nemytskii-Stepanov, and Sansone-Conti, mentioned in Chapter 6, we refer the reader to

J. J. Stoker. *Nonlinear Vibrations in Mechanical and Electrical Systems*, New York: Interscience (Wiley), 1950.

N. Minorsky. *Nonlinear Oscillations.* Princeton, N.J.: Van Nostrand, 1962.

For a more detailed study of particular equations, we suggest

M. L. Cartwright. *Forced Oscillations in Nonlinear Systems, Contributions to the Theory of Nonlinear Oscillations*, Volume I in the *Annals of Mathematical Studies.* Princeton, N.J.: Princeton Univ. Press, 1950 (pp. 149–241).

M. L. Cartwright. *Van der Pol's Equation and Relaxation Oscillations, Contributions to the Theory of Nonlinear Oscillations*, Volume II in the *Annals of Mathematical Studies.* Princeton, N.J.: Princeton Univ. Press, 1962 (pp. 3–18).

J. LaSalle. Relaxation Oscillations. *Quart. Appl. Math.* **7** (1949): 1–19.

For the subject of nonlinear oscillations, subharmonics, and related questions, we refer the reader to

J. K. Hale. *Ordinary Differential Equations.* New York: Interscience (Wiley), 1969.

J. K. Hale. *Oscillations in Nonlinear Systems.* New York: McGraw-Hill, 1963.

N. N. Bogoliubov and Y. A. Mitropolskii. *Asymptotic Methods in the Theory of Nonlinear Oscillations.* New York: Gordon and Breach, 1961; Paris: Gauthier Villars, 1962 (French transl.).

10

Stability

1 Stability of Rest Points

1.1 Concept of stability

In everyday language, the word stability has a meaning which is not very explicit. We may say that the weather, some government, or some social structure is "stable" or "unstable," and everyone will understand what is meant. Nevertheless, it would be quite hard to try to define in a mathematically precise way what these assertions really mean. In physics, where more precision is needed, a system is said to be in a stable state if the system tends to come back naturally to its initial state when there is a change in this state.

The mathematical formulation of this concept in its present form comes from Liapunov, who set down the basic definitions and clearly outlined the further development of the theory.

In the following definitions we are referring to a vector-differential equation of the type

$$\mathbf{x}' = \mathbf{f}(t, \mathbf{x}), \tag{10.1.1}$$

where $\mathbf{x}$ is an n-dimensional vector, t is a real variable interpreted as time, and the function $\mathbf{f}(t, \mathbf{x})$ satisfies conditions insuring existence, uniqueness, and continuous dependence of the solutions on initial conditions. The general solution of (10.1.1) is denoted by

$$\mathbf{x} = \mathbf{F}(t, t_0, \mathbf{x}_0). \tag{10.1.2}$$

We will first define different types of stability of a "rest point," that is, a point $\mathbf{x}_0$ such that

$$\mathbf{f}(t, \mathbf{x}_0) = 0 \quad \text{for every } t, \tag{10.1.3}$$

and therefore

$$\mathbf{F}(t, t_0, \mathbf{x}_0) = \mathbf{x}_0 \quad \text{for every } t, t_0. \tag{10.1.4}$$

To simplify the formulas, we assume that the rest point under consideration is the origin of coordinates, $\mathbf{x} = 0$. This obviously does not restrict generality.

1.2 Definitions

We assume that the differential equation (10.1.1) *is defined on the half-cylinder,* $\|\mathbf{x}\| \leqq A$, $t \geqq 0$. *We restrict our attention to this region. We also assume existence, uniqueness, and continuous dependence of the solution* (10.1.2) *of* (10.1.1) *on initial conditions. Finally we assume that* $\mathbf{x}_0 = 0$ *is a rest point, i.e., satisfies* (10.1.3) *and hence* (10.1.4).

(a) *The point* $\mathbf{x} = 0$ *is stable* [*in the sense of Liapunov* (*or Liapunov-stable*)] *if for every* $\varepsilon > 0$ *and* $t_0 \geqq 0$, *there is* $\delta > 0$ [$\delta = \delta(\varepsilon, t_0)$] *such that* $\|\mathbf{x}_0\| < \delta$ *implies that the solution* $\mathbf{F}(t, t_0, \mathbf{x}_0)$ *exists and satisfies* $\|\mathbf{F}(t, t_0, \mathbf{x}_0)\| < \varepsilon$ *for all* $t \geqq t_0$.

(b) *The point* $\mathbf{x} = 0$ *is uniformly stable if in the preceding definition it is possible to take* $\delta = \delta(\varepsilon)$ *independent of* t_0.

(c) *The point* $\mathbf{x} = 0$ *is asymptotically stable if it is stable* (a); *and*

(c′) *for every* $t_0 \geqq 0$ *there exists a* $\delta_0(t_0) > 0$ *such that* $\|\mathbf{x}_0\| < \delta_0$ *implies* $\mathbf{F}(t, t_0, \mathbf{x}_0) \to 0$ *for* $t \to +\infty$.

(d) *The point* $\mathbf{x} = 0$ *is equiasymptotically stable if it is stable* (a); *and*

(d′) *for every* $t_0 \geqq 0$ *there is a* $\delta_0(t_0) > 0$, *and for every* $\eta > 0$ *there is a* $T(t_0, \eta)$, *such that* $\|\mathbf{x}_0\| < \delta_0$, $t \geqq t_0 + T(t_0, \eta)$ *imply* $\|\mathbf{F}(t, t_0, \mathbf{x}_0)\| < \eta$. *In other words,* $\mathbf{F}(t, t_0, \mathbf{x}_0) \to 0$ *for* $t \to +\infty$ *uniformly in* $\|\mathbf{x}_0\| < \delta_0$.

(e) *The point* $\mathbf{x} = 0$ *is uniformly asymptotically stable if it is uniformly stable* (b); *and*

(e′) *there is a* $\delta_0 > 0$, *and for every* $\eta > 0$ *there is a* $T(\eta)$, *such that* $\|x_0\| < \delta_0$, $t_0 \geqq 0$, $t \geqq t_0 + T(\eta)$ *imply* $\|\mathbf{F}(t, t_0, \mathbf{x}_0)\| < \eta$. *In other words,* $\mathbf{F}(t, t_0, \mathbf{x}_0) \to 0$ *for* $t - t_0 \to +\infty$ *uniformly in* $\|\mathbf{x}_0\| < \delta_0$, $t_0 \geqq 0$.

(f) *The point* $\mathbf{x} = 0$ *is exponentially asymptotically stable if there exists* $\lambda > 0$ *and, for every* $\varepsilon > 0$, *a* $\delta(\varepsilon) > 0$, *such that* $t_0 \geqq 0$, $\|\mathbf{x}_0\| < \delta$ *imply* $\|\mathbf{F}(t, t_0, \mathbf{x}_0)\| < \varepsilon e^{-\lambda(t - t_0)}$ *for all* $t \geqq t_0$.

(g) *The point* $\mathbf{x} = 0$ *is unstable if it is not stable according to definition* (a).

The basic definitions are stability (a), asymptotic stability (c), and instability (g). Nevertheless, in practice the corresponding uniform properties (b) and (e) are more important. Equiasymptotic stability is between asymptotic and uniform-asymptotic stability. As a matter of fact, even more properties of this type can be inserted between those given, but this has little practical importance. We may only mention that properties (c′) and (e′) without stability (a) are sometimes called *quasi-asymptotic* and *uniform-quasi-asymptotic* stability. We will see that (d′) implies (a), so that "quasi-equi-asymptotic" stability is always equi-asymptotic. Finally, exponentially-asymptotic stability is the strongest and corresponds to the behavior of the systems with constant coefficients, where the solutions are precisely exponentials.

The difference between stability (a) and continuous dependence of the solutions on initial values should be especially noticed. Continuous dependence means, indeed, that for fixed t and t_0 we can find $\delta(\varepsilon)$ such that $\|\mathbf{F}(t, t_0, \mathbf{x}_0)\| < \varepsilon$ for every $\|\mathbf{x}_0\| < \delta$. Here in general δ will be a function $\delta(t, t_0, \varepsilon)$. If δ does not depend on t for all $t \geqq t_0$, we have stability. This means that *stability* (a) *is equivalent to* $\mathbf{F}(t, t_0, \mathbf{x}_0) \to 0$ *as* $\mathbf{x}_0 \to 0$, *uniformly in* $t \geqq t_0$.

Example 1

Let x be scalar, and $x' = \lambda x$. The solutions are $x = x_0 e^{\lambda t}$. The origin $x = 0$ satisfies: all definitions (a) to (f) if $\lambda < 0$ [see Figure 10.1,a]; definitions (a) and (b) if $\lambda = 0$ [see Figure 10.1,b]; definition (g) but none of definitions (a) to (f) if $\lambda > 0$ [see Figure 10.1,c].

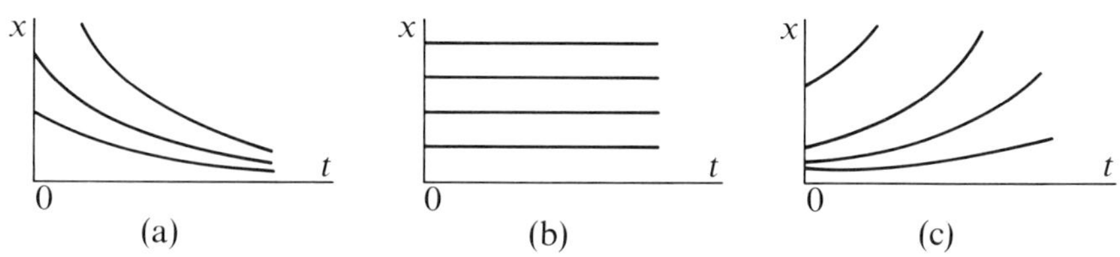

Figure 10.1

Example 2

A stable but not uniformly stable system is the following. Take the curves

$$x = x_0\left\{1 - (1 - \cos \pi t)\,\frac{2 - 2^{-t}}{4}\right\} = x_0\,\varphi(t).$$

They satisfy the differential equation

$$\frac{dx}{x} = \frac{\varphi'(t)}{\varphi(t)}\,dt$$

[see Figure 10.2]. It is easy to check that the point $x = 0$ is stable. Nevertheless, it is not uniformly stable, because by considering the solution for $t = 1, 3, 5, \ldots$ through $x(0) = 1$, we see that it assumes the values $x = 2^{-2}, 2^{-4}, 2^{-6}, \ldots$. However, for $t = 2, 4, 6, \ldots$ we see that it always returns to the value $x = 1$. Therefore no matter how small we take δ, there are always values of t_0 such that $x(t_0) < \delta$ but $x(t) = 1$ for some $t > t_0$.

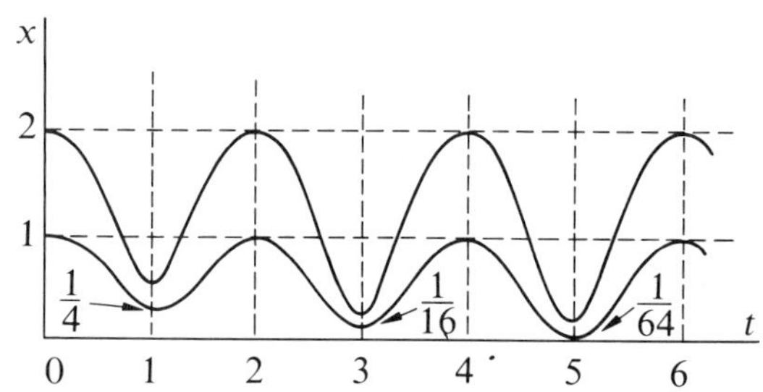

Figure 10.2

Example 3

The following system is asymptotically stable but not uniformly asymptotically stable. The differential equation

$$x' = -\frac{x}{t+1}$$

has the solution

$$x = x_0\,\frac{t_0+1}{t+1}$$

and the origin is asymptotically stable [it even satisfies (d)]. Nevertheless, if in definition (e′) we allow $\eta = \delta_0/2$, we should have

$$x(t_1) = x(t_0 + T) \leqq \tfrac{1}{2}x(t_0) = \tfrac{1}{2}\delta_0.$$

But $x(t_1) = \frac{1}{2}x(t_0)$ implies $(t_0 + 1)/(t_1 + 1) = 1/2$ or $t_1 = 2t_0 + 1$, and no T can work for all values of t_0.

Example 4

This system satisfies definition (c′) but not (a) [hence not (c)]. Let $\mathbf{x}$ be a point in the plane and the system be given in polar coordinates by

$$\theta' = \sin^2 \frac{\theta}{2}, \qquad r' = r(1 - r).$$

The trajectories are represented in Figure 10.3. The point $P(r = 1, \theta = 0)$ is a rest point to which all trajectories tend as $t \to +\infty$ (it is an *attractor*). On the other hand, it is obvious that it is unstable.

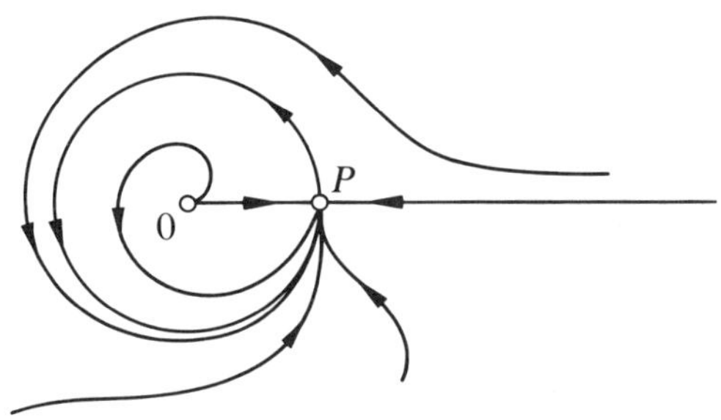

Figure 10.3

Example 5

It is more difficult to construct an example of a system satisfying (c) but not (d). Let x, y be coordinates in the plane and make the change of coordinates (which depends on t)

$$\begin{cases} x = re^{-t} \cos \theta \\ y = r \sin \theta. \end{cases}$$

For r and θ, we give the differential equations

$$r' = 0, \qquad \theta' = \sin \theta.$$

The solution is

$$r = \text{constant}, \quad \theta = \pm 2 \arctan e^{t-t_0}, \quad \text{or} \quad \theta = 0 \quad \text{or} \quad \theta = \pi.$$

In the original coordinates

$$\begin{cases} x = re^{-t} \dfrac{1 - e^{2(t-t_0)}}{1 + e^{2(t-t_0)}} \\[2ex] y = \pm 2r \dfrac{e^{t-t_0}}{1 + e^{2(t-t_0)}}, \end{cases} \tag{10.1.5}$$

or

$$\begin{cases} x = re^{-t} \\ y = 0. \end{cases}$$

In these formulas, t_0 corresponds to the maximum of $y(t)$, for which $y(t_0) = \pm r$ [see Figure 10.4]. The trajectories lie on the surface $r =$ constant.

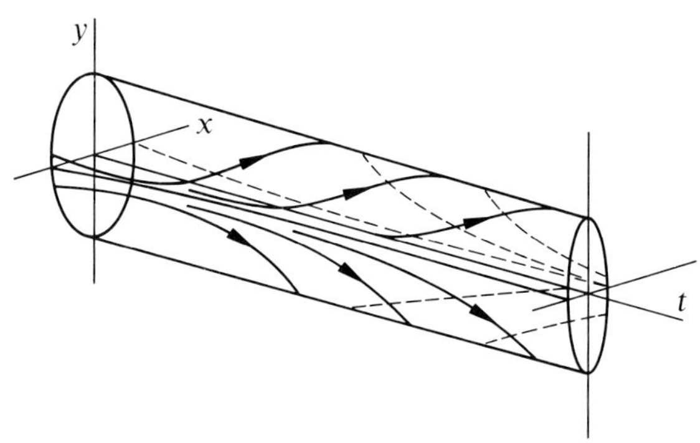

Figure 10.4

For any initial values x_i, y_i, and t_i we can find the constants r, t_0 of the trajectory (10.1.5) from

$$\begin{cases} r^2 = x_i^2 e^{2t_i} + y_i^2 \\[2ex] t_i - t_0 = -\arctan \left\{ \dfrac{x_i}{r} e^{t_i} \right\}. \end{cases} \tag{10.1.6}$$

Given $t_i \geqq 0$ and $\delta > 0$, we choose x_i, y_i such that $r^2 < \delta$ is some fixed value, while $t_0 - t_i$ is arbitrarily large. This can be done, according to (10.1.6), taking x_i positive sufficiently small. Then $y(t_0) = r$ for arbitrarily large values of t_0, and η in definition (d') cannot be given arbitrarily small. This proves that $x = y = 0$ is not equi-asymptotically stable.

1.3 Relations between the different types of stability

We will denote the different types of stability defined in Section 1.2 by the letter (a) to (g) of its definition. The following implications are immediate:

$$(\mathrm{b}) \Rightarrow (\mathrm{a}), \qquad (\mathrm{c}) \Rightarrow (\mathrm{a}), \qquad (\mathrm{e}) \Rightarrow (\mathrm{b}),$$
$$(\mathrm{e}') \Rightarrow (\mathrm{d}') \Rightarrow (\mathrm{c}'), \qquad (\mathrm{f}) \Rightarrow (\mathrm{e}) \Rightarrow (\mathrm{d}) \Rightarrow (\mathrm{c}).$$

The following lemmas are less obvious.

1.4 Lemma

For autonomous or periodic systems, (a) ⇒ (b).

Proof

This lemma is evident for autonomous systems ($\mathbf{x}' = \mathbf{f}(\mathbf{x})$). For periodic systems, let us assume that the period is 1. Let us assume also that (a) is true but that (b) is not. There exists then $\varepsilon > 0$ and sequences $\mathbf{x}_{0n} \to 0$, t_{0n} and t_n with $t_n > t_{0n}$ ($n = 1, 2, 3, \ldots$) such that

$$\|\mathbf{F}(t_n, t_{0n}, \mathbf{x}_{0n})\| > \varepsilon.$$

Due to periodicity, we may assume that all $t_{0n} \in [0, 1]$. This interval being compact, the sequence t_{0n} has a limit point and we may assume $t_{0n} \to t_0$.

By continuity of $\mathbf{F}(t, t_0, \mathbf{x}_0)$,

$$\mathbf{x}_{1n} = \mathbf{F}(0, t_{0n}, \mathbf{x}_{0n})$$

exists for n sufficiently large, and $\mathbf{x}_{1n} \to 0$. But since the origin is stable,

$$\mathbf{F}(t_n, t_{0n}, \mathbf{x}_{0n}) = \mathbf{F}(t_n, 0, \mathbf{x}_{1n}) \to 0 \quad \text{as} \quad n \to \infty,$$

contrary to our assumptions. This proves the lemma.

1.5 Lemma

For autonomous or periodic systems, (c) $\Rightarrow$ (e).

Proof

From (c) it follows that there exists $\delta_1 > 0$ such that $\|\mathbf{x}\| < \delta_1$ implies $\mathbf{F}(t, 0, \mathbf{x}) \to 0$ as $t \to +\infty$. By continuity of $\mathbf{F}$ there is $\delta_0 > 0$ such that $\|\mathbf{x}\| < \delta_0$, $0 \leqq t_0 \leqq 1$ imply that $\mathbf{F}(t, t_0, \mathbf{x}_0)$ exists for $0 \leqq t \leqq t_0$ and $\|\mathbf{F}(0, t_0, \mathbf{x}_0)\| < \delta_1$.

If the origin were not uniformly asymptotically stable, there would exist $\varepsilon > 0$ and sequences $\mathbf{x}_{0n}$, t_{0n}, T_n ($n = 1, 2, 3, \ldots$) such that $\mathbf{x}_{0n} < \delta_0$, $T_n \to +\infty$, $\|\mathbf{F}(t_{0n} + T_n, t_{0n}, \mathbf{x}_{0n})\| \geqq \varepsilon$. As in the previous lemma, we may assume $0 \leqq t_{0n} \leqq 1$ and that the sequences t_{0n} and $\mathbf{x}_{0n}$ have some limits t_0, $\mathbf{x}_0$. Then $\|\mathbf{x}_0\| \leqq \delta_0$, $0 \leqq t_0 \leqq 1$. Therefore $\mathbf{F}(0, t_{0n}, \mathbf{x}_{0n}) \to \mathbf{F}(0, t_0, \mathbf{x}_0) = \mathbf{x}_1$. As before, $\|\mathbf{x}_1\| < \delta_1$, and according to (c) there is an integer $N > 0$ such that $\|\mathbf{F}(N, t_0, \mathbf{x}_0)\| = \|\mathbf{F}(N, 0, \mathbf{x}_1)\| < 1/2\,\delta(\varepsilon)$, where $\delta(\varepsilon)$ is the function characterizing the uniform stability (which holds according to the preceding lemma).

Again by continuity of $\mathbf{F}$, $\mathbf{F}(0, t_{0n}, \mathbf{x}_{0n}) \to \mathbf{x}_1$; hence for sufficiently large n,

$$\|\mathbf{F}(N, t_{0n}, \mathbf{x}_{0n})\| = \|\mathbf{F}(N, 0, \mathbf{F}(0, t_{0n}, \mathbf{x}_{0n}))\| < \delta(\varepsilon),$$

and $\|\mathbf{F}(t, t_{0n}, \mathbf{x}_{0n})\| < \varepsilon$ for every $t \geqq t_{0n}$.

But this would contradict the hypothesis that for every n, no matter how large, there are values T_n as large as we please such that

$$\|\mathbf{F}(t_{0n} + T_n, t_{0n}, \mathbf{x}_{0n})\| \geqq \varepsilon.$$

This proves the lemma.

1.6 Lemma

Under the assumptions made in this chapter, (d′) $\Rightarrow$ (a); *hence* (d′) $\Rightarrow$ (d).

Proof

The proof is similar to the proofs of the two preceding lemmas. For every $t_0 \geqq 0$, $\varepsilon > 0$, there are $\eta(t_0) > 0$ and $T(\varepsilon, t_0)$ such that $\|\mathbf{x}_0\| < \eta$, $t \geqq t_0 + T$ imply $\|\mathbf{F}(t, t_0, \mathbf{x}_0)\| < \varepsilon$. By continuity of $\mathbf{F}$, on the other hand, there is $\delta_0 > 0$ such that $\|\mathbf{x}_0\| < \delta_0$, $t_0 \leqq t \leqq t_0 + T$ imply $\|\mathbf{F}(t, t_0, \mathbf{x}_0)\| < \varepsilon$. The smaller of the two values η, δ_0, satisfies the condition of $\delta(\varepsilon)$ of definition (a).

1.7 Other definitions

For practical applications, the above definitions have the inconvenience of being mathematical concepts of a "local" character. They only depend on—and also characterize—properties of the system in some neighborhood of the rest point. Nothing is said about the size of this neighborhood. Thus in Figure 10.5, a heavy sphere under the action of its own weight has a stable equilibrium at (a), is unstable at (b), while in (c) the equilibrium is mathematically stable but for practical purposes one may consider it unstable, as it is sufficient to move the sphere very little to make it roll away.

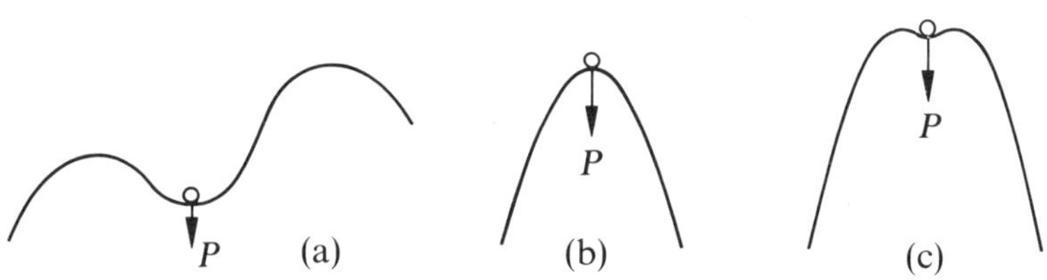

Figure 10.5

The asymptotically stable equilibrium, which is for practical purposes the most important, is only effective when the δ_0 neighborhood of definition (c′) is sufficiently large (and this depends, of course, of the possible disturbances which depend in turn on the physical conditions of the problem).

Definition

We define the region of attraction of the asymptotically stable rest point $\mathbf{x}_0$ *as the set of points such that all trajectories through them tend to* $\mathbf{x}_0$ *as* $t \to +\infty$.

If the region of attraction is the whole space, that is, all trajectories tend to $\mathbf{x}_0$ *as* $t \to +\infty$, *then* $\mathbf{x}_0$ *is said to be globally asymptotically stable.*

The practical importance of the notion of stability comes from the fact that every physical measurement is affected by some error. Therefore if we find that a system is in its equilibrium state, we can only be sure that the system will really remain there if the point is stable. Otherwise some very small error in the initial position could make it drift away from equilibrium.

The initial position is not the only thing normally affected by error, however. The differential equation itself has, physically speaking, only an approximate character. This may derive from lack of precision of the theory or simply because we do not know the exact coefficients entering in the differential equation. Hence the following definition.

Definition

The origin, assumed to be a rest point of equation

$$\mathbf{x}' = \mathbf{f}(t, \mathbf{x}),$$

is said to be totally stable (or stable with respect to perturbations), if for every $t_0 \geqq 0$ and every $\varepsilon > 0$ there is a $\delta > 0$ such that for every $\|\mathbf{x}_0\| < \delta$ and every differential equation

$$\mathbf{x}' = \mathbf{g}(t, \mathbf{x}) \tag{10.1.7}$$

satisfying $\|\mathbf{f}(t, \mathbf{x}) - \mathbf{g}(t, \mathbf{x})\| < \delta$ for all t and $\mathbf{x}$, the solution of this latter equation satisfies $\|\mathbf{G}(t, t_0, \mathbf{x}_0)\| < \varepsilon$ for all $t \geqq t_0$.

Notice that, in general, $\mathbf{x} = 0$ *is not* a solution of equation (10.1.7).

Problems

1. Find the set of values of the parameter p in the following differential equations for which the origin is asymptotically stable.

   (a) $x'' + (1 - p^2)x' + px = 0.$ (b) $x'' + px' + \dfrac{p^2}{1 - p^2}\, x = 0.$

   (c) $x''' + (1 - p)x'' + (1 + p)x' + (p^2 - p)x = 0.$

2. For the homogeneous system of degree m

$$x' = X(x, y), \qquad y' = Y(x, y),$$

   where $N(x, y) = xY(x, y) - yX(x, y)$ is never zero except at the origin, give conditions insuring stability and asymptotic stability (see also Section 7.2).

3. Prove that a homogeneous system for which the origin is an isolated critical point and which has some invariant rays (see Section 7.2.8) is stable if and only if all these invariant rays are trajectories tending to the origin.
4. Prove that a homogeneous system of even degree for which the origin is an isolated critical point cannot be stable.
5. Generalize the definition of stability to any invariant set (instead of just for a single rest point).

2 *Stability of Solutions*

2.1 *Stability of a solution*

The stability of a rest point, as considered in the preceding section, is a particular case. It is, indeed, the stability of the solution $\mathbf{x}(t) = \mathbf{x}_0$, and we can apply the same ideas to any solution $\mathbf{x}(t)$. So, we will say that the solution $\mathbf{u}(t)$ is stable if, for $\|\mathbf{u}(t_0) - \mathbf{x}(t_0)\|$ sufficiently small, $\|\mathbf{u}(t) - \mathbf{x}(t)\|$ remains small for all future t.

We will reduce the stability of a solution to the case of the stability of a point. This way the main concepts will remain the same.

Consider

$$\mathbf{x} = \mathbf{u}(t), \tag{10.2.1}$$

a particular solution of the differential equation (10.1.1) defined for all $t \geqq 0$. Set $\mathbf{u}_0 = \mathbf{u}(0)$. In order to compare $\mathbf{u}(t)$ with neighboring trajectories $\mathbf{x}(t)$, we introduce the new variable

$$\mathbf{y}(t) = \mathbf{x}(t) - \mathbf{u}(t). \tag{10.2.2}$$

$\mathbf{y}(t)$ satisfies the differential equation

$$\mathbf{y}'(t) = \mathbf{x}'(t) - \mathbf{u}'(t),$$

and from (10.1.1) it follows that

$$\mathbf{y}' = \mathbf{f}(t, \mathbf{y} + \mathbf{u}(t)) - \mathbf{f}(t, \mathbf{u}(t)) = \mathbf{g}(t, \mathbf{y}). \tag{10.2.3}$$

Notice that $\mathbf{u}(t)$ does not depend on the particular solution $\mathbf{x}(t)$.

The coordinate transformation (10.2.2) obviously transforms solutions of (10.1.1) into solutions of (10.2.3) and conversely. Thus, the solution $\mathbf{x} = \mathbf{u}(t)$ is transformed into $\mathbf{y} = 0$. Hence the origin is a rest point for equation (10.2.3). Compare also Figure 10.6. All definitions of stability of a rest point (a) to (g) can be applied, without change, to the stability of a trajectory, by the transformation (10.2.2).

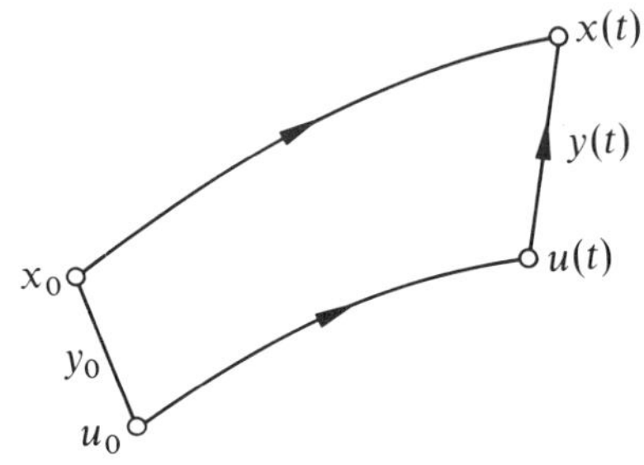

Figure 10.6

2.2 Orbital stability

This is a concept of geometric type introduced by Poincaré. Consider a system such as

$$\begin{cases} x' = y \\ y' = 0. \end{cases}$$

The trajectories are all horizontal lines but the velocity of $\mathbf{x}(t)$ on these lines is different from line to line. Two points $\mathbf{u}(t)$ and $\mathbf{x}(t)$ may be very close together at the beginning but moving on different lines they will get farther and farther apart with increasing t. Hence these trajectories are not Liapunov stable.

The same happens with the system

$$\begin{cases} x' = y(x^2 + y^2) \\ y' = -x(x^2 + y^2), \end{cases}$$

where the trajectories are concentric circles, and the angular velocity on each of them is proportional to the square of the radius. On the other hand, in both examples nearby trajectories remain near, if they are considered as curves. Thus, we have the following definition.

Definition

The trajectory $\mathbf{x} = \mathbf{u}(t)$ *is orbitally stable if for every* $t_0 \geqq 0$ *and every* $\varepsilon > 0$, *there is a* $\delta > 0$ *such that for* $\|\mathbf{x}_0 - \mathbf{u}_0\| < \delta$, *the solution* $\mathbf{x}(t) = \mathbf{F}(t, t_0, \mathbf{x}_0)$ *remains in the* ε*-neighborhood of the whole trajectory* $\{\mathbf{u}(t) \mid t \geqq t_0\}$. *That is,* $\|\mathbf{x}_0 - \mathbf{u}_0\| < \delta$ *implies that for each* $t \geqq t_0$, *there is a* $t' \geqq t_0$ *such that* $\|\mathbf{x}(t) - \mathbf{u}(t')\| < \varepsilon$.

2.3 *Variational equation*

Let us compare the solution $\mathbf{u}(t) = \mathbf{F}(t, t_0, \mathbf{u}_0)$ of equation (10.1.1) with the neighboring solutions $\mathbf{x}(t)$.

2.3a *Theorem*

Assume $\mathbf{f}(t, \mathbf{x})$ *is continuous and has continuous partial derivatives* $\partial f^i/\partial x^j$. *Assume the solution* $\mathbf{u}(t)$ *of* (10.1.1) *exists for* $t \geqq t_0$ *and let* $A(t)$ *be the jacobian matrix*

$$A(t) = \left[\frac{\partial \mathbf{f}(t, \mathbf{x})}{\partial \mathbf{x}}\right]_{\mathbf{x}=\mathbf{u}(t)} = \left(\frac{\partial f^i(t, \mathbf{x})}{\partial x^j}\right)_{\mathbf{x}=\mathbf{u}(t)}.$$

Then

$$Y(t, t_0, \mathbf{x}_0) = \frac{\partial \mathbf{F}(t, t_0, \mathbf{x}_0)}{\partial \mathbf{x}_0} = \left(\frac{\partial F^i(t, t_0, \mathbf{x}_0)}{\partial x_0^j}\right)$$

exists for $\mathbf{x}_0 = \mathbf{u}_0$ *and satisfies the differential equation*

$$Y' = \frac{\partial Y(t, t_0, \mathbf{x}_0)}{\partial t} = A(t)\,Y. \tag{10.2.5}$$

Proof

According to the differential equation,

$$x^i(t) - u^i(t) = x_0^i - u_0^i + \int_{t_0}^{t} [f^i(s, \mathbf{x}(s)) - f^i(s, \mathbf{u}(s))]\,ds.$$

By the mean value theorem,

$$f^i(s, \mathbf{x}(s)) - f^i(s, \mathbf{u}(s)) = \sum_j \frac{\partial f^i}{\partial x^j}(x^j(s) - u^j(s)),$$

where the partial derivatives $\partial f^i/\partial x^j$ have to be taken at some intermediate point.

Let $\mathbf{x}_0 - \mathbf{u}_0 = h\mathbf{v}$, where $\mathbf{v}$ is a fixed vector and $h \to 0$. Placing the following expression in the first equation and dividing by h, we obtain

$$\frac{x^i(t) - u^i(t)}{h} = v^i + \int_{t_0}^{t} \sum_j \frac{\partial f^i}{\partial x^j} \frac{x^j(s) - u^j(s)}{h} ds.$$

For fixed t, the quotient $\mathbf{x}(t) - \mathbf{u}(t)/h$ tends to the directional derivative

$$\left[\frac{d\mathbf{x}(t)}{dh}\right]_{\mathbf{x}(t)=\mathbf{u}(t)} = \left[\frac{d\mathbf{F}(t, t_0, \mathbf{x}_0)}{d\mathbf{x}_0}\right]_{\mathbf{x}_0=\mathbf{u}_0}$$

in the direction of $\mathbf{v}$. Denoting this directional derivative by $\mathbf{y}(t)$, we obtain

$$y^i(t) = \frac{dx^i(t)}{dh} = v^i + \int_{t_0}^{t} \sum_j \frac{\partial f^i}{\partial x^j} \frac{dx^j}{dh} ds.$$

Here, in the limit, the partial derivatives $\partial f^i/\partial x^j$ have to be taken at $h = 0$, i.e., for $\mathbf{x}(t) = \mathbf{u}(t)$. Therefore we may write in vector notation

$$\mathbf{y}(t) = \frac{d\mathbf{x}(t)}{dh} = \mathbf{v} + \int_{t_0}^{t} A(s) \frac{d\mathbf{x}(s)}{dh} ds = \mathbf{v} + \int_{t_0}^{t} A(s)\mathbf{y}(s)\, ds.$$

Hence $\mathbf{y}(t)$ satisfies the differential equation

$$\mathbf{y}'(t) = A(t)\mathbf{y}(t), \tag{10.2.6}$$

and the initial condition $\mathbf{y}(t_0) = \mathbf{v}$.

Now, take the following vectors for v:

$$\mathbf{v}_1 = \begin{pmatrix} 1 \\ 0 \\ 0 \\ \vdots \\ 0 \end{pmatrix}, \quad \mathbf{v}_2 = \begin{pmatrix} 0 \\ 1 \\ 0 \\ \vdots \\ 0 \end{pmatrix}, \quad \ldots, \quad \mathbf{v}_n = \begin{pmatrix} 0 \\ 0 \\ 0 \\ \vdots \\ 1 \end{pmatrix}.$$

The corresponding vectors $\mathbf{y}_j(t)$ $(j = 1, 2, \ldots, n)$ form the matrix $Y(t) = y^i_j(t)$. However, $y^i_j(t)$ is the derivative of $x^i(t)$ in the initial direction v_j and so

$$y^i_j(t) = \partial F^i(t, t_0, x_0)/\partial x^j_0.$$

Therefore the matrix $Y(t)$ so defined is identical with the matrix $Y(t, t_0, \mathbf{x}_0)$ of the statement of the theorem. From equation (10.2.6) we know that it satisfies (10.2.5).

Vector differential equation (10.2.6) [also matrix differential equation (10.2.5)] is called the "*variational equation*" of equation (10.1.1).

2.4 *The case of periodic solutions*

The most important case occurs when equation (10.1.1) is autonomous, that is, $\mathbf{x}' = \mathbf{f}(\mathbf{x})$, and when the solution $\mathbf{x} = \mathbf{u}(t)$ is periodic, say of period τ. Then the matrix $A(t)$ in (10.2.5–6) is also periodic of period τ and the variational equation is linear periodic. The same happens, of course, in the more general case in which equation (10.1.1) and the solution $\mathbf{x} = \mathbf{u}(t)$ are both periodic of period τ.

2.5 *Stability of transformations*

Given the periodic differential equation $\mathbf{x}' = \mathbf{f}(t, \mathbf{x})$, with $\mathbf{f}(t, \mathbf{x})$ periodic of period τ, one of the important problems is to establish if a periodic solution $\mathbf{x}(t)$ of the same period exists. For this it is convenient to consider the transformation T defined by

$$T[\mathbf{x}] = \mathbf{F}(\tau, 0, \mathbf{x}). \tag{10.2.7}$$

Then

$$T^2[\mathbf{x}] = T[T[\mathbf{x}]] = \mathbf{F}(2\tau, 0, \mathbf{x})$$

and so on for the successive powers $T^n[\mathbf{x}]$.

If $\mathbf{x}(t_0) = \mathbf{x}_0$ is a point of a periodic solution, then $\mathbf{x}(t_0 + \tau) = \mathbf{x}_0$. Hence

$$T[\mathbf{x}_0] = \mathbf{x}_0$$

and conversely. This means that $\mathbf{x}_0$ is a *fixed point* of the transformation T. For such a fixed point it is possible to give definitions of stability similar to those for solutions of differential equations. Let us state the two most important definitions.

Definitions

The fixed point $\mathbf{x}_0$ *of the transformation* $T[\mathbf{x}]$ *is stable if for every* $\varepsilon > 0$ *there is a* $\delta > 0$, *such that* $\|\mathbf{x} - \mathbf{x}_0\| < \delta$ *implies* $\|T^n[\mathbf{x}] - \mathbf{x}_0\| < \varepsilon$ *for every positive integer* n.

The fixed point $\mathbf{x}_0$ is asymptotically stable, if it is stable and furthermore there is some $\eta > 0$ such that $\|\mathbf{x} - \mathbf{x}_0\| < \eta$ implies $T^n[\mathbf{x}] \to \mathbf{x}_0$ as $n \to +\infty$.

There is no need to stress the similarity of these definitions with those given previously.

Problems

1. For differential systems (a)–(e) which are written in the form of polar coordinates, do the following: write the system in cartesian coordinates; write the variational equations; and find out which of the solutions are Liapunov stable and which are orbitally stable.

   (a) $r' = 0,\ \theta' = 1.$ (b) $r' = 0,\ \theta' = r.$

   (c) $r' = 0,\ \theta' = \dfrac{1}{r}.$ (d) $r' = r,\ \theta' = 1.$

   (e) $r' = -r,\ \theta' = 1.$

2. For the following linear transformations $T\colon R^2 \to R^2$ given by $(x_2, y_2) = T[(x_1, y_1)]$, the origin is a fixed point. Find the characteristic multipliers μ and the correspondent eigenvectors (x, y) such that $T[(x, y)] = (\mu x, \mu y)$.

   (a) $\begin{cases} x_2 = 2x_1 \\ y_2 = \frac{1}{2}y_1 \end{cases}$ (b) $\begin{cases} x_2 = x_1 + y_1 \\ y_2 = x_1 - y_1 \end{cases}$

   (c) $\begin{cases} x_2 = y_1 \\ y_2 = x_1 \end{cases}$ (d) $\begin{cases} x_2 = y_1 \\ y_2 = -x_1 \end{cases}$

   (e) $\begin{cases} x_2 = \frac{1}{2}x_1 + \frac{1}{2}y_1 \\ y_2 = \frac{1}{2}x_1 - \frac{1}{2}y_1 \end{cases}$ (f) $\begin{cases} x_2 = y_1 \\ y_2 = 0 \end{cases}$

   (g) $\begin{cases} x_2 = x_1 + y_1 \\ y_2 = 0 \end{cases}$ (h) $\begin{cases} x_2 = \frac{1}{2}x_1 - \frac{1}{2}y_1 \\ y_2 = \frac{1}{2}x_1 - \frac{1}{2}y_1. \end{cases}$

3. Show that for the linear transformation

$$\begin{cases} x_2 = ax_1 + by_1 \\ y_2 = cx_1 + dy_1 \end{cases}$$

   the origin is asymptotically stable if and only if the characteristic multipliers μ_1, μ_2 are of absolute value less than one. Show also that if for both characteristic multipliers, $|\mu_i| \leqq 1$, then the origin is stable except in the case when the matrix of the transformation has canonical form

$$\begin{pmatrix} \mu & 1 \\ 0 & \mu \end{pmatrix}.$$

4. Compare the definition of orbital stability with the generalization of the definition of stability given in Problem 5 (page 311).

3 Direct Method of Liapunov

3.1 Idea of the method

If the general solution of a differential equation is known, it is relatively easy to establish the stability or instability of a given rest point. Unfortunately, in many cases it is practically and even theoretically impossible to obtain the solution of a given differential equation in closed form. Therefore it is important to have other methods which allow one to assert the stability or instability of a given rest point, without the need of obtaining the general solution.

Liapunov developed two methods to do this in his fundamental work on stability. The first one is based on the series expansions of the solutions of a differential equation, from which we can obtain conclusions regarding their asymptotic behavior and stability.

The so-called *second method of Liapunov*, also called the *direct method*, compares the given (in most cases vector) differential equation, with a simpler scalar differential equation for a so-called *Liapunov function.* From the behavior of this comparison equation it is possible to draw conclusions about the stability of the given one.

The basic idea of this method comes from physics and particularly from mechanics, where the total energy of a system is such a Liapunov function. For example, assume that for a mechanical system, $x(t)$ represents the position, $y = x'$ the velocity, and let the equation of motion be acceleration $= y' = f(x, y)$. Assume that the forces producing this acceleration derive from a *potential energy* $V(x, y)$. The kinetic energy may be given by $k(y^2/2)$, where k is a constant. If the system is *dissipative*, i.e., dissipates energy, the time derivative

$$\frac{d}{dt}\left[V(x, y) + \frac{ky^2}{2}\right]$$

can never be positive, i.e., the total energy $W(x, y) = V(x, y) + ky^2/2$ can never increase. Now, if $(x_0, y_0 = 0)$ is a rest point of the system, and if $W(x, y)$ has at (x_0, y_0) a strict minimum, this means that the system can never leave the point (x_0, y_0) without an additional supply of some finite amount of energy. So, infinitesimal disturbances cannot bring the system out of the equilibrium position—this means stability. Essentially this is a theorem of Lagrange,who states that the stable equilibrium positions correspond to minima of the potential energy (the kinetic energy being zero at rest).

3.2 Definite functions

Let, as usual, $\mathbf{x}$ be the "state" vector and t the "time." We will consider scalar functions $v(t, \mathbf{x})$, defined in some half-cylinder $R^+ \times S_a$, where R^+ is the half-line $[0, +\infty)$ and S_a is the n-ball in R^n of radius a and center at the origin. We will also assume that $v(t, \mathbf{x})$ is continuously differentiable.

We shall say that $v(t, \mathbf{x})$ is positive definite if

(a) $v(t, 0) = 0$ for all $t \geqq 0$, and

(b) to every $\varepsilon > 0$ there corresponds $\alpha(\varepsilon) > 0$ such that $\|\mathbf{x}\| > \varepsilon$ implies $v(t, \mathbf{x}) > \alpha$ for all $t \geqq 0$.

By continuity of $v(t, \mathbf{x})$, we have $\lim \alpha(\varepsilon) = 0$ as $\varepsilon \to 0$. It is easy to see that we can choose $\alpha(\varepsilon)$ to be nondecreasing and that we may always assume it to be continuous.

Further, we see that condition (b) can be stated as follows:

> *there is a function $\alpha(\varepsilon)$ continuous and nondecreasing, $\alpha(0) = 0$, $\alpha(\varepsilon) > 0$ for $\varepsilon > 0$, such that $v(t, \mathbf{x}) \geqq \alpha(\|\mathbf{x}\|)$.*

Notice that (b) is stronger than $v(t, \mathbf{x}) > 0$ for $\|\mathbf{x}\| > 0$; for example the function $v = x/(t + 1)$ satisfies this last condition but not (b).

A negative definite function is defined similarly.

Consider the total derivative of $v(t, \mathbf{x})$, that is, the derivative of $v(t, \mathbf{x}(t))$ along any trajectory $\mathbf{x}(t)$ of the differential equation (10.1.1). This derivative is

$$v' = \frac{dv}{dt} = \frac{\partial v}{\partial t} + \frac{\partial v}{\partial \mathbf{x}} \mathbf{f}(t, \mathbf{x}), \tag{10.3.1}$$

as $d\mathbf{x}/dt = \mathbf{f}(t, \mathbf{x})$. Here $\partial v/\partial \mathbf{x}$ is to be considered a row vector $[\partial v/\partial x^1, \ldots, \partial v/\partial x^n]$ and $\mathbf{f}(t, \mathbf{x})$ as a column vector $[f^1(t, x^1, \ldots, x^n), \ldots, f^n(t, x^1, \ldots, x^n)]^T$. Notice that the value of dv/dt can be computed as a function of the point $(t, \mathbf{x})$, without knowing the explicit equation of the trajectories $\mathbf{x}(t)$, and hence without having to integrate the differential equation.

Theorem 3.3 *Sufficient conditions for stability*

> *Given the differential equation (10.1.1), defined on $R^+ \times S_a$, if there exists a function $v(t, \mathbf{x})$ also defined on $R^+ \times S_a$, positive definite and such that at each point $(t, \mathbf{x})$ the total derivative*
>
> $$\frac{dv(t, \mathbf{x})}{dt} \leqq 0, \tag{10.3.2}$$
>
> *then the origin is stable.*

Proof

According to (10.3.2), on every trajectory $\mathbf{x}(t)$ of (10.1.1),

$$v(t_1, \mathbf{x}(t_1)) = v(t_0, \mathbf{x}(t_0)) + \int_{t_0}^{t_1} \frac{dv}{dt}\, dt \leqq v(t_0, \mathbf{x}(t_0)) \qquad (10.3.3)$$

holds for $t_1 \geqq t_0$. It follows immediately from property (a) that the origin is a rest point for (10.1.1).

Let $\alpha(\varepsilon)$ be the function defined above and assume that $t_0 \geqq 0$ and ε, $0 < \varepsilon < a$, are given. Then $v(t, \mathbf{x}) < \alpha(\varepsilon)$ implies $\|\mathbf{x}\| < \varepsilon$. By continuity of $v(t, \mathbf{x})$ at $(t_0, 0)$ there is $\delta > 0$ such that $\|\mathbf{x}_0\| < \delta$ implies $v(t_0, \mathbf{x}_0) < \alpha(\varepsilon)$. Then, by (10.3.3), $v(t, \mathbf{F}(t, t_0, \mathbf{x}_0)) < \alpha(\varepsilon)$ for all $t \geqq t_0$ and the theorem is proved.

Example 6

Consider the differential equations $x' = y$, $y' = -x$. The function $v = x^2 + y^2$ is a Liapunov function. Indeed, it is positive definite and

$$\frac{dv}{dt} = 2xx' + 2yy' = 2(xy - yx) = 0.$$

Hence the origin is stable, as we have seen already in Chapter 4.

3.4 Theorem

Under the hypothesis of Theorem 3.3, *if* $v(t, \mathbf{x}) \to 0$ *as* $\mathbf{x} \to 0$ *uniformly in* $t \geqq 0$, *then the origin is uniformly stable.*

Indeed, in the proof of the preceding theorem, we see that by virtue of this additional assumption we can take δ independent of t_0.

The additional condition of this theorem is stated by some authors (for example *Massera*), by saying that "$v(t, \mathbf{x})$ has *an infinitely small upper bound.*"

3.5 Theorem

Given equation (10.1.1) *defined on* $R^+ \times S_a$, *if there exists a Liapunov function* $v(t, \mathbf{x})$, *defined and positive definite in the same region, such that* $v(t, \mathbf{x}) \to 0$ *as* $\mathbf{x} \to 0$ *uniformly in* $t \geqq 0$, *and its total derivative* $v'(t, \mathbf{x})$ *is negative definite, then the origin is uniformly asymptotically stable.*

Proof

According to the preceding theorem, the origin is uniformly stable. We must still prove that there is $\delta_0 > 0$ and, for every $\eta > 0$, $T(\eta)$ such that $\|\mathbf{x}_0\| < \delta_0$, $t_0 \geqq 0$, $t \geqq t_0 + T(\eta)$ imply $\|\mathbf{F}(t, t_0, \mathbf{x}_0)\| < \eta$.

Take δ_0 such that $\|\mathbf{x}_0\| < \delta_0$, $0 \leqq t_0 \leqq t$ imply $\|\mathbf{F}(t, t_0, \mathbf{x}_0)\| < \varepsilon_0$, where $\varepsilon_0 < a$ is such that in the region $\|\mathbf{x}\| < \varepsilon_0$ the Liapunov function is bounded $[v(t, \mathbf{x}) < M]$. This is possible because $v(t, \mathbf{x}) \to 0$ uniformly in $t \geqq 0$. For the rest of the proof, δ_0, ε_0, and M will be fixed.

Given $\eta > 0$, $\eta < \varepsilon_0$, there exists $\delta(\eta) > 0$ such that $\|\mathbf{x}_0\| < \delta(\eta)$ implies $\|\mathbf{F}(t, t_0, \mathbf{x}_0)\| < \eta$ for all $t \geqq t_0$. Let $-m(\eta) < 0$ be an upper bound of $v'(t, \mathbf{x})$ for the points $(t, \mathbf{x})$ with $\delta(\eta) \leqq \|\mathbf{x}\| \leqq \varepsilon_0$, $t \geqq 0$, and let $T(\eta) > M/m$.

We will prove that $T(\eta)$ satisfies the conditions required by the uniform asymptotic stability. We need only prove that every solution with $t_0 \geqq 0$, $\|\mathbf{x}_0\| < \delta_0$, enters during the interval $[t_0, t_0 + T(\eta)]$ into the cylinder $\|\mathbf{x}\| < \delta(\eta)$. Assuming that this takes place at $t = t_1 < t_0 + T(\eta)$, it follows that $\|\mathbf{F}(t_1, t_0, \mathbf{x}_0)\| < \delta(\eta)$. Hence $\|\mathbf{F}(t, t_0, \mathbf{x}_0)\| < \eta$ for all $t \geqq t_1$, and in particular for all $t \geqq t_0 + T(\eta)$ (see Figure 10.7).

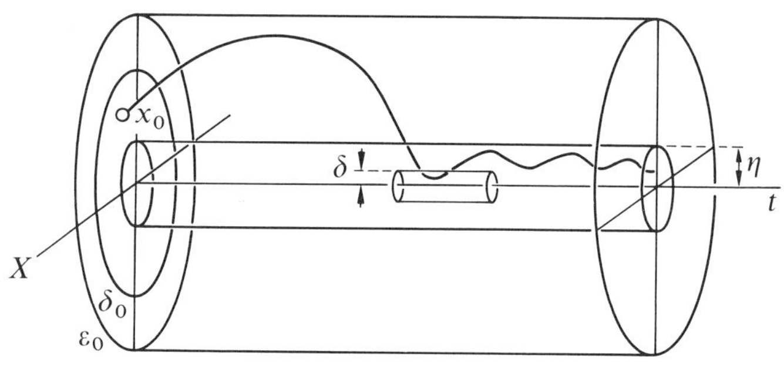

Figure 10.7

To prove that in the interval $[t_0, t_0 + T(\eta)]$ the solution enters the cylinder $\|\mathbf{x}\| < \delta(\eta)$, let us assume for a moment that this is not so. Then, for all t with $t_0 \leqq t \leqq t_0 + T(\eta)$ we would have $\|\mathbf{x}(t)\| \geqq \delta(\eta)$ and therefore $v'(t, \mathbf{x}) \leqq -m$. But then

$$v(t_0 + T(\eta), \quad \mathbf{x}(t_0 + T(\eta))) = v(t_0, \mathbf{x}_0) + \int_{t_0}^{t_0+T(\eta)} v'(t, \mathbf{x}(t))\, dt$$

$$\leqq M - mT(\eta) \leqq M - m\frac{M}{m} \leqq 0.$$

We know that this is impossible; hence the theorem is proved.

Example 7

Consider the equation $x'' + kx' + x = 0$ written as the system

$$x' = y, \qquad y' = -x - ky, \quad (k > 0).$$

As a Liapunov function we take $v = x^2 + y^2 + 2axy$, leaving the constant a to be determined later. The derivative is

$$v' = -2[(k-a)y^2 + akxy + ax^2].$$

In order that this expression has a constant sign, $a^2k^2 - 4a(k-a)$ must be negative, or

$$0 < a < \frac{4k}{k^2+4}.$$

For any value of a which satisfies this condition, v' will be negative definite if $k > 0$. To see this it is best to make $x = 0$ and observe that

$$\frac{4k}{k^2+4} = \frac{k}{1+(k/2)^2} < k.$$

It follows that for $k > 0$ it is always possible to choose a satisfying this condition; hence the origin is uniformly asymptotically stable.

3.6 ***Theorem***

If $\mathbf{f}(t, \mathbf{x})$ *and* $v(t, \mathbf{x})$ *are defined on* $R^+ \times S_a$, $v(t, \mathbf{x}) \to 0$ *uniformly in* $t \geqq 0$ *as* $\mathbf{x} \to 0$; *if, furthermore, for a given* t_0 *and arbitrarily small values of* $\mathbf{x}$, $v(t_0, \mathbf{x})$ *has positive values, and if the derivative* $v'(t, \mathbf{x})$ *is positive definite, then the origin is unstable.*

Proof

Let $\mathbf{x}_0$ be such that $v(t_0, \mathbf{x}_0) = v_0 > 0$. As v tends uniformly to zero with $\mathbf{x}$, one may assume that $v(t, \mathbf{x})$ is bounded on $\|\mathbf{x}\| \leqq a$ (changing the value of a, if this were not the case). For the same reason there is an $\varepsilon > 0$ such that $v(t, \mathbf{x}) > v_0$ implies $\|\mathbf{x}\| > \varepsilon$. As v' is nonnegative along the trajectory $\mathbf{x}(t)$ which starts at $(t_0, \mathbf{x}_0)$, we will have $v(t, \mathbf{x}(t))$ nondecreasing; hence $\|\mathbf{x}(t)\| > \varepsilon$ for $t \geqq t_0$. On the other hand $v'(t, \mathbf{x})$ is positive definite; hence, for $t \geqq t_0$ we have $v'(t, \mathbf{x}(t)) \geqq k > 0$. Therefore, if the trajectory remains in the cylinder $\|\mathbf{x}\| \leqq a$ forever, we must have $v(t, \mathbf{x}(t)) \to +\infty$, which is impossible because $v(t, \mathbf{x})$ was bounded there. Hence the trajectory must leave the cylinder and the origin is unstable.

Example 8

Consider the equation of the last example, this time with $k < 0$. Using the same Liapunov function $v(x, y)$, now with $(4k/k^2 + 4) < a < 0$, all conditions are satisfied in order to prove that the origin is unstable.

3.7 Theorem

If the differential equation (10.1.1) *is defined for all* $\mathbf{x} \in R^n$, $t \geqq 0$, *and there is a Liapunov function* $v(t, \mathbf{x})$ *also defined on* $R^+ \times R^n$ *and such that*

(a) $v(t, \mathbf{x})$ *is positive definite,*
(b) $v(t, \mathbf{x}) \to 0$ *as* $\mathbf{x} \to 0$ *uniformly in* $t \geqq 0$,
(c) $v(t, \mathbf{x}) \to +\infty$ *as* $\mathbf{x} \to \infty$, *uniformly in* $t \geqq 0$,
(d) $v'(t, \mathbf{x})$ *is negative definite,*

then the origin is globally asymptotically stable.

Proof

Conditions (a), (b), and (d) insure that the origin is asymptotically stable (Theorem 3.6). We must still prove that all solutions tend to the origin.

For any trajectory, take $(t_0, \mathbf{x}_0)$ as an arbitrary point of it. According to condition (c), there exists a constant M such that $\|\mathbf{x}\| > M$ implies $v(t, \mathbf{x}) > v(t_0, \mathbf{x}_0)$. The considered solution cannot escape to infinity, as $v'(t, \mathbf{x}) \leqq 0$ implies $v(t, \mathbf{x}(t)) \leqq v(t_0, \mathbf{x}_0)$ for $t \geqq t_0$; hence the solution must remain in the cylinder $\|\mathbf{x}\| \leqq M$. Therefore, $\mathbf{x}(t)$ exists forever and the proof of Theorem 3.6 will show that $\mathbf{x}(t)$ must tend to the origin.

3.8 Theorem

If $\mathbf{f}(t, \mathbf{x})$ *and* $v(t, \mathbf{x})$ *are defined on* $R^+ \times S_a$, *where* $v(t, \mathbf{x})$ *is positive definite and its derivative* $v'(t, \mathbf{x})$ *negative definite, and if the partial derivatives* $\partial v/\partial x^i$ *are bounded, then the origin is totally stable (that is, with respect to perturbations), and this even uniformly in the sense that the* δ *does not depend on* t_0.

Proof

Notice that the hypothesis implies that $v(t, \mathbf{x}) \to 0$ uniformly in $t \geqq 0$, since $v(t, \mathbf{x}) = |v(t, \mathbf{x}) - v(t, 0)| \leqq M \|\mathbf{x}\|$. Notice also that now we have to consider two differential equations, the basic one $\mathbf{x}' = \mathbf{f}(t, \mathbf{x})$ and the perturbed one $\mathbf{x}' = \mathbf{g}(t, \mathbf{x})$. The solutions we will denote, respectively, by $\mathbf{F}(t, t_0, \mathbf{x}_0)$ and $\mathbf{G}(t, t_0, \mathbf{x}_0)$. We must consider the derivative of $v(t, \mathbf{x})$ along solutions of both equations, and will denote them by $v'_{\mathbf{f}}$ and $v'_{\mathbf{g}}$.

Let:

$\varepsilon > 0$ be arbitrary;

$\eta_1(\varepsilon) > 0$ be such that $\|\mathbf{x}\| \geqq \varepsilon$ implies $v(t, \mathbf{x}) \geqq \eta_1(\varepsilon)$;

$\delta_1(\varepsilon) > 0$, $\delta_1(\varepsilon) < \varepsilon$, be such that $\|\mathbf{x}\| < \delta_1(\varepsilon)$ implies $v(t, \mathbf{x}) < \eta_1(\varepsilon)$;

$\left\| \dfrac{\partial v}{\partial x} \right\| \leqq M$ in $\|\mathbf{x}\| \leqq a$, $t \geqq 0$;

$\eta_2(\varepsilon) > 0$ be such that $\|\mathbf{x}\| > \delta_1(\varepsilon)$ implies $v'_f \leqq -\eta_2(\varepsilon)$;

$0 < \delta_2(\varepsilon) < \dfrac{1}{M} \eta_2(\varepsilon)$;

$\delta(\varepsilon) = \min \{\delta_1(\varepsilon), \delta_2(\varepsilon)\}$.

Assume that t_0, $\mathbf{x}_0$, and $\mathbf{g}(t, \mathbf{x})$ satisfy the conditions $\|\mathbf{x}_0\| < \delta(\varepsilon)$, $t_0 \geqq 0$, $\|\mathbf{g}(t, \mathbf{x}) - \mathbf{f}(t, \mathbf{x})\| < \delta(\varepsilon)$ for all $\|\mathbf{x}\| \leqq a$, $t \geqq 0$. We will then prove that $\|\mathbf{G}(t, t_0, \mathbf{x}_0)\| < \varepsilon$ for all $t \geqq t_0$. Let us indeed assume that this is false. There exists then some interval (t_1, t_2), $t_0 \leqq t_1 < t_2$ such that $\mathbf{x}_1 = \mathbf{G}(t_1, t_0, \mathbf{x}_0)$, $\mathbf{x}_2 = \mathbf{G}(t_2, t_0, \mathbf{x}_0)$, $\|\mathbf{x}_1\| = \delta_1(\varepsilon)$, $\|x_2\| = \varepsilon$, and $\delta_1(\varepsilon) \leqq \|\mathbf{G}(t, t_0, \mathbf{x}_0)\| \leqq \varepsilon$ for all $t_1 \leqq t \leqq t_2$. Then

$$v(t_1, \mathbf{x}_1) \leqq \eta_1(\varepsilon)$$

and

$$\begin{aligned} v'_{\mathbf{g}} = \frac{\partial v}{\partial t} + \frac{\partial v}{\partial \mathbf{x}} \mathbf{g} = v'_f + \frac{\partial v}{\partial \mathbf{x}} (\mathbf{g} - \mathbf{f}) &\leqq -\eta_2(\varepsilon) \\ &+ M\delta(\varepsilon) < -\eta_2(\varepsilon) + \eta_2(\varepsilon) = 0. \end{aligned}$$

Hence

$$v(t_2, \mathbf{x}_2) < \eta_1(\varepsilon), \qquad \text{or } \|\mathbf{x}_2\| < \varepsilon,$$

contrary to our assumption. Consequently, we have proved the theorem.

Example 9

Example 7 also applies to this case. It is easy to check that all conditions are satisfied and that the Liapunov function given there establishes the total stability of the origin.

3.9 Necessary conditions for stability

The preceding theorems provide *sufficient conditions* to insure stability, asymptotic stability, and so on. There is interest in inverting these theorems giving *necessary conditions* of the same type. Consider for example the case of ordinary Liapunov stability. In order to determine if the origin is stable, we may seek a Liapunov function satisfying the conditions of Theorem 3.3. If we find such a function, we have proved that the origin is stable. But there is the logical possibility that such a function does not exist, yet nevertheless the origin is stable. If this could happen then the question of stability could not be completely decided by this method. But if we can prove the converse theorem—that is, that if the origin is stable, then there is necessarily some Liapunov function with the appropriate properties—then there remains no doubt in this case. If the origin is stable, it is only a question of skill to find the (actually, any) Liapunov function with the indicated properties.

In general it is much more difficult to give *necessary and sufficient conditions.* If we are only interested in obtaining a sufficient condition, we may ask for stronger assumptions than are really necessary in order to make the proof easier [for example, conditions of differentiability]. But if we do this, it will be impossible to find the converse theorem. Indeed, for some cases there may not exist a Liapunov function satisfying these overly restrictive conditions. A necessary and sufficient condition is quite critical; it is indeed an *equivalent* condition to the one under consideration.

Here we will restrict ourselves to giving only one converse theorem, the one corresponding to ordinary stability.

3.10 Theorem

If the origin is a stable rest point of the differential equation (10.1.1), *then there is a Liapunov function* $v(t, \mathbf{x})$, *positive definite, such that* $v'(t, \mathbf{x}) \leqq 0$.

Proof

We will construct the Liapunov function. For this let us consider, in $(t, \mathbf{x})$ space, the "tube" of trajectories $\{(t, \mathbf{x}) \mid \mathbf{x} = \mathbf{F}(t, 0, \mathbf{x}_0), t \geqq 0, \|\mathbf{x}_0\| = r\}$ for a given $r > 0$. If r is sufficiently small, by virtue of the stability and continuous dependence on the initial conditions, these trajectories are defined for all $t \geqq 0$, and constitute a hypersurface V_r with the following properties:

(a) For each $T \geqq 0$, the intersection of V_r with the hyperplane $t = T$ is an $n - 1$ dimensional topological sphere (that is homeomorphic to the sphere $\|\mathbf{x}\| = 1$ in R^n). This topological sphere separates the hyperplane $t = T$ into two regions, the bounded one containing the origin $\mathbf{x} = 0$.

(b) For each $\varepsilon > 0$ there is a $\delta > 0$ such that $r < \delta$ implies V_r is contained in the cylinder $\|\mathbf{x}\| < \varepsilon$.

(c) The vectors $\mathbf{f}(t, \mathbf{x})$ and therefore the trajectories of (10.1.1) "do not leave" the interior of V_r through its surface; indeed, the trajectories lie on V_r; hence the vectors $\mathbf{f}(t, \mathbf{x})$ are always tangent to this surface (see Figure 10.8).

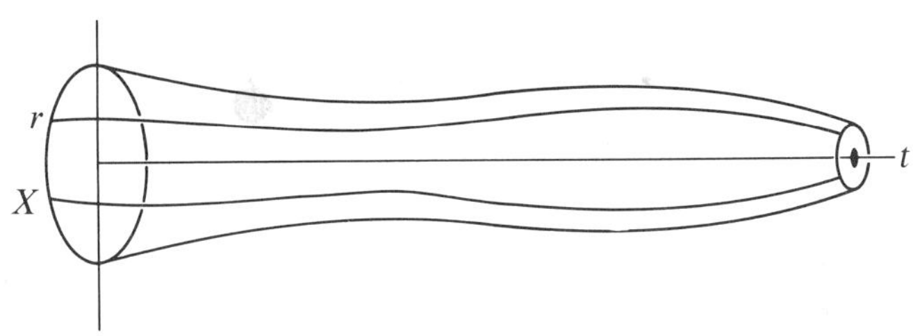

Figure 10.8

We may take the function

$$v(t, \mathbf{x}) = \|\mathbf{F}(0, t, \mathbf{x})\| = \|\mathbf{x}_0\| \tag{10.3.3}$$

as Liapunov function. This function is indeed positive definite. If $\|\mathbf{x}\| > \varepsilon$, by (b) we have $\|\mathbf{x}_0\| > \delta$. According to (c), $v'(t, \mathbf{x}) \leqq 0$ (actually $= 0$).

Comparing this with Theorem 10.3.3, we find the following differences:

The Liapunov function constructed here is not necessarily defined on some cylinder $\|\mathbf{x}\| \leqq r_0$, but only on the interior of a "tube" V_{r_0} for some r_0 (the cross section of this tube may tend to zero for $t \to +\infty$).

The constructed $v(t, \mathbf{x})$ is continuous by continuous dependence of $\mathbf{F}(t, t_0, \mathbf{x}_0)$ on the initial data. But in order to insure its differentiability, we must make the additional assumption on the differential equation, that $F(.,.,.)$ depends differentiably on the initial data.

Problems

1. With the aid of the Liapunov function $v = x^2 + 2y^2$, show that the origin is stable for the system

$$\begin{cases} x' = -x^3 - 2y^2 \\ y' = xy - y^3. \end{cases}$$

2. Consider the system

$$\begin{cases} x' = -k_1 x - b_1 \varphi(\sigma) \\ y' = -k_2 y - b_2 \varphi(\sigma) \\ \sigma' = c_1 x + c_2 y - \rho\varphi(\sigma), \end{cases}$$

where $k_1 > 0$, $k_2 > 0$, b_1, b_2, c_1, c_2, and ρ are given constants and $\varphi(\sigma)$ is a continuous function defined for all σ, such that $\varphi(0) = 0$, $\sigma\varphi(\sigma) > 0$ for all $\sigma \neq 0$, and the integrals $\int_0^{\pm\infty} \varphi(\sigma)\, d\sigma$ diverge.

With the aid of the Liapunov function

$$v(x, y, \sigma) = x^2 + y^2 + \int_0^\sigma \varphi(\sigma)\, d\sigma$$

show that

$$\rho > \frac{(b_1 - c_1/2)^2}{2k_1} + \frac{(b_2 - c_2/2)^2}{2k_2}$$

is a sufficient condition insuring the stability of the origin. (Lefschetz, *Stability of Nonlinear Control Systems*, pp. 14–15.) This kind of stability for every function $\varphi(\sigma)$ satisfying the above conditions is called "*absolute stability*."

---

4 *Stability of Particular Systems*

4.1 Linear systems

Given the linear system

$$\mathbf{x} - A(t)\mathbf{x} = \mathbf{f}(t), \tag{10.4.1}$$

in order to examine the stability of the solution

$$\mathbf{x} = \mathbf{u}(t), \tag{10.4.2}$$

we consider the difference

$$\mathbf{y} = \mathbf{x} - \mathbf{u}(t). \tag{10.4.3}$$

This satisfies the differential equation

$$\mathbf{y}' - A(t)\mathbf{y} = 0,$$

and $\mathbf{x} = \mathbf{u}(t)$ corresponds to the trivial solution $\mathbf{y} \equiv 0$.

Therefore, *the stability properties of all solutions of* (10.4.1) *are the same and equal to the stability properties of the homogeneous equation.* Furthermore, *these stability properties are independent of the right-hand side* $f(t)$.

4.2 *Theorem*

For the homogeneous linear system

$$\mathbf{x}' = A(t)\mathbf{x}, \tag{10.4.4}$$

the stability of the origin is equivalent to the boundedness of all solutions for $t \geqq 0$.

Proof

If the origin is stable, there are $\varepsilon > 0$, $\delta > 0$ such that $\|\mathbf{x}_0\| < \delta$ implies $\|\mathbf{F}(t, 0, \mathbf{x}_0)\| < \varepsilon$ for all $t \geqq 0$. From the linearity of the equation it follows that every solution $\mathbf{x}(t)$ of (10.4.4) is a constant multiple of some solution with initial value $<\delta$. Hence $\|\mathbf{x}(t)\| < M\varepsilon$ for all $t \geqq 0$ and some constant M.

Conversely, all solutions of (10.4.4) are linear combinations of n linearly independent ones. We can therefore give a single constant M such that for all solutions, $\|\mathbf{x}(t)\| < M\|\mathbf{x}(0)\|$ is true for $t \geqq 0$. Hence in the definition of stability, we may take $\delta = \varepsilon/M$ if $t_0 = 0$. In case $t_0 \neq 0$, the same reasoning is applicable with $M = M(t_0)$.

4.3 Linear systems with constant coefficients

In the case of constant coefficients, the solution of the homogeneous equation is explicitly known, and is a linear combination of real or complex exponentials $e^{\lambda t}$, possibly multiplied by polynomials in t (see Chapter 4). We may therefore state:

If all characteristic roots λ_i of the matrix A of the differential equation

$$\mathbf{x}' = A\mathbf{x}$$

have negative real parts, then the system is stable and even asymptotically stable.

If some characteristic root λ_i of the matrix A has a positive real part, then the system is unstable.

If some characteristic roots λ_i have zero real parts, while for all others the real part is negative, then the system will be stable provided the roots with zero real part are simple or, if multiple, the corresponding canonical form is purely diagonal.

The ordinary stability is the limiting case between asymptotic stability and instability. For practical purposes, asymptotic stability is the most interesting case, from which ordinary stability follows as a limiting case.

The question of asymptotic stability for linear systems with constant coefficients is therefore to insure that a certain polynomial (the characteristic polynomial of the matrix A)

$$P(z) = a_0 z^n + a_1 z^{n-1} + \cdots + a_n \tag{10.4.5}$$

has all its roots with negative real parts.

For the case of real coefficients, a necessary condition for this is that all coefficients of $P(z)$ have the same sign. This results from considering the decomposition of $P(z)$ into linear and quadratic factors. These factors are of the form $(z + \alpha)$ if $z = -\alpha$ is a real root, and

$$(z + \beta + i\gamma)(z + \beta - i\gamma) = z^2 + 2\beta z + \beta^2 + \gamma^2$$

if $z = -\beta \pm i\gamma$ are complex roots. Thus only positive coefficients appear in the product.

4.4 Routh-Hurwitz condition

We form the following determinants:

$$D_1 = a_1, \qquad D_2 = \begin{vmatrix} a_1 & a_0 \\ a_3 & a_2 \end{vmatrix}, \qquad D_3 = \begin{vmatrix} a_1 & a_0 & 0 \\ a_3 & a_2 & a_1 \\ a_5 & a_4 & a_3 \end{vmatrix}, \ldots,$$

$$D_n = \begin{vmatrix} a_1 & a_0 & 0 & 0 & 0 & \cdots & \cdot \\ a_3 & a_2 & a_1 & a_0 & 0 & \cdots & \cdot \\ a_5 & a_4 & a_3 & a_2 & a_1 & \cdots & \cdot \\ \cdot & \cdot & \cdot & \cdot & \cdot & \cdot & \cdot \\ a_{2n-1} & a_{2n-2} & a_{2n-3} & a_{2n-4} & a_{2n-5} & \cdots & a_n \end{vmatrix}$$

where a_i should be replaced by zero if $i < 0$ or $i > n$. Let us assume that $a_0 > 0$. *In order that all the roots of* (10.4.5) *have negative real parts, it is necessary and sufficient that all these determinants be positive.*

The proof of this is somewhat lengthy and will not be given here.

Example 10

In order that the polynomial

$$z^3 + a_1 z^2 + a_2 z + a_3$$

have all its roots with negative real parts, it is necessary and sufficient that

$$a_1 > 0, \qquad a_1 a_2 - a_3 > 0, \qquad \begin{vmatrix} a_1 & 1 & 0 \\ a_3 & a_2 & a_1 \\ 0 & 0 & a_3 \end{vmatrix} = a_1 a_2 a_3 - a_3^2 > 0.$$

This is therefore the condition for

$$x''' + a_1 x'' + a_2 x' + a_3 x = 0$$

to be asymptotically stable.

For high-order systems, this condition is quite lengthy to apply because of the high-order determinants which must be evaluated. In such a case the next criterion may be easier.

4.5 Sturm sequences

Given $P(z)$ by (10.4.5), assume $a_0 = 1$ and let $Q(z) = a_1 z^{n-1} + a_3 z^{n-3} + \ldots,$ whose last term is $a_{n-1}z$ or a_n (according to the parity of n). We apply the Euclidean algorithm to $P(z)/Q(z)$ and obtain the sequence

$$P = Qq_1 + p_2, \qquad Q = p_2 q_2 + p_3, \qquad p_2 = p_3 q_3 + p_4, \ldots$$

where

$$q_1 = b_1 z + 1, \qquad q_2 = b_2 z, \qquad q_3 = b_3 z, \ldots .$$

A necessary and sufficient condition for $P(z)$ to have all its roots with negative real parts, is that all numbers b_1, b_2, ... be positive.

4.6 Application of complex integrals

If the differentiable Jordan curve $\mathscr{C}$ is the boundary of the domain $\mathscr{D}$ in the complex plane, then a well-known formula of the theory of analytic functions of a complex variable states that as the variable z describes $\mathscr{C}$, the total increase in the argument of $P(z)$ is 2π times the number of zeros minus the number of poles of $P(z)$ in $\mathscr{D}$:

$$N - P = \frac{1}{2\pi} \int_{\mathscr{C}} d(\arg P(z)).$$

In our case $P(z)$ is a polynomial and has no poles. Therefore the condition that $P(z)$ have no roots in the complex half-plane corresponding to positive real parts is equivalent to asking that this integral be zero when taken along a curve $\mathscr{C}$, as represented in Figure 10.9. Here $\mathscr{C}$ is composed of a segment of the imaginary axis and a semicircle of radius R, and the integral should be zero for all values of R.

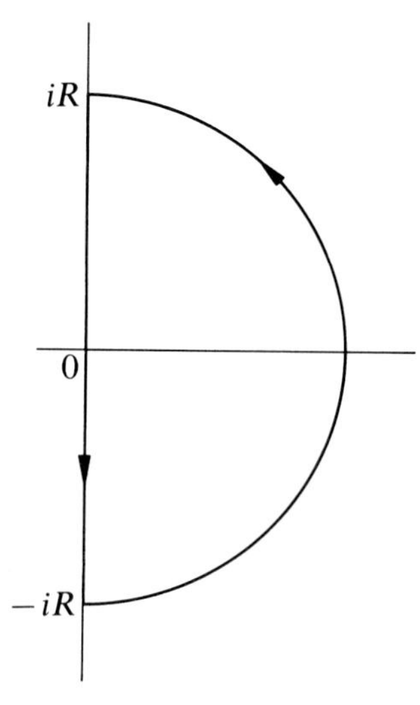

Figure 10.9

In order to evaluate this integral, let us consider the semicircle. R can be taken large enough so that $P(z) \sim z^n$, so that we may put

$$\Delta \arg P(z) = \Delta \arg z^n = n\, \Delta \arg z = n\pi.$$

For the integral along the imaginary axis we write $z = i\omega$; hence $P(z) = P(i\omega)$. We let ω vary from $+\infty$ to $-\infty$. Observe that $P(i\omega)$ is the reciprocal of the transfer function (see Chapter 5).

Example 11

Let $P(z) = z^3 + z^2 + 4z + 1$. Then $P(i\omega) = -i\omega^3 - \omega^2 + 4i\omega + 1$. Writing $P(z) = u + iv$, we have the parametric representation

$$\begin{cases} u = 1 - \omega^2 \\ v = 4\omega - \omega^3, \end{cases}$$

and this curve is represented graphically in Figure 10.10. We have to close this curve with something like a very large circle, in such a way that on it $\Delta \arg P(z) = 3\pi$. This has been done in Figure 10.10, where we see that the origin is not enclosed, algebraically speaking, by this curve. Indeed, if we take a half-line through the origin and count the intersections of it with the curve $P(i\omega)$, taking $+1$ for each intersection

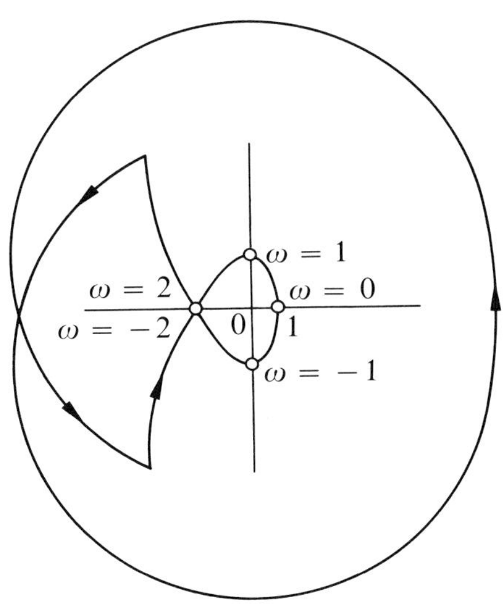

Figure 10.10

where $P(i\omega)$ goes counterclockwise and -1 for each intersection where $P(i\omega)$ goes clockwise, the total result is zero. Hence, for the entire closed curve $\mathscr{C}$, $\Delta \arg P(z) = 0$ and there are no roots with positive real parts. That there are no roots with zero real parts is obvious from the fact that the curve $P(z)$ does not pass through the origin.

4.7 Nyquist criterion

A particularly important case is the following. Consider the "feedback" system represented schematically in Figure 10.11. A linear system, characterized by its transfer function $\Phi(i\omega)$ has as input the primary input $x(t)$ minus its own output $y(t)$. We ask for the relation between $x(t)$ and $y(t)$.

Assuming all magnitudes oscillate harmonically with angular frequency ω, we may disregard the common factor $e^{i\omega t}$ as explained in Chapter 5, and write the relations

$$y = \Phi \cdot (x - y) = \Phi \cdot x - \Phi \cdot y;$$

hence

$$(1 + \Phi)y = \Phi x.$$

We recall that from the operational point of view this is really a differential equation in which the left side is the homogeneous part and the right side is the forcing term (input). Solving for y we get

$$y = \frac{\Phi}{1 + \Phi} x.$$

The stability of this solution is determined by the homogeneous part of the differential equation, that is, $(1 + \Phi)y = 0$. There will be asymptotic stability if the curve $1 + \Phi(i\omega)$ does not "enclose" the origin (algebraically speaking). Therefore we may say that there will be asymptotic stability if the curve $\Phi(i\omega)$, for $-\infty < \omega < +\infty$, does not enclose the point $z = -1$. This is known as Nyquist's stability criterion. Notice that the transfer function $\Phi(i\omega)$ is $1/P(i\omega)$, where $P(d/dt)y(t) = f(t)$ is the open loop differential equation of the system.

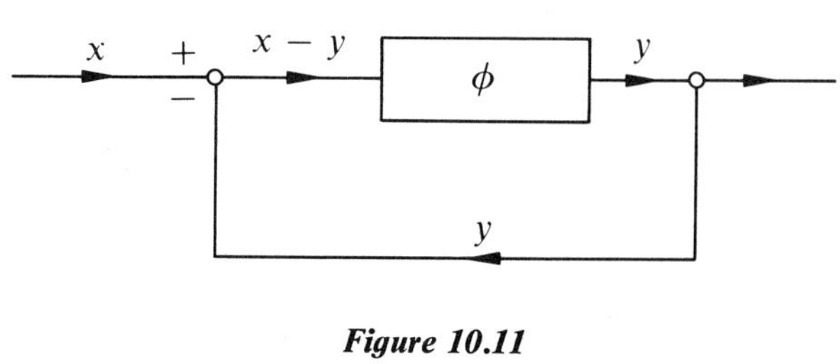

Figure 10.11

4.8 *Liapunov functions for systems with constant coefficients*

For some problems it is important to be able to construct explicitly a Liapunov function for a given differential system.

Problem: *Given the differential system* (10.4.4), $\mathbf{x}' = A\mathbf{x}$, *where* A *is a constant matrix of elements* a^i_j *and characteristic roots* $\lambda_1, \ldots, \lambda_n$, *and given a form (homogeneous polynomial)* $u(x^1, x^2, \ldots, x^n)$ *of degree* m, *find another form* $v(x^1, x^2, \ldots, x^n)$ *also of degree* m, *such that*

$$\sum \frac{\partial v}{\partial x^i} x^{i'} = \sum \frac{\partial v}{\partial x^i}(a^i_1 x^1 + a^i_2 x^2 + \cdots + a^i_n x^n) = u. \quad (10.4.6)$$

4.9 *Theorem*

If there is no system of nonnegative integers $m_1, m_2, \ldots, m_n$ *satisfying*

$$m_1 + m_2 + \cdots + m_n = m \quad (=\text{degree of } u)$$

and

$$m_1\lambda_1 + m_2\lambda_2 + \cdots + m_n\lambda_n = 0, \quad (10.4.7)$$

then the above problem has exactly one solution.

Proof

To start with let us try to determine a form v such that

$$v' = \sum \frac{\partial v}{\partial x^i}(a^i_1 x^1 + \cdots + a^i_n x^n) = \lambda v. \quad (10.4.8)$$

In the case $m = 1$, write

$$v = p_1 x^1 + p_2 x^2 + \cdots + p_n x^n. \quad (10.4.9)$$

Substituting in (10.4.8), the conditions

$$p_1 a^1_j + p_2 a^2_j + \cdots + p_n a^n_j = \lambda p_j \quad (i = 1, 2, \ldots, n)$$

are obtained. These conditions can be satisfied if and only if the determinant

$$D(\lambda) = \det(A - \lambda E) = 0. \tag{10.4.10}$$

Therefore, if this equation has n distinct roots $\lambda_1, \ldots, \lambda_n$, there will be n distinct forms

$$v_i = \sum p_k^i x^k \quad (i = 1, 2, \ldots, n) \tag{10.4.11}$$

such that

$$v_i' = \lambda_i v_i. \tag{10.4.12}$$

In the case $m > 1$, let N be the number of terms a homogeneous polynomial of degree m has; this number will now be the number of unknown coefficients $q_1, \ldots, q_N$ of the form $v(x^1, \ldots, x^n)$ we should determine. As in the preceding case, from the condition $v' = \lambda v$ we get the equations

$$q_1 A_j^1 + q_2 A_j^2 + \cdots + q_N A_j^N = \lambda q_j \quad (i = 1, 2, \ldots, N). \tag{10.4.13}$$

The coefficients A_j^i are linear combinations of the preceding a_j^i's. The condition for existence of a solution of (10.4.13) is

$$D_m(\lambda) = \det((A_j^i) - \lambda E) = 0. \tag{10.4.14}$$

Liapunov proved that the characteristic roots of (10.4.14) are obtained from the characteristic roots of (10.4.10) as linear combinations:

$$\lambda = m_1\lambda_1 + m_2\lambda_2 + \cdots + m_n\lambda_n, \tag{10.4.15}$$

where $m_1, m_2, \ldots, m_n$ are nonnegative integers of sum $= m$.

This can be seen easily if all the roots $\lambda_1, \ldots, \lambda_n$ are distinct. In this case, all the forms of degree m obtained as product of the v_i's of (10.4.12),

$$v = v_1^{m_1} v_2^{m_2} \cdots v_n^{m_n} \tag{10.4.16}$$

are distinct and satisfy condition (10.4.8) by virtue of

$$v' = v\left(m_1 \frac{v_1'}{v_1} + \cdots + m_n \frac{v_n'}{v_n}\right) = v(m_1\lambda_1 + \cdots + m_n\lambda_n).$$

On the other hand, the number of terms of a homogeneous polynomial of degree m, terms which are of the type $(x^1)^{m_1}(x^2)^{m_2} \cdots (x^n)^{m_n}$, is precisely the number of sets of nonnegative integers $m_1, \ldots, m_n$ such that the sum $m_1 + \cdots + m_n = m$. This number N is therefore the degree of equation (10.4.14), and the N linearly independent solutions are the forms (10.4.16).

Let us now return to our original problem: given $u(x^1, \ldots, x^n)$, we must determine $v(x^1, \ldots, x^n)$ satisfying (10.4.6). The coefficients of u are given, and for the determination of the coefficients of v we have, by substitution in (10.4.6), a system of linear algebraic equations which differ from the system (10.4.13) in the right-hand side, where now the coefficients of u appear. According to our hypothesis (10.4.7), $\lambda = 0$ cannot be a characteristic root of the system (10.4.13). Hence the coefficients of v are uniquely determined by those of u and the theorem is proved (at least for distinct roots).

Notice that the condition (10.4.7) is satisfied in the case when all characteristic roots $\lambda_1, \ldots, \lambda_n$ have negative real parts, that is, when system (10.4.4) is asymptotically stable.

As an application to stability theory, assume that the system (10.4.4) is asymptotically stable. Let us give arbitrarily, as $u(\mathbf{x})$, a negative definite form. We then determine $v(\mathbf{x})$ according to the last theorem. Then $v(\mathbf{x})$ must be positive definite for the following reasons:

(a) $v(\mathbf{x})$ cannot take negative values, because in that case we could apply Theorem 10.3.8 (taking $-v(x)$ and $-u(x)$), and system (10.4.4) would be unstable.

(b) $v(\mathbf{x})$ cannot take the value zero except for $\mathbf{x} = 0$, because in such a case its derivative $v' = u$, being negative, would indicate that $v(\mathbf{x})$ also takes negative values.

(c) $v(\mathbf{x})$ does not depend on t, so that its continuity and the fact that it is zero only at the origin makes it positive definite.

We thus have proved that if the system (10.4.4), with constant coefficients, is asymptotically stable, we can always find Liapunov functions, as shown in Theorem 10.3.6. As function $u(\mathbf{x})$ an arbitrary form of degree m can be given (but m must be even so that $u(\mathbf{x})$ can be negative definite). For simplicity, in practical applications a form of degree 2 is usually chosen as $u(\mathbf{x})$.

Problem

1. Let the constant matrix A be such that $\mathbf{x}' = A\mathbf{x}$ is asymptotically stable. Let $B(t)$ be continuous for $t \geqq 0$ and such that $\lim B(t) = 0$ as $t \to +\infty$. Prove that the origin is asymptotically stable for the differential equation

$$\mathbf{x}' = [A + B(t)]\mathbf{x}.$$

(Brauer and Nohel, *Qualitative Theory of Ordinary Differential Equations,* p. 156.) *Hint*: prove first that all solutions exist for $t > 0$. Next write the equation as $\mathbf{x}' - A\mathbf{x} = B(t)\mathbf{x}$ and apply the formula for variation of constants. Take into account that $\|e^{At}\| < Ke^{-rt}$ for $r > 0$ and some K. Multiplying all terms by e^{rt} it is then possible to apply Gronwall's inequality. Thus, all solutions tend to zero. That the origin is stable follows from a similar proof giving a uniform bound for the solutions in any finite time interval.

References

The origin of Liapunov's direct method, an historic landmark, was first published as a paper in Russian in 1892, then translated into French in 1907 and reproduced as

A. A. Liapunov. *Problème Général de la Stabilité du Mouvement*, in the *Annals of Mathematics*. Princeton, N.J.: Princeton Univ. Press, 1949.

A good introduction to this subject is given in

J. LaSalle and S. Lefschetz. *Stability by Liapunov's Direct Method.* New York: Academic Press, 1961.

This method is also discussed in some detail in

L. Cesari. *Asymptotic Behavior and Stability Problems in Ordinary Differential Equation,* in the series *Ergebnisse der Mathematik und ihrer Grenzgebiete.* Berlin: Springer-Verlag, 1963.

S. Lefschetz. *Differential Equations, Geometric Theory*, New York: Interscience (Wiley), 1957.

W. Hahn. *Theorie und Anwendung der Direkten Methoden von Liapunov*, in the series *Ergebnisse der Mathematik und ihrer Grenzgebiete.* Berlin: Springer-Verlag, 1959.

W. Hahn. *Stability of Motion*, in the series *Grundlehren der Mathematiksche Wissenschaft.* New York: Springer-Verlag, 1967.

Further applications of the Liapunov direct method, particularly *absolute stability*, are given in

S. Lefschetz. *Stability of Nonlinear Control Systems.* New York: Academic Press, 1965.

For the proofs concerning the Routh–Hurwitz criterion and the Sturm sequences, we refer the reader to

F. R. Gantmacher. *The Theory of Matrices*, Volume II, Chapter XV. New York: Chelsea, 1960.

M. Marden. *The Geometry of the Zeros of a Polynomial*, in *Mathematics Surveys*, Volume 3. Providence, R.I.: American Mathematical Society, 1949.

For related subjects, particularly good exercises on stability and asymptotic behavior of solutions, see

F. Brauer and J. A. Nohel. *Qualitative Theory of Ordinary Differential Equations.* New York: Benjamin, 1969.

Answers

Chapter 1

1.4. $x''x' - x''x + x'^2 - x'x = (x'' + x')(x' - x)$.

1.5. $y' = -\dfrac{x'}{x^2} = -\dfrac{1+x^2}{x^2} = -y^2 - 1$.

1.6. $x(t) = t^2 - 8t + 15$.

1.7. $x(t) = t^3 + t$.

1.8. $x(t) = \sin t - \cos t$.

1.9. $x(t) = e^{(t^2+c)}$.

1.10. $x(t) = (t^3 + c)e^t$.

2.2. (a) $y' = y \cot x$. (b) $y' = x/y$.
(c) $y' = x^2/y^2$. (d) $y' = 4y^{3/4}$.

(e) $y' = \dfrac{y \ln y}{x \ln x}$. (f) $y' = \dfrac{y^2 - x^2}{2\,xy}$.

(g) $y' = \dfrac{y^2 - x^2 + 1}{2xy}$. (h) $y' = \dfrac{y^2 - x^2 + a^2}{2xy}$.

2.6. (a) $y^2 - x^2 = c$. (b) $y = \sqrt{2(c - x)}$.

3.1. (a) $y = kx + c$. (b) $y = ce^{kx}$.
(c) $y = cx^k$. (d) $kx^2 - y^2 = c$.
(e) $y = a + c(x - b)$. (f) $\cos y = \cos x + c$.

3.2. $y = ce^x - x - 1$.

3.3. $y = 4x - 2\,\dfrac{ce^{4x} - 1}{ce^{4x} + 1}$.

3.4. $\dfrac{(x+y)^2}{2} - \dfrac{(x-y)^3}{3} = c.$

3.5. $y = x \tan \ln cx.$

3.6. $y^2 + x^2 = cx^4.$

3.7. (a) $y = (x^3 - 1)(c + \ln (x^3 - 1)).$
(b) $y = (x + c) \cos x.$

(c) $x = \dfrac{1}{2y} + \dfrac{c}{y^3}.$

3.8. (a) $y = z^{-1}, \; z = e^{x^2/2}\left(-\int^x e^{-s^2/2} \sin s \, ds + c\right).$

(b) $y = z^2, \; z = ce^{-x} + \sin x.$

3.9. The equation for z is:

$$z' = -(2p(x)y_1(x) + q(x))z - p(x).$$

3.10. $y = x^2 + (-1 + ce^{-x})^{-1}.$

3.11. $a_2 x^2 - b_1 y^2 - 2a_1 xy = c.$

3.12. $\sin y \tan x + x = c.$

3.13. $e^{xy} + \sin x + \cos y = c.$

4.1. (a)–(iii) $y = 1 - \dfrac{(x+c)^2}{4};$

–(iv) $y = 1;$ (v) Yes.

(b)–(iii) $y = 1 - \left(\dfrac{3}{2}\right)^{2/3} (x + c)^{2/3};$

–(iv) $y = 1;$ (v) No.
(c)–(iii) $(x - c)^2 + y^2 = 1;$
–(iv) $y = \pm 1;$ (v) Yes.

4.2. (a) $y = cx - \dfrac{2}{3} c^{3/2}; \quad y = \dfrac{x^3}{3}.$

(b) $y = cx + \sqrt{1 + c^2}; \quad x^2 + y^2 = 1.$

(c) $y = cx - 2\sqrt{c}; \quad y = -\dfrac{1}{x}.$

5.1. (a) $x = x_0 + \dfrac{g}{k}\left(t + \dfrac{e^{-kt}}{k}\right) - \dfrac{v_0 e^{-kt}}{k}.$

(b) $x = c_1 t^2 + c_2.$

(c) $x = \dfrac{t}{2}\sqrt{x^2 + c} + \dfrac{c}{2} \ln (x + \sqrt{x^2 + c}) + c_2.$

(d) $x = c_1 \int e^{-t^2} \, dt + c_2.$

(e) $t = c_1 + \int \frac{dx}{\sqrt{2k \cos x - c_2}}$.

(f) $x = c_1 e^t + c_2 e^{-t}$.

(g) $x^2 = \frac{1}{c_1} + c_1(t - c_2)^2$.

5.2. (a) $x'' = 2x' - 2x$. (b) $x''' = -3x'$.
(c) $x'' = 2xx' + x' - 2x^2$. (d) $x'' = x'e^x$.

5.3. $x' = -k_1 x$; $y' = k_1 x - k_2 y$; $x(0) = x_0$, $y(0) = 0$.

$$x = x_0 e^{-k_1 t}; \quad y = x_0 \frac{k_1}{k_2 - k_1}(e^{-k_1 t} - e^{-k_2 t}), \text{ if } k_1 \neq k_2;$$

or $y = x_0 kte^{-kt}$, if $k_1 = k_2 = k$.

Chapter 2

1.1. (a) $x' = tx^2$; $x(0) = 1$. (b) $x' = b + x$; $x(1) = a + b$.
(c) $x' = (-t^{-1} + t^{-2})x$; $x(1) = 2$. (d) $x' = 2t^{-1}x$; $x(2) = 0$.

1.2. Differentiate the double integral two times; evaluate this integral by interchanging the order of integration.

2.1. $|f_n(t_2) - f_n(t_1)| \leqq K|t_2 - t_1|$.

2.2. Given $\varepsilon > 0$, for some N, $n > N$ implies $|f_n(t)| < \varepsilon$. For $f_1, \ldots, f_N$, there is a common bound on the derivative, as in the previous problem.

2.3. Assuming $\lim f_{n_i}(t) = g(t)$ almost everywhere, it follows from Egoroff's theorem (H. L. Royden. *Real Analysis*, p. 72) that the convergence is uniform on a set of measure $2\pi - \varepsilon$. This is impossible because two functions $f_n(t)$, $f_m(t)$, $n \neq m$, do not differ in less than δ on a set of measure $2\pi - \varepsilon$, even if we take n, m very large.

2.4. $\lim f_n(t) = 0$ for $0 \leqq t < 1$; $= 1$ for $t = 1$.

3.1. (i) existence; (ii) uniqueness.
(a) (i) and (ii) if $t^2 \neq x^2$.
(b) (i) $x \leqq 0$ or $x \geqq 1$; (ii) $x < 0$ or $x > 1$.
(c) (i) everywhere; (ii) everywhere, except at $x = 0$ and at $x = 1$.
(d) (i) and (ii) $0 < x < \pi$, $2\pi < x < 3\pi, \ldots$.
(e) (i) everywhere, except at $x = 0$, $t \leqq 0$.
(ii) everywhere except at $x = 0$, $t < 1$.

3.2. (a) $|x - y| \leqq 0.1e^{2|t|}$. (b) $|x - y| \leqq 0.01te^{t/2}$, $t > 0$.

3.3. Lipschitz constant is t.

3.4. $f'(x) = \lim \frac{\Delta f}{\Delta x} = 0$.

4.1. (a) $x = t + \frac{t^2}{2} + \frac{t^3}{3!} + \cdots = e^t - 1$. (b) $x = 1 + t + t^2 + \cdots = \frac{1}{1-t}$.

(c) $x = 1 - t^2 + \frac{t^4}{2} - \frac{t^6}{3!} + \cdots = e^{-t^2}$.

5.1. (a) $(-\infty, 1/2]$. (b) $(\text{arc cot}\,(a-1) - \pi, \text{arc cot}\,(a-1))$.
(c) $(-\infty, 0)$. (d) No solution.
(e) No solution.

6.1. It is a linear equation $y' = A(t)y$ with continuous $A(t)$.

7.1. $$\begin{pmatrix} x \\ y \end{pmatrix}' = \begin{pmatrix} x^2 + y^2 \\ x^2 - y^2 \end{pmatrix}; \quad \begin{pmatrix} x(0) \\ y(0) \end{pmatrix} = \begin{pmatrix} 1 \\ 0 \end{pmatrix}.$$

7.2. $$x' = \frac{x}{\sqrt{x^2 + y^2}}, \qquad y' = \frac{y}{\sqrt{x^2 + y^2}}.$$

7.4. (a) $x = 1 + \frac{t^2}{2} + \frac{t^4}{4!} + \cdots$.

(b) With $x_n = 1 + \int_0^t y_{n-1}\,dt$ and $y_n = \int_0^t 2x_n^3\,dt$,

we obtain $x_2 = 1 + t^2 + \frac{t^4}{2} + \frac{t^6}{5} + \frac{t^8}{28}$.

(c) With $x_{-1} = y_{-1} = 0$ and $x_0 = 1, y_0 = -1$,

$$\Delta_n x = x_n - x_{n-1} = \int \Delta_{n-1} y\,dt,$$

$$\Delta_n y = y_n - y_{n-1} = -2 \int (\Delta_{n-1} x + \Delta_{n-1} y)\,dt,$$

we obtain $x = 1 - t + \frac{t^3}{3} - \frac{t^4}{6} + \frac{t^5}{30} - \cdots$.

8.2. Consider $v(t) = \ln u(t)$; hence $v'(t) = kv(t)$. Then apply Gronwall's lemma.

8.3. See Coddington and Levinson. *Ordinary Differential Equations*, p. 60, problem 1.

8.4. In these problems, integrate the differential equation under the assumption that the sign function is constant $= +1$ or $= -1$. Then discuss how the solutions match on the boundary where the sign function changes. This can be done by drawing the solutions $x(t)$, $y(t)$ in the x, y-phase plane.

8.5. (a) (i) $x_0 = 0$; $x = \pm\left(\frac{t - t_0}{2}\right)^2$ for $t \geqq t_0$, $x = 0$ for $t \leqq t_0$.

(ii) $x_0 \neq 0$, $t - t_0 > -2\sqrt{|x_0|}$.

(b) (i) $x_0 = 0$; $x = \pm\left(\frac{t - t_0}{2}\right)^2$ for $t \leqq t_0$, $x = 0$ for $t \geqq t_0$.

(ii) $x_0 \neq 0$, $t - t_0 < 2\sqrt{|x_0|}$.

(c) (i) $x_0 = 0$; $x = 0,\ x = \left[\frac{2}{3}\,|t_0 - t|\right]^{3/2} \operatorname{sgn}(t_0 - t)$.

(ii) $x_0 \neq 0,\ t - t_0 < -\frac{3}{2}\,x_0^{2/3}$ if $x_0 > 0$;

$t - t_0 > -\frac{3}{2}x_0^{2/3}$ if $x_0 < 0$.

(d) (i) $(t, x) = (0, 0)$; $x = t_0^2,\ x = 0$.
(ii) $t > 0$ or $t < 0$, provided $0 \leqq x_0 \leqq t_0^2$.

Chapter 3

1.8. (a) $\begin{pmatrix} \cos\alpha & -\sin\alpha \\ \sin\alpha & \cos\alpha \end{pmatrix}$. (b) $\begin{pmatrix} 1 & 0 \\ 0 & -1 \end{pmatrix}$. (c) $\begin{pmatrix} 0 & 1 \\ 1 & 0 \end{pmatrix}$.

1.9. $\begin{pmatrix} 2 & & & \\ & 2 & & \\ & & 2 & \\ & & & 2 \end{pmatrix}$, $\begin{pmatrix} 2 & & & \\ & 2 & & \\ & & 2 & 1 \\ & & & 2 \end{pmatrix}$, $\begin{pmatrix} 2 & 1 & & \\ & 2 & & \\ & & 2 & 1 \\ & & & 2 \end{pmatrix}$,

$\begin{pmatrix} 2 & & & \\ & 2 & 1 & \\ & & 2 & 1 \\ & & & 1 \end{pmatrix}$, $\begin{pmatrix} 2 & 1 & & \\ & 2 & 1 & \\ & & 2 & 1 \\ & & & 2 \end{pmatrix}$.

1.10. $\lambda = \cos\alpha \pm i\sin\alpha$, real for $\alpha = 0$ and $\alpha = \pi$.

1.14. (a) $\lambda = 1,\ x = \begin{pmatrix} 1 \\ 1 \\ 1 \end{pmatrix}$; $\lambda = 2,\ x = \begin{pmatrix} 1 \\ 0 \\ 1 \end{pmatrix}$; $\lambda = 3,\ x = \begin{pmatrix} 1 \\ 2 \\ 2 \end{pmatrix}$.

(b) $\lambda = 2,\ x = \begin{pmatrix} 1 \\ 1 \\ 1 \end{pmatrix}$; $\lambda = 2,\ x = \begin{pmatrix} 1 \\ 0 \\ 1 \end{pmatrix}$; $\lambda = -1,\ x = \begin{pmatrix} 1 \\ 2 \\ 2 \end{pmatrix}$.

(c) $\lambda = 1,\ x = \begin{pmatrix} 1 \\ 1 \\ 1 \end{pmatrix}$; $\lambda = 1,\ x = \begin{pmatrix} 1 \\ 2 \\ 2 \end{pmatrix}$.

(d) $\lambda = 1,\ x = \begin{pmatrix} 1 \\ 1 \\ 1 \end{pmatrix}$.

1.15. (a) $\begin{pmatrix} 2 & 1 \\ 0 & 2 \end{pmatrix}$; (b) $\begin{pmatrix} 5 & 0 \\ 0 & -1 \end{pmatrix}$; (c) $\begin{pmatrix} 3 & 0 \\ 0 & -3 \end{pmatrix}$;

(d) $\begin{pmatrix} 0 & -2 \\ 2 & 0 \end{pmatrix}$; (e) $\begin{pmatrix} 3 & -2 \\ 2 & 3 \end{pmatrix}$; (f) $\begin{pmatrix} 1 & 1 \\ -1 & 1 \end{pmatrix}$.

1.16. (a) $\begin{pmatrix}1 & & \\ & 2 & \\ & & 3\end{pmatrix}$; (b) $\begin{pmatrix}2 & & \\ & 2 & \\ & & -1\end{pmatrix}$; (c) $\begin{pmatrix}1 & & \\ & 1 & 1\\ & & 1\end{pmatrix}$;

(d) $\begin{pmatrix}1 & 1 & \\ & 1 & 1\\ & & 1\end{pmatrix}$.

2.1. (a) $-\infty < t < \infty$. (b) $t \neq k\pi$, $k = 0, \pm 1, \pm 2, \ldots$.
(c) $-\infty < t < \infty$. (d) $t \neq 0$.
(e) $t \neq -1, 0, 1$. (f) $t \neq k\pi$, $k = 0, \pm 1, \pm 2, \ldots$.

2.2. $\|\mathbf{x}(2)\| \leqq \|\mathbf{x}(0)\| e^{10}$.

2.3. (a) Linearly independent in [1, 2], dependent in the other cases.
(b) Linearly dependent in all intervals.

2.4. Express $\mathbf{y}_1$ and $\mathbf{y}_2$ as linear combinations of $\mathbf{x}_1$ and $\mathbf{x}_2$.

2.5. $X(t)$ is a fundamental matrix in (a)-(c)-(d).

2.6. (a) $\begin{pmatrix}-t^{-1} & 2t^{-3}\\ -t & 4t^{-1}\end{pmatrix}$. (b) $\begin{pmatrix}t^{-1} & 0\\ t^{-2} & -t^{-1}\end{pmatrix}$.

(c) $\begin{pmatrix}\tan t & 0\\ 0 & \cot t\end{pmatrix}$. (d) $\begin{pmatrix}1 & 0\\ \dfrac{e^{-t}-e^{t}}{2} & 1\end{pmatrix}$.

4.2. Integrate $yL[x] - xM[y]$ term by term by parts and observe that only the boundary terms remain, while the terms with integrals cancel.

5.1. $x = t^2 - t - t \ln t$.

6.1. Linearly independent are (a) and (b).

6.2. Take the solutions $1, t, e^t$.

6.3. $t^2x'' - 2tx' + 2x = 0$.

Chapter 4

1.1. (a) $\begin{pmatrix}e^{-t} + e^{-3t}\\ -3e^{-t} - 5e^{-3t}\end{pmatrix}$. (b) $e^{-t}\begin{pmatrix}2\\ -2\end{pmatrix}$.

(c) $e^{-2t}\begin{pmatrix}1 - t\\ -3 + t\end{pmatrix}$. (d) $e^{t}\begin{pmatrix}1\\ t\end{pmatrix}$.

(e) $\begin{pmatrix}5\cos 3t - 5\sin 3t\\ 4\cos 3t + 2\sin 3t\end{pmatrix}$. (f) $-e^{-2t}\begin{pmatrix}\cos t + \sin t\\ 2\cos t\end{pmatrix}$.

1.2. (a) $\begin{pmatrix}e^t - e^{-t}\\ 1 + e^t\\ -1 + e^t\end{pmatrix}$. (b) $\begin{pmatrix}-5e^t\\ 5e^t + 6e^{3t}\\ -e^t - 10e^{3t} + 8e^{6t}\end{pmatrix}$.

(c) All orbits (for $r = \pi^{-1}, (2\pi)^{-1}, (3\pi)^{-1}, \ldots$) are stable from the inside, unstable from the outside.

(d) Orbits for $r = \pi^{-1}, (3\pi)^{-1}, \ldots$ are unstable, orbits for $r = (2\pi)^{-1}, (4\pi)^{-1}, \ldots$ are stable.

Chapter 9

1.1. (a) Soft ($|x| < \pi$). (b) Hard $\left(|x| < \frac{\pi}{4}\right)$.

(c) Soft ($|x| < \pi$). (d) Hard $\left(|x| < \frac{\pi}{2}\right)$.

(e) Soft ($|x| > 1$).

1.2. (a) $T = 4 \int_0^A \frac{dx}{\sqrt{2(\cos x - \cos A)}} \cdot \lim_{A \to 0} T = 2\pi.$

(b) $T = 4 \int_0^A \frac{dx}{\sqrt{2 \ln \frac{\cos x}{\cos A}}} \cdot \lim_{A \to 0} T = 2\pi.$

(c) $T = 4 \int_0^A \frac{dx}{\sqrt{x_0'^2 - x^2 - \frac{2}{3} x^3}}$, where $A^2 + \frac{2}{3} A^3 = x_0'^2$.

$\lim_{A \to 0} T = 2\pi.$

(d) $T = 4 \int_0^A \frac{dx}{\sqrt{x_0'^2 - x^2 - \frac{x^4}{4}}}$, where $A^2 + \frac{A^4}{4} = x_0'^2$.

$\lim_{A \to 0} T = 2\pi.$

(e) $T = 4 \int_0^A \frac{dx}{\sqrt{x_0'^2 - x + \ln(1 + x)}}$, where $A - \ln(1 + A) = x_0'^2$

$\lim_{A \to 0} T = 2\pi.$

(f) $T = 4 \int_0^A \frac{dx}{\sqrt{x_0'^2 - \frac{1 - e^{-x^2}}{2}}}$, where $\frac{1 - e^{-A^2}}{2} = x_0'^2$.

$\lim_{A \to 0} T = 2\pi.$

(g) $T = 4\int_0^A \frac{dx}{\sqrt{x_0'^2 - \frac{x^4}{4}}}$, where $\frac{A^4}{4} = x_0'^2$.

$\lim_{A\to 0} T = \infty$ (obtained by series expansion of

$$A^2\sqrt{1-\left(\frac{x}{A}\right)^4} = A^2\left(1-\frac{1}{2}\left(\frac{x}{A}\right)^4+\cdots\right)\Bigg).$$

1.3. $T = \pi\left(\frac{1}{\sqrt{k_1}} + \frac{1}{\sqrt{k_2}}\right)$.

1.4. $T = \sqrt{\frac{8A}{k_1}}\left(1+\frac{k^1}{k_2}\right)$, where A is the amplitude of oscillation on the side $x > 0$.

1.5. In (9.1.9), $\dim x = \text{inch}$, $\dim x'' = \dim F = \text{inch sec}^{-2}$, $\dim a = \text{sec}^{-2}$, $\dim b = \text{inch}^{-2}\,\text{sec}^{-2}$. In (9.1.12), $\dim A = \text{inch}$.

1.6. $\omega^2 = k^2 - \frac{k^2}{8}A - \frac{F}{A}$.

3.1. For large μ, $T \cong 2\mu \ln 2$.

3.2. For large μ, $T \cong 2\mu k \ln\left(1+\frac{2}{k}\right) \cong 4\mu$.

Chapter 10

1.1. (a) $0 < p < 1$. (b) $0 < p < 1$.

(c) $-\frac{1}{2} < p < 0$ (Use Hurwitz' condition; see 10.4.4–5.)

1.2. Assume $X(x, y)$ and $Y(x, y)$ continuous, $K = \int_0^2 \frac{Z(\cos\theta, \sin\theta)}{N(\cos\theta, \sin\theta)}\,d\theta$.

Then $KN(\cos\theta, \sin\theta) \leqq 0$ is equivalent to stability, $KN(\cdots) < 0$ equivalent to asymptotic stability.

2.1. (a) $\begin{cases} x' = -y \\ y' = x. \end{cases}$

The matrix of the variational equation A is $A = \begin{pmatrix} 0 & -1 \\ 1 & 0 \end{pmatrix}$. All trajectories are Liapunov stable.

(b) $\begin{cases} x' = -y\sqrt{x^2+y^2} \\ y' = x\sqrt{x^2+y^2} \end{cases}$

$$A = \begin{pmatrix} -r\sin rt\cos rt & -r(1+\sin^2 rt) \\ r(1+\cos^2 rt) & r\sin rt\cos rt \end{pmatrix}$$

All trajectories are orbitally stable.

(c) $$\begin{cases} x' = -\dfrac{y}{r} \\ y' = \dfrac{x}{r} \end{cases}$$

$$A = \begin{pmatrix} \dfrac{1}{r}\sin\dfrac{t}{r}\cos\dfrac{t}{r} & -\dfrac{1}{r}\cos^2\dfrac{t}{r} \\ \dfrac{1}{r}\sin^2\dfrac{t}{r} & -\dfrac{1}{r}\sin\dfrac{t}{r}\cos\dfrac{t}{r} \end{pmatrix}$$

All trajectories are orbitally stable.

(d) $$\begin{cases} x' = x - y \\ y' = x + y \end{cases} \qquad A = \begin{pmatrix} 1 & -1 \\ 1 & 1 \end{pmatrix}$$

All trajectories are unstable.

(e) $$\begin{cases} x' = -x - y \\ y' = x - y \end{cases} \qquad A = \begin{pmatrix} -1 & -1 \\ 1 & -1 \end{pmatrix}$$

All trajectories are stable.

2.2. (a) $\mu_{1,2} = 2, \dfrac{1}{2}$; $(1, 0), (0, 1)$.

(b) $\mu_{1,2} = \sqrt{2}, -\sqrt{2}$; $(1, \sqrt{2} - 1), (1, -\sqrt{2} - 1)$.
(c) $\mu_{1,2} = 1, -1$; $(1, 1), (1, -1)$.
(d) $\mu_{1,2} = i, -i$; no real eigenvectors.

(e) $\mu_{1,2} = \dfrac{\sqrt{2}}{2}, -\dfrac{\sqrt{2}}{2}$; $(1, \sqrt{2} - 1), (1, -\sqrt{2} - 1)$.

(f) $\mu_{1,2} = 0$; only eigenvector $(1, 0)$. (g) $\mu_{1,2} = 0, 1$; $(1, -1), (1, 0)$.
(h) $\mu_{1,2} = 0$; only eigenvector $(1, 1)$.

3.1. $\dfrac{dv}{dt} = -2x^4 - 4y^4$.

Index

Abel's theorem, 91
Absolutely continuous function, 36
Adjoint equation, 94
Admittance, 160
Alexandroff, P., 281
Almost periodic function, 152
Almost periodic solution, 151
Alternating current circuits, 161
Analog computers, 32
Analytic systems, 217
Angular frequency, 151
Aperiodic case, 150
Apostol, T. M., 123
Approximate methods, 31
Approximate solution, 38
Approximation, linear, 229
Approximation theorem, 45
Arbib, M. A., 168
Arzelá, C., 69
Arzelá-Ascoli lemma, 40
Ascoli, G., 40
Associated matrix equation, 87
Astatic, 149
Attractor, 305
Autonomous systems, 143, 170

Bang-bang, 64
Bellman, R., 240
Bendixson, I., 240
  theorem, 276
Bernoulli's equation, 14
Besicovitch, A. S., 168
Bieberbach, L., 105, 281
Birkhoff, G. D., 200
Black box, 140
Bode diagram, 166
Bogoliubov, N. N., 300
Bohr, H., 152, 168
Bouncing ball, equation of, 290
Brauer, F., 336

Canonical form
  of a matrix, 77
  real, 123
Cantor, G., 41
Capacitor, 147, 162
Carathéodory, C., 65
Carathéodory's theorem, 64
Cartwright, M. L., 300
Cartwright-Littlewood equation, 294
Cauchy-Peano theorem, 37
Center, 208
  for homogeneous systems, 213, 215
Cesari, L., 240, 336
Characteristic
  equation, 74, 115
  multipliers, 316
  roots, 74, 107
    complex, 123
Churchill, R. V., 136
Circuits, alternating current, 161
Clairaut's equation, 23
Coddington, E. A., 65, 68, 340
Collatz, L., 33
Compactification of the euclidean
    plane, 267
Complementary term, 158
Conservative systems, 282
  free oscillations in, 283
Conti, R., 200, 240, 280, 300
Continuation of solutions, 53
Continuation theorem, 53, 65
Continuous interest, 9
Contraction operator, 51
Control, 139
Convergence theorem, 49

Convolution, 129
Critical case, 150
Critical point, 144, 171, 173, 189, 201
  classification, 207
  isolated (in homogeneous systems), 214
Cylinder, equations on the, 242

Damping, critical, 151
Degenerate linear systems, 208, 209
Delta function, 130
  derivative of, 132
  Laplace transform of, 132
Dense set, 41
Dependence of solutions on initial data and parameters, 54
Diagonal sequence, 41
Diffeomorphic, locally, 241
Differential equations
  approximate methods of solution, 31
  approximate solution, 38
  autonomous, 170
  in the complex domain, 103
  definition, 1
  degree, 2
  elementary methods of solution, 2
  explicit, 3
  first order, 5
  generalized solution, 35
  general solution, 1
  graphical methods, 31
  higher order (special methods), 25
  homogeneous, 11
  homogeneous in x, x', x'', 27
  implicit, 19
  linear, 70
    homogeneous, 82
    nonhomogeneous, 96
  on manifolds, 241
  numerical methods, 31
  order, 2
  particular solution, 1
  periodic, 173
  qualitative methods, 32
  second order, 282
  singular solution, 21
  systems of, 2, 28
    explicit form, 3
Dirac delta function, 130
Direction field, 5, 171
Dirichlet function, 35
Discriminant curve, 21
Dissipative system, 283, 317
Distributions, 133
Doetsch, G., 136
Domain, definition, 5
Duffing's equation, 285, 288
  with damping, 294, 298
Dynamical system, 188

Eigenvalue, 73
Eigenvector, 74, 107
Elliptic sector, 228
Equivalent dynamical systems, 189
Euler, L., 137
Euler equation, 98
Euler formula, 154
Euler-Poincaré characteristic, 265
Evolution, 55
Exact equations, 15
Existence
  interval of, 53
  theorems, 34, 61, 83
Exponential of a matrix, 118

Falb, P. L., 168
Fan, 228
Fasol, K. H., 168
First integral (of a differential equation), 283
Fixed point, 51, 179, 250, 315
Focus, 208
  for analytical systems, 226
  for homogeneous systems, 215
Fomin, S. V., 51
Forced motion, 143
Forced oscillations, 157, 285, 294
Fourier, J. B., 137
Fourier integral, 166
Fourier series, 166
Fractional part of x, 249
Free motion, 143
Frequency
  angular, 151
  natural, 151
Frequency response, 157
Friction, negative, 151
Friction coefficient, 147
Function
  almost periodic, 152
  delta (*see* Delta function)
  Liapunov (*see* Liapunov function)
  negative definite, 318
  positive definite, 318
Fundamental matrix, 87
  principal, 88
Fundamental system, 87

Gain, 166
Gantmacher, F. R., 105, 336
Gauss' sphere, 267
Generalized solution, 35
General solution, 1
Graphical methods, 31
Gronwall's lemma, 44

Hahn, W., 336
Hale, J. K., 300
Hard spring, 284
Harmonic (component), 180
Harmonic oscillator, 149

Harmonic response, 165
Hartman, P., 67, 240
Heaviside, O., 126
Hille, E., 105
Holbrook, J. G., 136
Homoemorphism, 179
Homogeneous equation, 11
Homogeneous systems, 211
Homotopic curves, 258
Homotopy, 258
Hopf, H., 281
Hurwitz, A., 328
Hyperbolic sector, 228
Hysteresis, 298, 300

Ideal point, 267
Impedance, 160
Implicit function theorem, 19, 184
Impulse function, 130
Ince, E. L., 33
Index, 257
  of a curve, 257
  of the elementary critical points, 262
  of a point, 261
  of the sphere, 265
  of a surface, 263
Inductance, 147
Inductor, 162
Infinitely small upper bound, 319
Infinitesimal, 221
Initial condition, 8
Input, 139
Integer part of x, 249
Integral of motion, 283
Integrating factor, 16
Interest, continuous, 9
Interval of existence, 36, 53
Invariant ray, 216
Invariant set, 191

Jordan canonical form, 77, 119
Jordan curve, 257

Kalman, R. E., 168
Kamke, E., 33, 69
Kaplan, W., 136
Kneser, H., 69
Kneser's theorem, 67
Kolmogorov, A. N., 51
Krabbe, G., 136
Kronecker's delta (formula 3.1.4), 71

Lagrange, J. L., 137
Lagrange stability, 197
Lakshmikantham, V., 68
Laplace transform, 126
  definition, 128
  table, 136
LaSalle, J. P., 300, 336
Lebesgue measure zero, set of, 35
Leela, S., 68
Lefschetz, S., 105, 200, 217, 240, 280, 300, 326, 336
Lefschetz' lemma, 183
Levinson, N., 65, 68, 340
Liapunov, A. A., 301, 334, 336
  direct method, 317
  second method, 317
Liapunov function, 232, 317
  for systems with constant coefficients, 333
Lienard's equation, 290
Limit cycle, 274
Limit point, 194
Limit set, 194
  configuration in the plane, 279
  for two-dimensional systems, 274
Linear approximation, 229
Linear equation
  first order, 12
  of order n, 99
    with constant coefficients, 115
Linear independence, 84
Linear normed space, 51
Linear systems, 141
  autonomous, two-dimensional, 201
  degenerate, 208, 209
  perturbed, 229
Linear transformation, 73, 75
Linearity theorem, 84
Liouville's transformation, 91
Lipschitz condition, 44

MacDuffee, C. C., 105
Manifold, differentiable, 241
Marden, M., 336
Massera, J. L., 319
Matrix
  canonical form, 77
  characteristic equation, 74
  characteristic roots, 74
  companion, 100
  conjugate, 72
  eigenvector, 74
  exponential of, 80
  function of, 79
  inverse, 72
  logarithm of, 80
  nonsingular, 72
  norm of, 72
  singular, 72
  transpose, 72
Matrix algebra, 70
Matrix equation, associate, 87
Matrix solution (of a system of differential equations), 118
Metric space, 169
Mikusinski, J., 126, 136
Minorsky, N., 300
Mitropolskii, Y. A., 300

Model, mathematical, 137
Motion, 170
  forced, 143
  free, 143
Murphy, G. J., 168

Natural frequency, 151
Nemytskii, V. V., 200, 240, 280, 300, 346
Newton, I., 137
Node, 207
  in analytic systems, 226
  improper, 207
  isotropic, 207
  proper, 207
Nohel, J. A., 336
Nonconservative systems, 290
Norm of a vector, 59
Numerical methods, 31
Nyquist criterion, 332
Nyquist locus, 166

Operational method, 126
Orbit, 170
Orthogonal vectors, 60
Oscillations, 147
  complex notation, 154
  forced, 157, 285, 294
  normal mode, 153
Output, 139

Parallel connection, 161
Particular solution, 1
Path polygon, 280
Peano, G., 69
Pendulum, equation of, 242
Periodic equations, 173
Periodic solutions, 147, 179
Perturbation, 139
Phase space, 169
Poincaré, H., 200, 251, 257, 312
Poincaré-Bendixson theorem, 276
Poincaré formula for the index, 259
Poincaré theorem, 281
Poincaré theorem on index, 264
Point
  critical, 144, 171, 173, 189, 201
    classification, 207
  of exterior contact, 259
  fixed, 51, 179, 250, 315
  of interior contact, 259
  limit, 194
  regular, 173
  of rest, 144, 173, 189
  singular, 282
Poisson stability, 200
Projective plane, 267
  differential equations on the, 269
  trajectories on the, 267

Qualitative methods, 32

Reduction of order, 92
Region of attraction, 309
Regular arc, 186
Regular point, 173
Reissig, R., 200
Relaxation oscillations, 295
Relay control, 64
Resistance, negative, 151
Resistor, 147, 162
Resonance, 157
Restoring force, 284
Rest point, 144, 173, 189
Riccati equation, 14
Riemann sphere, 267
Rotation number, 249
Routh-Hurwitz criterion, 328, 336
Royden, H. L., 36, 68, 188, 339

Saddle point, 208
Sansone, G., 200, 240, 280, 300
Scalar product, 60
Schwartz, L., 133, 136
Sector
  elliptic, 228
  hyperbolic, 228
Separable variables, 9
Sequence, diagonal, 41
Series connection, 161
Similar matrices, 75
Singer, I. M., 280
Singular point, 282
Singular solution, 21
Smith, O. J. M., 168
Soft spring, 285
Solution (*see also* differential equation)
  continuation of, 53, 62
  dependence on initial conditions
    and parameters, 66
  general, 1
  generalized, 35
  maximal, 67
  minimal, 67
  particular, 1
  periodic, 147, 179
Solution curve, 5
Soroka, W. W., 33
Space
  of events, 170
  linear normed, 51
  metric, 169
  phase, 169
  state, 169
Spring
  hard, 284
  soft, 285
Spring constant, 147
Stability, 145, 301
  absolute, 326, 336
  asymptotic, 146, 302
    of a transformation, 316

Stability (*continued*)
of critical points, 230
definition, 302
equiasymptotic, 302
exponentially asymptotic, 302
globally asymptotic, 309
Lagrange, 197
of linear systems, 326
with constant coefficients, 328
orbital, 312
Poisson, 200
quasi-asymptotic, 303
with respect to perturbations, 310
of a rest point, 302
of solutions, 311
total, 310
of transformations, 315
uniform, 302
uniform asymptotic, 302
uniform quasi-asymptotic, 303
State space, 169
State variable, 139
Step function, 132
Stepanov, V. V., 200, 240, 280, 300, 346
Stoker, J. J., 300
Sturm sequences, 329, 336
Subharmonic (component), 180, 288
Successive approximations, 48, 62
Superposition principle, 142
System
astatic, 149
fundamental, 87
Systems
autonomous, 170
two-dimensional, 201
conservative, 282
coupled, 152
of differential equations, 59
dissipative, 283, 317
dynamical, 188
Systems (*continued*)
homogeneous, 211
nonconservative, 290
time invariant, 142

Thorpe, J. A., 280
Titchmarsh, E. C., 221
Topological transformation, 179, 189
Torus, equations on the, 245
Towing curve, 6
Trajectory, 170
Trajectory tube, 186
Transfer function, 164, 332
Transversal cells and arcs, 181
Tsien, H. S., 168

Undetermined coefficients, method of, 115
Uniformly bounded sequence, 41
Uniqueness theorems, 34, 42, 46, 47, 62, 65, 84
Unstability, 146, 302

Van der Pol's equation, 295
Vandermonde, determinant, 109
Variational equation, 58, 313
Variations of constants, 97, 125
Vector field, 171
Vector function, 59
derivative of, 61
integral of, 61
limit of, 60
Vector operations, 59
Vinograd theorem, 190

Willers, F. A., 33
Wronskian, 101

Z-sector, 219, 221
combination of, 228
description, 225